The Ark and Beyond

Convening Science: Discovery at the Marine Biological Laboratory

A Series Edited by Jane Maienschein

I am delighted to serve as editor for the new University of Chicago Press series Convening Science: Discovery at the Marine Biological Laboratory. These books will "highlight the ongoing role the Marine Biological Laboratory plays in the creation and dissemination of science, in its broader historical context, as well as current practice and future potential." Each volume is anchored at the MBL, including work about the MBL and MBL science and scientists, work by those scientists, work that begins with workshops or research or courses at the MBL, collaborations made possible by the MBL, and so on. Books by, about, with, for, inspired by, and otherwise related to the MBL will capture the spirit of discovery by the community of MBL scientists and students. Some will be monographic, while others are collaborative coherent collections.

We look forward to discovering new ideas and approaches that find their way into volumes of the series. I first did summer research, with a small NSF grant, as a graduate student in 1976, which led to my first edited volume inspired by this special place. Many others have been similarly inspired, and this new series invites us to bring together our works into a collection of reflections on the MBL's role in promoting discovery through its exceptional role in convening science at the seaside.

JANE MAIENSCHEIN
Series Editor
University Professor of the Center for Biology and Society
Arizona State University
Fellow, Marine Biological Laboratory

The Ark and Beyond

The Evolution of Zoo and Aquarium Conservation

EDITED BY BEN A. MINTEER,
JANE MAIENSCHEIN,
AND JAMES P. COLLINS

with a Foreword by
GEORGE RABB

The University of Chicago Press

CHICAGO AND LONDON

The University of Chicago Press, Chicago 60637
The University of Chicago Press, Ltd., London

Published 2018
Printed in the United States of America

27 26 25 24 23 22 21 20 19 18   1 2 3 4 5

ISBN-13: 978-0-226-53832-7 (cloth)
ISBN-13: 978-0-226-53846-4 (paper)
ISBN-13: 978-0-226-53863-1 (e-book)

DOI: https://doi.org/10.7208/chicago/9780226538631.001.0001

Library of Congress Cataloging-in-Publication Data

Names: Minteer, Ben A., 1969– editor. | Maienschein, Jane, editor. | Collins, James P., editor.
Title: The ark and beyond : the evolution of zoo and aquarium conservation / edited by Ben A. Minteer, Jane Maienschein, and James P. Collins.
Other titles: Convening science.
Description: Chicago ; London : The University of Chicago Press, 2018. | Series: Convening science
Identifiers: LCCN 2017042866 | ISBN 9780226538327 (cloth : alk. paper) | ISBN 9780226538464 (pbk. : alk. paper) | ISBN 9780226538631 (e-book)
Subjects: LCSH: Zoos. | Aquariums. | Conservation biology.
Classification: LCC QL76 .A75 2018 | DDC 590.73—dc23 LC record available at https://lccn.loc.gov/2017042866

♾ This paper meets the requirements of ANSI/NISO Z39.48–1992 (Permanence of Paper).

For George Rabb

Contents

PART 3 ZOO AND AQUARIUM CONSERVATION TODAY: VISIONS AND PROGRAMS • 105

PART 4 CARING FOR NATURE: WELFARE, WELLNESS, AND NATURAL CONNECTIONS • 177

Foreword

Shortly after being asked to write this foreword, I was sidelined by upsetting news: the tree frog Toughie, the last known representative of *Ecnomiohyla rabborum*, had died at the Atlanta Botanical Garden. The extinction of a species is always hard to understand and difficult to come to terms with. It is one of those moments when we wish we could turn back time, make other decisions, and choose options that would lead to a viable outcome. In this case the extinction of Rabbs' fringe-limbed tree frog was personal to me, since Joseph Mendelson and his colleagues had named the species for me and my wife, Mary. We had lost Toughie, we had lost our species, and it seems we are continuing to lose the fight against the chytrid fungi that are decimating amphibians worldwide.

In a time when some are questioning society's need for zoos, Toughie's story perfectly illustrates the need for zoos, aquariums, and botanical gardens to be conservation organizations, the subject of this wide-ranging volume. The last few individuals of Rabbs' fringe-limbed tree frog had been collected in 2005 from the canopy of a cloud forest in central Panama when it was realized the species was in peril from a chytrid fungal epidemic. Efforts to breed the final survivors failed owing to our lack of knowledge. Researchers discovered that in the wild, tadpoles in their tree-hole abodes would nibble on the father frog's back, gathering essential nutrition. For this species, such knowledge came one step too late: the last female died in 2009, rendering extinction a certainty. Experts at the Atlanta Botanical Garden and Zoo Atlanta cared for Toughie and his fellow frogs and extended their lives as far as possible, but in this case vital knowledge of the species did not come in time.

There are many other species in states of peril, as documented by the Red List of endangered species of the IUCN, the International Union for Conservation of Nature. There is much more research to be done, more care to be given, and more education to impart. The contributions in this volume illustrate some of the critical work being done in and by zoological institutions around the globe to prevent the extinction of more species and to educate the public about their conservation.

Here, forty-eight authors document and illustrate the transformation of zoos and aquariums from entertaining menageries to conservation institutions. They describe the history of such places, current undertakings in conservation for several species such as gorillas in the wild as well as in institutional facilities, and future prospects for such institutions to change further to more effectively counter the increasing tide of extinctions.

The beginning of conservation activities by American institutions is rightly attributed to William Hornaday of the New York Zoological Society, who orchestrated the salvage of the American bison. Nowadays, as the Wildlife Conservation Society, this organization has projects and programs in more than sixty countries around the world. Here readers can also learn about the origins, missions, and operations of more recent institutions with distinctive conservation programs, such as the Arizona-Sonora Desert Museum, the Monterey Bay Aquarium, and the Phoenix Zoo.

This volume speaks of the conflicts in the operating agendas of many institutions as they grapple with providing for the welfare of the individual animals kept in zoos and aquariums and for the welfare/conservation of their species in the wild. The science involved in adequately providing for both individual and species welfare is reviewed in several chapters, and the ethical questions of keeping wild animals in captivity are explored in other sections of this book. Clearly shown is the cooperative nature of meaningful conservation activities, involving not just fellow zoological institutions but wildlife societies, government agencies, professional organizations, and caring communities. The One Plan Approach to species conservation, in situ and ex situ, of the Conservation Breeding Specialist Group of the Species Survival Commission/IUCN that is described in the chapter by Kathy Traylor-Holzer and her coauthors obviously depends on such cooperation.

One finds interesting contrasts given on the prospective futures for the physical structures that house the animal ambassadors in these institutions. On one hand, there is advocacy for returning aquariums from giant immersion experiences for people back to smaller containments that allow more intimate connections between visitors and the animals displayed. In regard to the simulation of the natural environments of the terrestrial species a zoo chooses to keep, there is advocacy for deconstructing enclosures so the spaces become actual sanctuaries. If space is limited, there are nevertheless utopic conceptions such as the Zootopia in Denmark where the visitors will not be discernible to the animals at all, as described in the last chapter by Ben Minteer. Such settings will also convey the importance of the integrity of ecosystems to the flourishing of most species.

Many authors of the thirty chapters explicitly or implicitly welcome other views on the challenges of these institutions' becoming more substantial conservation organizations. In this regard, readers might find enlightening perspectives from twenty-two other authors who have responded very briefly to the question posed by the Center for Humans and Nature: "How can zoos and aquariums foster cultures of care and conservation?" The answers come from conservation leaders in zoos, a sensitive architect, a champion for the seas, and a compassionate ecologist. And in this volume it is good to have the affirmative perspective of Rick Barongi, a principal in assembling the third version of the World Conservation Strategy of the World Association of Zoos and Aquariums.

My own institution has fostered knowledge of the genetics of small populations, but it also has seen to the development of conservation psychology as a means to influence and better the attitudes and behavior of visitors and people generally in respect to a sustainable relationship with the natural world and its diversity of life forms. Here Susan Clayton, coauthor of the first text on conservation psychology, and Khoa D. Le Nguyen explore the potentials for this approach.

In sum, it is encouraging to have the authors in this volume inform us on how to deal with the manifest threats to the existence of other species. However, the ultimate threat leading to the extinction vortex for populations and species is us, the human species. Further, most members of our species have become more separated from the natural world in urban concentrations, and thereby appreciation of and concern for the diversity of life have been diminished. Thus I hope there will be a following complementary volume on environmental conservation as behavior expected of all peoples. A means to effect such an embrace of responsibility is to inform and educate the visitors to our institutions so that they not only become practitioners of conservation in their own lives, but also spread the concern for the existence of all other life to friends and neighbors and acquaintances. Visitors to zoos, aquariums, botanical gardens, nature centers, and natural history museums make up a tenth of the people on the planet. The reformulated ark envisioned by many in this book can certainly also embark on changing the outlook and behavior of zoo visitors and supporters to achieve a sustainable relationship with the natural world and all its species. A concluding chapter by Adrián Cerezo and Kelly Kapsar dwells on this enormous charge in the context of sustainable development, as outlined in Agenda 2030 of the United Nations. For zoos and aquariums, I see this as a charge to transform their visitors and communities into responsible and respectful global citizens who care for all life.

Finally, I say many thanks for the first step in this charge given by Ben Minteer, Jane Maienschein, and James Collins in assembling and editing the thoughtful essays in this volume. It is the beginning of moving beyond the ark!

George Rabb
President Emeritus
Chicago Zoological Society / Brookfield Zoo

INTRODUCTION

Zoo and Aquarium Conservation: Past, Present, Future

Ben A. Minteer, Jane Maienschein, and James P. Collins

ZOOS AND AQUARIUMS CONFRONT THE CONSERVATION CHALLENGE

We are living, most biodiversity scientists and conservationists will tell you, in a time of profound ecological change. The mounting pressures imposed by habitat destruction and fragmentation, human population growth, overexploitation, spread of infectious diseases, and rapid climate change threaten scores of wild species around the globe (Thomas et al. 2004; Collins and Crump 2009; Hoffmann et al. 2010; Dirzo et al. 2014; Urban 2015). It's a situation so severe that the distinguished biologist E. O. Wilson recently called for setting aside no less than half the planet for biodiversity protection (Wilson 2016). Wilson's proposal no doubt sets a high-water mark for bold responses to global biodiversity loss, but it also signifies just how serious the challenge has become for many concerned about the viability of threatened species and ecosystems in the coming decades (Kolbert 2014).

Can zoos and aquariums play any role in tackling a challenge of this magnitude, let alone a significant one? Many scientists and leaders within and outside the zoo and aquarium communities believe they can, especially if they make biodiversity conservation a top institutional priority (see, e.g., Zimmermann et al. 2007; Conde et al. 2011a; Barongi et al. 2015). Although the sense of a looming global extinction crisis has clearly magnified these calls, they too are not new. Zoos and aquariums have for decades been encouraged to play a more significant role in the global effort to combat species extinction, to protect habitats, and in general to conserve biodiversity at a range of scales, from tens of hectares, to large national parks, to immense ocean reserves (see, e.g., Bendiner 1981; Norton et al. 1995; Hancocks 2002; Fa, Funk, and O'Connell 2011).

This agenda has been met with both skepticism and encouragement. Critics often argue that zoos and aquariums in fact do very little for biodiversity conservation and that any promotion as legitimate conservation organizations is a cynical appeal to justify an anachronistic and exploitative kind of institution. Supporters respond by asserting that zoos and aquariums have been serious contributors to conservation programs for more than a century, with an impressive track record of saving species that would be extinct without their efforts (see Tullis 2014 for an overview of the argument).

Such categorical views may appeal to a desire for simple and unqualified assessments, but they do not get us far in understanding zoos and aquariums as conservation actors. As the chapters in this volume will show, the story is much more complex.

ROOTS AND LEGACIES

As many zoo historians have written (including those in this volume), the early menageries from which the modern zoo emerged were anything but "centers of conservation" (see, e.g., Hoage and Deiss 1996; Reid and Moore 2014). Nevertheless, a concern with wildlife preservation, not just the protection of animals for public display in captivity, is a significant part of the history of the modern, professionally run zoo that developed in the nineteenth century. The incipient wildlife protection goals of early zoos, however, emerged at best only fitfully and partially. In European zoos in the eighteenth and early nineteenth centuries, for example, the interest lay primarily in the exotic qualities of certain animal species and in utilitarian applications of animal husbandry (Guerrini and Osborne, this volume; Ritvo, this volume). Still, by the early nineteenth century many European zoological gardens (like their later American counterparts) were viewed, especially by the scientific societies that founded them, as cultural hubs of nature study and public learning as well as recreational landscapes (Hochadel 2005).

In the United States, an identifiable concern with wildlife protection within zoos began to coalesce only in the late nineteenth century, coinciding with the early American conservation movement and the establishment of many of the nation's first zoos (see Barrow, this volume; Henson, this volume). Natural history, scientific study and education, and wildlife preservation all became tightly linked in the original aspirational missions of many US zoological institutions during the period. Through zoological study and education, for example, early American zoos such as New York's Bronx Zoo, which opened to the public in late 1899, hoped to encourage interest in exotic and native animals and a concern for wildlife protection within a largely urban visitor base (Stott 1981, 58).

Captive breeding and reintroduction quickly emerged as the primary technique in zoos' early wildlife preservation efforts, with the notable example of the recovery of the American bison in the early twentieth century, where again the Bronx Zoo was a leader (Barrow; Rothfels; Henson; and Kisling, all this volume). Zoos' scientific and conservation ambitions and capacities became more sophisticated and extensive as the century wore on. Animal husbandry, zookeeping, and zoo biology, for example, became more professionalized in the later twentieth century, supported by the growth of scientific journals, conferences, and professional societies (Kisling 2000b). Some of the largest and best-equipped zoos, such as the one at San Diego, eventually moved to the cutting edge of research and conservation science, developing impressive new genetic and reproductive techniques (see, e.g., the chapters by Ryder; Friese; and Tubbs, all this volume). The best of these researchers contributed to advances in basic biology as they sought to conserve species using an ever-improving understanding in areas such as physiology, developmental biology, genetics, ecology, and host-pathogen interactions in zoo and aquarium species. Basic discoveries were applied to conserving animals, and also to educating the public about the importance of zoos as distinctive institutions where this sort of work could be done. Some zoos developed into "boundary institutions" using translational research to move the best ideas and most advanced techniques that modern science had to offer from laboratory or field experiments to conserving Earth's biodiversity. Although expensive, it was also the sort of discovery-based work that attracted the public's interest and donations.

Not surprisingly, this period marked the high tide of the zoo as a surrogate "Noah's ark" (see, e.g., Durrell 1976) geared to creating captive assurance populations as a hedge against extinction in the wild. The ark model was solidified in the 1980s by the creation of Species Survival Plans (SSPs) by (what is now) the Association of Zoos and Aquariums (AZA) as an overarching strategy for recovering and managing threatened and endangered species held in zoos (Grow, Luke, and Ogden, this volume; Kisling, this volume). The Amphibian Ark (Mendelson, this volume) is an excellent example of this sort of thinking. As researchers in the 1980s and 1990s began to realize how many amphibian species were declining, some to extinction (Collins and Crump 2009), zoos, aquariums, and other conservation-related institutions created ex situ facilities to house and breed threatened species in the hope that at some future date the frogs can be reestablished in their natural habitats (Reid and Zippel 2008).

Breeding zoo animals to recover select threatened species and to reinforce in situ populations isn't the only way for zoos to contribute to species protection in the wild, however, and a suite of activities soon flourished under the general banner of "conservation." Over the past several decades, for example, many zoos expanded their capacity to protect endangered species by a range of tactics and methods, such as developing integrative species conservation programs integrating ex situ and in situ populations into a hybrid metapopulation management system (Redford, Jensen, and Breheny 2012; see also Traylor-Holzer, Leus, and Byers, this volume). Zoos also began to participate in collaborative field conservation, including significant community-based conservation initiatives (Lukas and Stoinski; Allard and Wells; and Cerezo and Kapsar, all this volume). As a result, zoos have been an important part of the institutional and intellectual context in which the science of conservation biology began to take shape in the late 1970s and early 1980s (Meine, Soulé, and Noss 2006).

Public aquariums, too, became more active participants in aquatic field conservation and science during this same period, with a growing number of calls over the past decade for them to play a more significant role in helping to sustain marine ecosystems (e.g., Penning et al. 2009; Tlusty et al. 2013; see also the chapters by Knapp and by Spring, this volume). The aquarium conservation story is less well known and documented than that of zoos, and although their conservation agenda overlaps to a considerable extent with that of zoos, aquariums' own history, as well as their technological and design requirements, also presents a distinct set of conservation challenges and opportunities deserving more research (see, e.g., Brunner 2005; also Muka, this volume; Linquist, this volume).

Zoo-based visitor education, a more indirect approach to conservation than breeding and reintroduction or participation in field projects, has long been a staple of zoo and aquarium programs (as the chapters in this volume by Henson and by Palmer, Kasperbauer, and Sandøe illustrate). The education-conservation link rests on the expectation that the better visitors understand the behavior, natural history, and status of threatened wildlife, the more likely they will be to engage in pro-conservation/environmental behavior. Although the evidence supporting this claim for the link's effectiveness is mixed and often hotly contested (e.g., Falk et al. 2007; Marino et al. 2010; Jensen 2014), many within and outside the zoo and aquarium community suggest that under the right conditions these institutions can have an appreciable impact on biodiversity knowledge and conservation actions, especially if such programs foster a sense of connection between visitors and animals (Rabb and Saunders 2005; Moss, Jensen, and Gusset 2015; also Grajal, Luebke, and Kelly, this volume; Clayton and Le Nguyen, this volume).

Finally, scores of zoos and aquariums have increasingly become patrons of field conservation, devoting part of their operating budgets to funding in situ wildlife programs. According to the AZA, its accredited zoos and aquariums "fund over 2500 conservation projects in more than 100 countries and spend on average $160 million on conservation initiatives annually" (https://www.aza.org/conservation-funding; also Grow, Luke and Ogden, this volume; Allard and Wells, this volume). Even though these numbers sound impressive—especially for a field (conservation) notorious for feeling cash-strapped—as Grazian (2015) points out, this effort is largely carried by a small set of larger and comparatively more resource-rich zoos. Furthermore, conservation expenditures currently account for only a very small percentage of most zoological institutions' budgets. The most recent conservation vision of WAZA, for example called for a minimum of 3 percent of zoos' and aquariums' annual operating budgets to be devoted to conservation, which would mark a considerable increase for some (Barongi et al. 2015). Many within the zoo and aquarium communities are therefore pressing for even more investment and for these institutions to contribute even more of their budgets to conservation in the future (see, e.g., the chapters by Barongi and by Lukas and Stoinski, this volume).

TENSIONS AND TRADE-OFFS

To the degree that zoos and aquariums become more seriously engaged in conservation in their visions, programs, and budgets, they will have to contend with the constraints of their own institutional histories and capacities (as noted by Henson, this volume, and by Monfort and Christen, this volume). But they will also have to come to grips with a complex and rapidly evolving ethical and social context (see, e.g., the contributions in this volume by Palmer, Kasperbauer, and Sandøe; Norton; Clay; and Maple and Segura). Of course, zoological institutions have for decades prompted discussions about a range of animal welfare and animal rights issues, from the basic question of the moral acceptability of keeping animals in captivity (e.g., Jamieson 1985, 1995; Regan 1995; Gruen 2014) to more specific ethical arguments and debates over practices such as captive breeding, manipulative zoo- and aquarium-based research, wild animal acquisition, habitat enrichment, and commercialization of wildlife (see, e.g., Norton et al. 1995; Kreger and Hutchins 2010; Maple and Perdue 2013).

In the past several years purported ethical violations by zoos and aquariums have been litigated in the news and popular media as a series of controversial cases have reignited a debate about the place of zoological parks in modern society, from the breeding and keeping of orcas at SeaWorld to the culling of captive animals for population management (e.g., Greene, this volume). As we write this introduction in summer 2016, the decision by the Cincinnati Zoo to shoot Harambe the Gorilla after a child entered his enclosure has provoked an international uproar (e.g., McPhate 2016; Walters 2016), yet it remains unclear what lessons we have learned from that unfortunate episode. Collectively, these recent cases have energized a growing chorus of critics calling for the end of what they believe to be an unethical and unnecessary form of animal exploitation in the twenty-first century (e.g., Wallace-Wells 2014; Zimmermann 2015; Bekoff 2016).

It is too early to tell how this newly emboldened critique of zoological parks will play out. As the discussions in this volume remind us, zoos and aquariums have weathered many storms over the decades while managing to remain incredibly popular among the public. Nevertheless, as they strive to become more serious partners in the study, recovery, management, and preservation of wildlife "outside the enclosure," they will have to make difficult decisions and consider complex trade-offs regarding the values and interests that have traditionally shaped their mission, including decisions that will affect not only their identity, but also their bottom line (Cohen 2013). This means that the fuller ethical evaluation of zoo and aquarium conservation requires more than navigating and balancing animal welfare concerns (as important as these are). It also requires understanding and managing an emerging set of value-laden and ethical questions about our responsibility to conserve biodiversity, preserve wildness, and achieve sustainability across a spectrum of rapidly changing ex situ and in situ contexts (Minteer and Collins 2013; also Norton; Monfort and Christen; Cerezo and Kapsar; and Linquist, all this volume).

The SeaWorld orca case in particular has underscored perhaps the most difficult and controversial of these value-driven challenges: balancing zoological parks' long-standing entertainment and recreation interests with the growing commitment to animal welfare, scientific research, and biodiversity conservation (Conway 2011; Grazian 2015).[1] Again, this challenge to balance the entertainment and scientific aims of zoological parks is far from new. As Henson (this volume) writes, zoos have been struggling with the task of harmonizing their public and scientific goals since the late nineteenth century. As far back as the early nineteenth century (when the modern version of the institution emerged in Europe), zoos were forced to realize that they could not survive as purely scientific institutions for research and education. They had to build and sustain their popularity with an often fickle public, typically more interested in seeing new and exhilarating animal attractions than in learning about natural history or about wildlife conservation (Hochadel 2005, 39–40).

It's a pragmatic concession, however, that has made some trouble, reflected in the widely voiced criticism that philistine entertainment values, more than "higher" educational, scientific, or conservation goals, define the zoo and aquarium today. Yet at the same time some zoological parks are embracing rather than fleeing from their public entertainment function, arguing that more ambitious, immersive, and challenging designs can lure more visitors *and* display a more respectable animal and environmental ethic (see, e.g., Minteer, this volume). Regardless, as Ivanyi and Colodner (this volume) remind us (by their example of the Arizona-Sonora Desert Museum), public popularity and a strong zoo conservation identity are by no means mutually exclusive.

THE ARK AND BEYOND

It's clear that zoos and aquariums continue to evolve and mature in complex ways as modern conservation actors, all while being buffeted by the shifting winds of institutional, scientific, and societal change. As a result, the expanding conservation role for zoological parks raises a set of intriguing and difficult questions and challenges as we think more deeply about the history of zoos and aquariums, what they are today, and where they might—and perhaps should—be going. This book wrestles with these questions and challenges.

The questions and challenges concern the trajectory of zoological parks themselves, but they also concern zoo- and aquarium-centered scholarship, especially a set of important discussions in conservation history, the history and philosophy of the life sciences, environmental and animal ethics/studies, and conservation biology. For example:

- How and why did the conservation mission emerge and evolve in modern zoos and aquariums, and how does its development situate these institutions in the wider history, philosophy, and practice of science and conservation?

- "Conservation" has been used to describe a wide array of practices within and by zoos and aquariums, from breeding to release in situ and conserving species (and habitats) in the field, to conducting ex situ animal research that may inform field conservation, to promoting pro-conservation attitudes and behavior through education. How have these different practices developed at various times and in various places? Are all these interpretations of conservation equally compatible? Which, in the end, are the most important?
- What do we mean by "animal welfare" and "animal wellness," and how do these commitments and responsibilities relate to the conservation values and goals of zoos and aquariums today as well as to a wider set of conservation values applied to animals in their natural habitats? Can zoos make good on their promise to be leaders in both conservation and animal care while also catering to public entertainment and recreation interests?
- What is "wild" or "natural"—and how do we know? Can zoos ever be wild in any meaningful sense? To what extent does it matter whether zoos or aquariums replicate natural conditions (which they cannot do without qualifications)? How do narratives about acclimatization, husbandry, "enculturation" of both animals and humans, and wildlife protection all fit into this mix?
- What are the implications for conservation philosophy, science, and policy when zoos and aquariums become more engaged in field conservation projects—and as they take on an enhanced role as breeding centers and as "conservation arks" for threatened wild populations? Is the ark metaphor even useful today given the practical limitations of zoos and aquariums (space, resources)—limitations that become more glaring in the face of a global biodiversity crisis?

To try to understand these questions better, and to do our best to answer them, we created a "thinking community" (as the great conservationist Aldo Leopold might have put it) around the story of zoo and aquarium conservation, a community that would itself grow and evolve as we moved through the various stages of our project. The first event was held at the Marine Biological Laboratory (MBL) in Woods Hole, Massachusetts, in May 2014 as part of the MBL-Arizona State University (ASU) History of Biology seminar series. That meeting focused primarily on the *historical* foundations of zoo and aquarium conservation, with some consideration of contemporary management and scientific issues. In fall 2015 we held a companion zoo and aquarium conservation symposium at ASU in Tempe and the Phoenix Zoo that effectively "flipped" the earlier gathering, emphasizing the participation of zoo and aquarium leaders while bringing along many of the historians from the earlier Woods Hole seminar and expanding the group to include ethicists, biologists, and social scientists. By the time we began compiling the chapters collected in this volume, our "thinking community" had swelled to four dozen contributors representing an impressive breadth of zoological and academic institutions (and an equally wide array of scholarly and professional fields and traditions).

The overarching theme of the project, the theme that bound us all together, was the meaning and significance of zoo and aquarium conservation as an idea and as a set of practices, and how an understanding of its complex traditions, challenges, and opportunities could be absorbed into broader narratives and discussions in conservation history, environmental ethics, the history of the life sciences, and conservation biology. The primary focus was on the US story, though the international context figures prominently in several of the contributions.

An early model for the present volume was the groundbreaking collection *Ethics on the Ark* (Norton et al. 1995), which drew together an eclectic group of zoo professionals, philosophers, activists, and biologists to consider a range of ethical, scientific, and management issues confronting the modern zoo (see the chapters in this volume by Norton and by Maple and Segura). A lot of water has flowed under the conservation bridge since the mid-1990s; the time therefore seemed right for a new assembly of voices. We also sought to expand the discussion by emphasizing the historical dimension, especially because zoos and aquariums have been nearly invisible in the standard conservation narratives.[2] And we wanted *The Ark and Beyond* to incorporate key perspectives from the social sciences (especially psychology and sociology), as well as a set of questions about zoo ethics and values that acknowledged, but also went beyond, the usual animal welfare debates.

THE ROAD AHEAD

We've arranged the chapters that follow into six clusters. The first two parts focus on the historical dimension of zoo and aquarium conservation, beginning with the rise of the modern zoo in Europe, which had only a tenuous link to what we would consider conservation today (though the roots of later conceptual and practical tensions in zoo conservation are clearly visible). Chapters in part 2 explore the emergence of an identifiable conservation agenda in the early American zoos and aquariums of the late nineteenth and early twentieth centuries, concluding with the growth and maturation of this agenda late in the twentieth century. One of the more intriguing conclusions that emerges from these discussions is that there is no single, unified narrative of zoo and aquarium conservation, but rather a tangle of converging and diverging ideas and practices surrounding the circulation, breeding, reintroduction, and preservation of wildlife in and by zoological parks. As a result, although there is historical evidence supporting the claims that zoos have been championing conservation since the late nineteenth century, with a few notable exceptions this commitment was neither consistent nor all that significant for species protection until the later part of the twentieth century.

The next part moves to a consideration of today's zoo and aquarium conservation landscape, including emerging agendas motivating major zoo and aquarium bodies such as the World Association of Zoos and Aquariums (WAZA), the US-based Association of Zoos and Aquariums (AZA), and the Conservation Planning Specialist Group of the International Union for Conservation of Nature (IUCN) Species Survival Commission (SSC). This third part also includes a series of in-depth accounts of particular programs at zoological parks known for their conservation programs and their influence on species survival and environmental sustainability more generally. Taken together, the chapters in this part demonstrate the diversity and scope of the contemporary zoo and aquarium conservation vision while emphasizing the importance of integrating zoological parks into a larger network of institutions working to recover, protect, and manage biodiversity across the landscape.

As we mentioned above, the ethical and societal context of the modern zoological park has figured prominently in discussions of the value of zoos and aquariums today, especially in light of significant concerns about animal welfare. Chapters in part 4 explore the welfare-conservation relation across a number of dimensions, revealing the complexity and ongoing challenges of running an ethically accountable institution premised on the public display of animals for human enjoyment. Contributions examine the nuances of the human connection to zoo animals and reflect on evolving ethical responsibilities surrounding animal wellness and the intricacies of the animal welfare-conservation interface. The part closes with an argument for zoological parks to promote a wider environmental ethic of ecological resilience and sustainability.

The scientific dimension of zoo and aquarium conservation is the subject of the penultimate part of the volume, which draws together a series of chapters examining one of the more fascinating and innovative frontiers of zoo biology: the use of advanced genetic and reproductive technologies to help recover and conserve threatened animal populations—and perhaps even to "revive and restore" species through genetic engineering and synthetic biology (aka "de-extinction"). The role of zoological parks in the "genetic rescue" of species, especially at the more interventionist end of the spectrum, raises a host of intriguing issues, including ethical and philosophical questions about the appropriate use of these technologies and what they might have in store both for zoo and aquarium animals and for the future of endangered species in the wild. Zoological research supporting critical conservation efforts is also highlighted, as well as the myriad scientific and institutional questions confronting zoos and aquariums as they address one of the more urgent and formidable biodiversity challenges today: the amphibian extinction crisis.

The final part of the book considers several alternative pathways and models for zoos and aquariums moving forward, starting with a cautionary argument for zoos to solidify what they historically do well (caring for and managing zoo animal populations) rather than overcommitting themselves as "full service" conservation organizations. It offers defenses of place-based zoological institutions rooted in natural history and regional culture as well as arguments to rethink the societal role and ecological footprint of zoos and aquariums in the age of sustainability. The discussion closes with a pair of chapters probing the biological, aesthetic, and cultural implications of enhancing wildness and naturalness in the zoo environment, including whether we really want zoos to be red in tooth and claw—and whether radical efforts to make zoos more immersive offer a more naturalistic alternative to the traditional visitor experience or just a more ingeniously designed one.

No single volume on a subject as complex and context-dependent as zoo and aquarium conservation, even one as expansive as ours, could persuasively claim to be complete or exhaustive. Still, in *The Ark and Beyond* we've worked hard to provide as informed and as rich a treatment of the story of zoo and aquarium conservation as possible. We believe the book captures some of the very best thinking about zoos and conservation available today.

If we were asked (in closing this introduction) to condense a long list of project goals into a single ambition for this book, we'd say it is to show

how a carefully chosen set of critical and diverse perspectives on the evolution and character of zoological parks can help us better understand them as conservation actors in a time of rapid ecological change, rising expectations, and deepening societal skepticism. We also hope to demonstrate the great value of assembling a diverse and thoughtful group of academics and practitioners to engage in a challenging, and not always easy, conversation about the legacy, limitations, and hopes of one of our more popular public institutions as it continues to change and transform in this century. That involves two goals rather than one, we realize. But you need to have pairs on the ark.

PART I

Protoconservation in Early European Zoos

CHAPTER ONE

Animals in Circulation: The "Prehistory" of Modern Zoos

Anita Guerrini and Michael A. Osborne

INTRODUCTION

Humans have collected and displayed nonhuman animals for at least twenty-five hundred years. Some of this history is recorded in the works of Gustave Loisel and Vernon N. Kisling (Loisel 1912; Kisling 2000a).[1] The prehistory of modern zoos encompasses Roman arenas, Hannibal's elephants, and pet monkeys. What these animals have in common, and what they share with the inhabitants of many later menageries and zoos, is that they are classified as exotic: foreign rather than local. These animals were also for the most part wild rather than domesticated, but their essential quality was their foreignness. Dogs and cats as well as apes and monkeys served as pets for ancient Greeks and Romans, but the latter enjoyed much higher status. Exotic animals represented power, both political and social.

At the same time, such animals also held what we would now call scientific significance. Alexander the Great collected animals during his military campaigns to gain prestige, but he also sent many to his former tutor Aristotle for analysis. If there is not a concept of conservation in the modern ecological sense in such collecting, there is a concept of these animals as rare and worthy of study.

The intertwining identities of exotic animals as status objects and as scientific objects continued through premodern history, establishing an instrumentalist perspective on animals that persists today. In other words, animals held value insofar as they were useful and beneficial to humans, whether as pets, food, or transportation or as objects of research. Exotic animals held particular value. As such, their preservation and conservation demanded particular attention, and menageries provided the conditions for their survival in the alien environments of Europe. By the twelfth century, many European monarchs had menageries filled with the spoils of the Crusades and the gifts of diplomatic exchange. The menagerie at the Tower of London began in 1235 with a gift of three leopards from the Holy Roman Emperor Frederick II to the English king Henry III. Frederick himself had three menageries. The value of these animals was in their rarity and foreignness, and also in their provenance as diplomatic gifts. The greatest prize in Henry III's menagerie was an elephant presented to him in 1255 by the king of France (Hahn 2003, 13–14). Early modern zoos continued to be the affair of royalty and aristocracy. Neither public education nor conservation in the modern sense of species survival was central to their aims. As we will see below, this changed at the end of the eighteenth century as the modern zoo developed out of its aristocratic origins. Our examples are drawn mainly from France but illustrate the wider European development of zoos. The "prehistory" of the modern zoo reveals underlying contradictions and tensions that continue to figure in modern discourse on zoos as well as in other human uses of animals. The modern, conservation-oriented zoo is a product of its history, both recent and not so recent.

THE STATUS OF ANIMALS IN EARLY MODERN EUROPE

In *The Parts of Animals* Aristotle commented, "In all things of nature there is something of the marvelous," adding, "We should venture on the study of every kind of animal without distaste; for each and all will reveal to us something natural and something beautiful" (Aristotle 1968). He studied not only the exotic animals Alexander brought back, but every animal he came across, no matter how humble or mundane. He made little distinction between wild and domesticated.

The medieval rediscovery of Aristotle's works led to the revival of natural history. Albertus Magnus's thirteenth-century *On Animals* included his own observations as well as Aristotle's. Humanist naturalists of the sixteenth and seventeenth centuries followed Albertus in considering rarity to be one among many reasons to study animals. In the 1550s, Konrad Gessner's encyclopedic four-volume *History of Animals* (history here meaning natural history) included the rare and exotic such as the rhinoceros and the camel (Albertus Magnus 1999; Gessner 1551–58), but he also described native wild animals such as foxes and domesticated ones such as dogs. Gessner included few New World animals: the guinea pig appeared in the first volume, the armadillo in a supplement. By the end of the sixteenth century, the animal trade between Europe and the rest of the world, intimately linked to exploration and colonization, brought an influx of new animals to Europe. These novel types raised questions of classification and the overall order of nature that Gessner did not address. How did new animals fit into a system that had been thought to be full? These new animals soon made their way to royal menageries.

While naturalists continued to emphasize the medical and culinary usefulness of animals (following the ancient Roman Pliny), they also began once more to recognize the value of animals as scientific objects. Historians differ on the consequences of these changes for animals themselves. In his landmark *Man and the Natural World* (1983), cultural historian Keith Thomas identified the early modern era as a time of transition in English ideas toward a new recognition of animal cognition and sensibility (Thomas 1983). Other historians have seen the mind-body division drawn by philosopher René Descartes as inaugurating an unprecedented reign of cruelty over animals, particularly in the context of science. Pointing to their lack of speech as evidence, Descartes maintained that animals did not possess a mind or soul as humans did and therefore resembled lifelike machines. Their actions were merely instinctual, and they could not experience pain cognitively. While the new science led to a great number of animal experiments, few experimenters believed animals experienced pain less keenly than humans. They experimented despite the "beast machine" notion, not because of it, and many philosophers refuted these ideas altogether.[2]

Nonetheless, in this period animals remained instruments toward human ends, whether food, entertainment, or scientific knowledge. Although cruelty to animals became increasingly frowned on over the course of the eighteenth century, concern was less for animals than for the effects of cruelty on the human soul and moral character. The English artist William Hogarth neatly summarized this point of view, derived from the thirteenth-century theologian Thomas Aquinas, in his series of engravings titled *The Four Stages of Cruelty*. Hogarth's protagonist begins his career of crime as a child by torturing a dog and ends on the anatomists' table as an executed murderer. German philosopher Immanuel Kant kept Hogarth's engravings before him in 1780 when he wrote about the moral responsibility of humans toward animals, concluding that they had none, but that cruelty could damage human moral sensibility. Less than a decade later, however, political philosopher Jeremy Bentham asserted that animals, like humans, were capable of happiness and therefore deserving of equal moral consideration with humans. Whether they could speak was irrelevant; it was plain to him that they could suffer (Guerrini 2003, 63–66).

VERSAILLES AND THE ZOO AS LABORATORY

When King Louis XIV began in the 1660s to build his magnificent estate at Versailles, outside Paris, he included a menagerie in his plans. The octagonal menagerie building was among the first to be completed, in 1664. There had long been a royal menagerie at Vincennes, just south of Paris, to supply the royal tables. A 1694 definition of *ménagerie* referred to "a place built next to a country house to fatten animals, poultry, etc." But alongside the cows and sheep at Vincennes there were also lions and elephants, many of them employed in animal combats. In the 1670s, a "valiant cow" fought off a lion and a wolf in a staged battle (Loisel 1912, 2:99).

The king's vision for Versailles included fierce wild animals that exemplified royal power (the lion, Alexander the Great's symbol, was a particular favorite) as well as smaller and less fierce but still exotic animals. Many of these came from new French colonies, while others were diplomatic gifts. Among them, several species of exotic birds modeled courtly behavior. Versailles was preeminently a place of social interactions, and animals played a central role not only in the menagerie, but also in the Labyrinth, built in the 1670s. Here, among fountains with life-sized animal sculptures, courtiers could enjoy cultured conversation (Guerrini 2015, 173–77; Mabille and Pieragnoli 2010).

Animals circulated into and out of Versailles. Louis XIV's chief minister Colbert employed a man to travel the world and collect a large number of animals; something like one hundred ostriches came to Versailles between 1687 and 1694. The collection, transport, and maintenance of these animals employed dozens. Artists drew the animals, poets described them in verse, and the octagonal central building, from which one could see the entire zoo—a panopticon—hosted lavish dinners (Mabille and Pieragnoli 2010).

Two years after the menagerie opened, Colbert inaugurated the Paris Academy of Sciences. The physician Claude Perrault soon initiated a program of human and animal dissection. Among the first animals to be dissected was a lion that died at Vincennes in June 1667, followed over the next two decades by numerous animals from Versailles. This program had multiple objectives; comparative anatomy was only one of them. Academicians regularly observed animals at Versailles and Vincennes in life as well as in death and debated their behavior and diet. They speculated on the differences between native and exotic animals and on similar animals from widely differing places. They struggled to place animals within familiar classificatory and etymological categories. They measured them and drew their internal and external parts. Some of this work appeared in the 1670s in the form of two lavish volumes, funded by the king, on the natural history of animals. These volumes glorified the king and the exotic animals in his possession while also raising serious scientific issues of anatomy and classification (Guerrini 2015, 128–64).

Was conservation among the academicians' goals? If conservation meant keeping animals alive in captivity, yes. But a broader vision of animals' lives in their own environments emerged only fitfully. Although the images of the animals in Perrault's *Memoirs for a Natural History of Animals* depicted them in natural environments—unlike earlier works of natural history—the environments were not their native ones, but Versailles itself or idealized classical landscapes. Animals from widely differing locales, such as the Canadian and Sardinian deer and Old and New World monkeys, appeared together. Their status as wild animals in confinement seldom merited comment, as opposed to their value as exotic exemplars of royal power.

Perrault recognized the uniqueness of these specimens. Uniqueness was in part a matter of circumstance: many of the species at Versailles were single specimens (such as the elephant), with at most two or three in residence at any time. Their individuality led Perrault and his contemporaries to emphasize their differences rather than their similarities. Resisting any concept of biological species, and with an imperfect understanding of extinction, Perrault could not develop an idea of conservation in the modern sense. These animals circulated in one direction: from natural environments to unnatural ones (Guerrini 2015, 160–62).

We know less about the relation between the animals at the Tower Menagerie in London and the scientific programs of the Royal Society. Animals were dissected at meetings of the society, and descriptions of some of them appeared in its journal, the *Philosophical Transactions*, but there was no organized program of research such as existed in Paris. Nehemiah Grew's catalog of the society's "Musaeum," published in 1681, included many exotic animals, either skeletal or stuffed. Grew seldom indicated their provenance, but it is likely that many of them originated in the royal menagerie; others came from private collectors or directly from abroad. The flying squirrel, for example, "was sent from *Virginia*, its breeding place" (Grew 1681, 20). The absolutist state of Louis XIV allowed the close relationship between his zoo and the Paris Academy of Sciences, a relationship that was much envied but difficult to duplicate in other lands.

EARLY MODERN ZOOS AND THE EMERGENCE OF CONSERVATION

Nonetheless, royal menageries such as those at Versailles and at the Tower of London, and a proliferation of private menageries at aristocratic estates from the end of the seventeenth century onward, played important scientific roles in the early modern era (MacGregor 2014; Robbins 2002). They greatly broadened the range of known (and seen) animals. Gessner had included in his natural history many animals he had never seen, and some of his illustrations were obviously drawn from verbal descriptions rather than direct experience (Kusukawa 2010). Perrault and his fellow academicians were scrupulous in their observation and representation of animals, and they influenced contemporary representation. As Grew's catalog shows, menagerie animals held direct connections to natural history collections. The king's elephant, which had figured in paintings and tapestries celebrating the majesty of Louis XIV, died in 1681. Her bones still reside at the Paris Museum of Natural History. Seventy years after she died, the Comte de Buffon, director of the royal botanical garden in Paris, used these same bones in the 1750s to tell a very different story about elephants in his *Natural History* (Guerrini 2012). Grew's was only one of many catalogs that detailed the numbers of exotic animals in public and private collections.

As animals circulated in premodern Europe, so too did their multiple uses and meanings over time and place and to various audiences. Concepts of species, population, and extinction developed, employing observations made in zoos and observations of museum specimens that often originated in zoos. In these ways and in the expansion of European consciousness afforded by the view of exotic animals, premodern zoos contributed to the eventual recognition by scientists and the public of the necessity for conservation.

ANIMAL PRACTICES UNDER THE REVOLUTIONARY SHADOW

We lack a comprehensive history of animal practices across Europe in the nineteenth century, but we know that significant differences in regional and national diets, scientific views, and sensibilities abounded. All three influenced the formation of menageries, zoos, and vivaria. No firm criteria differentiated these sorts of animal collections. The Versailles menagerie had been lavish and expansive, while the early menagerie at the Paris Museum of Natural History appeared to be merely a series of rooms displaying animals. The "zoological gardens" of the nineteenth century connoted somewhat more attention to animals' needs, and vivaria specifically reared animals for scientific study. The history of nineteenth-century French zoos and state-sponsored animal collections, and the collection of exotic animals, intersects with the needs of industrial society. The Industrial Revolution and a gradual movement toward rational agricultural policies in England and the Low Countries conferred greater economic value on domesticated animals. It also inspired a search for appropriate animals for naturalization, acclimatization, and domestication. Even long-domesticated animals found new uses.

Notions of the exotic and foreign, themselves concepts in transition as modern nations took form in the aftermath of the Napoleonic Wars, could be applied indiscriminately. Interest in behavior, diet, and morphology, as well as the aesthetic value and productive capacities of exotic animals, was widespread. Harriet Ritvo's chapter in this volume gives one example of the development of nineteenth-century zoo and conservation practices. The French experience provides another. The introduction of merino sheep in France in the 1780s provides a case study of utilitarian aims and conservation goals. Government interest in improving the quality of French wools led Buffon's collaborator Louis-Jean-Marie Daubenton to apply his knowledge of the comparative anatomy of quadrupeds to adapting Spanish merino sheep and other exotic breeds from as far away as Tibet to French soil and climate. Daubenton's many experimental sites included Versailles and the royal farm at Rambouillet, and Spanish merinos at Rambouillet were protected during the Revolution. Although merino sheep ate more than French breeds, Daubenton demonstrated that they also produced finer wool and could therefore enhance France's pastoral economy. Circulating throughout France, merinos were shown at agricultural fairs, and Daubenton read several papers on sheep at the Paris Academy of Sciences. His investigations led to an instruction manual for shepherds and owners, printed and distributed by the Crown, and even a school for shepherds. These efforts to develop best practices of exotic animal care likely spared Daubenton from the fate of the chemist Lavoisier and other savants who died at the guillotine. One might term this conservation, but of a decidedly utilitarian bent (Daubenton 1801).

The French Revolution affected how animals were sourced and how they circulated. As we have seen, menageries enjoyed a vogue among early modern elites (Robbins 2002; Hahn 2003, 143–64; MacGregor 2014; Loisel 1912). Animals might be sourced in a variety of ways, but chartered Crown companies such as the French East India Company provided fairly consistent links with India and Africa and transported animals until the Revolution (Lacroix 1978). During and after the Napoleonic era, the French state threw its financial support behind scientists and science, and this support encompassed the circulation of exotic animals and their study. State sponsorship of science, robust in the German-speaking states as well, further differentiated France from the scientific environment across the Channel, where gentlemen-naturalists such as Charles Darwin and Charles Lyell constituted a different professional motif. In his 1830 *Reflections on the Decline of Science in England*, the English mathematician and polymath Charles Babbage wrote with envy of post-Napoleonic France and its state-sponsored posts for science, and its astounding diversity of technical and scientific schools (Babbage 1830). The Paris Museum of Natural History became France's central node for animal studies.

ANIMALS AS TECHNOLOGY

Concerted study of the human-animal relationship went hand in hand with instrumentalist perspectives on the animal economy. If the kings were gone, the legacies of the Versailles menagerie and Buffon's royal botanical garden, now rechristened the Paris Museum of Natural History, persisted. Five months after its own foundation in 1793, the museum became home to one to Europe's first public menageries. Populated by an initial core of performing animals collected by Revolutionary authorities from the streets of Paris, the menagerie became a pivot of the new biology (Burkhardt 1997). The museum would later house a public aquarium and maintain breeding colonies of exotic animals for exchange and scientific study. One example of this network of exchange is the museum's axolotl salamander colony, started in the 1850s, which has provided animals for study throughout Europe and beyond for more than 150 years (Reiss, Olsson, and Bossfeld 2015).

Britain had led the way in the reform of agriculture in the eighteenth century by means of enclosures, selective breeding, and "improving" landowners. Despite the efforts of the physiocrats and other reformers, French agriculture remained largely traditional. Moreover, France entered the nineteenth century with a much reduced colonial presence, having lost much of its earlier empire to England, and for a time circulation of exotic animals through France dwindled. But as its empire expanded in the nineteenth century into Africa and Southeast Asia, exotic animals resumed their travels to France. Naturalists investigated exotic animal physiology and morphology and experimented with acclimatization and domestication to improve French agriculture and industry (Osborne 1994). France, in contrast to England, lacked a robust culture of cattle rearing, so several acclimatization experiments concerned exotic quadrupeds such as llamas, vicunas, yaks, and others thought to be fit candidates for incorporation into French agriculture.

Babbage had been prescient in his assessment of post-Revolutionary French science. The structure of education for animal husbandry changed with the development of zootechny, a term that first appeared in French in the 1830s, not long after the foundation of the Zoological Society of London (Desmond 1985). Zootechny, the techniques of domestication, breeding, maintenance, and improvement of domestic animals, was championed by the zoologist Isidore Geoffroy Saint-Hilaire and a host of acclimatizers. It continued Daubenton's program of research in several particulars. The French state supported it with successive academic chairs at the Institut national agronomique at Versailles and later at the Conservatoire des arts et métiers. The zoologist Émile Baudiment was the incumbent of both professorial chairs; at his death in 1863 he left three unpublished books, including one on merino sheep (Drouin 1994).

NEW BIOLOGY AND NEW ZOOS

Buffon had compared animals to slaves, yet by the nineteenth century Jean-Baptiste Lamarck's transformism had reconceived the human-animal relationship as a partnership rather than one of slavery under human imperium. Advanced civilizations, added Isidore Geoffroy Saint-Hilaire, treated animals better than did primitive peoples (Blanckaert 1992). Buffon had also asserted that living forms necessarily, and irreversibly, degenerated over time, especially when domestication or other experiences in the animal lineage were ancient. Buffon viewed New World animals as particularly degenerate and feeble compared with those of Europe, a theory Thomas Jefferson vigorously refuted in his *Notes on the State of Virginia* (1780). This notion of decline nonetheless persisted (Dugatkin 2009).

In contrast to an inevitable degeneration of types discerned by Buffon, beginning about 1800 Lamarck and the zoologist Étienne Geoffroy Saint-Hilaire proposed that animal form was highly malleable. Étienne's son Isidore and Isidore's son Albert carried his research program forward into the middle and late decades of the nineteenth century. The three men posited that degeneration might be reversible and improvement possible within limits yet to be determined. This notion continued a tenuous lineage of Revolutionary ideas and much rhetoric on the topic of regeneration of the French nation (Spary 2000). The three generations were associated with the founding of the three zoos of Paris (the Museum ménagerie, the Jardin d'acclimatation, and the modern zoo at Vincennes) (Osborne 1996).

In contrast to Lamarck's concept of limitless variability, Isidore Geoffroy Saint-Hilaire thought that variability of the organism was limited but could be amplified dramatically by domestication, acclimatization, and naturalization. "Wildness" became a biological category. Both zoologists believed that diet and the external environment changed the animal economy, but Isidore Geoffroy Saint-Hilaire thought the process might be managed and accelerated. Indeed, Isidore's father, Étienne, a founder of experimental teratology, had once poured hot wax on fertilized chicken eggs to see how it might alter embryogenesis and possibly cause a new species to emerge. The verb "to acclimatize," used by Buffon as early as 1775, gained increased usage by naturalists in the 1830s. Imbued with multiple meanings depending on the national and colonial context, it might or might not have transformist implications (Osborne 2000; see also Aragón 2005; Ritvo, this volume).

The connections between conservation and acclimatization become evident in the thought of Isidore Geoffroy Saint-Hilaire, who argued for and set in place an apparatus for the conservation and protection of wild animals well before he founded the largest and most vital of all acclimatization societies, the Société zoologique d'acclimatation (hereafter the society) in 1854. By this time France was rapidly industrializing, and Geoffroy Saint-Hilaire called for new and more efficient uses of animals, both domesticated and wild. It was not enough, he argued, to continue rearing common domestic animals when the new industrial working classes were hungry and ate too little meat. Wild species needed to be conserved and studied for their contributions to human agriculture, industry, and diet. This instrumentalist idea of animals as resource remains prominent in modern concepts of animal conservation, preservation, and bioprospecting. Geoffroy Saint-Hilaire called for an applied natural history directed primarily at exotic animals with the goal of acclimatizing and ultimately domesticating them (Osborne 1992).

The society's showpiece, the Jardin d'acclimatation, opened in 1860 on a thirty-seven-acre plot in the city's Bois de Boulogne. The design was inspired by the open layout of the London Zoo and incorporated into Baron von Haussmann's plans to renovate Paris. Many of the Jardin's animals found their way to France as gifts to the Emperor Napoléon III and his family and were housed, provisionally, at the museum ménagerie. Jardin founders sought to avoid a circus-style display of animals, envisioning instead a multifunctional institution focused on public instruction about exotic animals and their products, laboratory investigations of acclimatization theory and practice, and exotic animal commerce. Daubenton's activities inspired the Jardin's programs, and in 1864 its managers erected a statue of him with a merino ram. The institution frequently displayed animals from the French colonies and French spheres of influence, including sheep from Algeria and Morocco and Angora goats formerly the property of the defeated Algerian patriot Abd-el-Kader. Great Britain experienced an aquarium craze in the 1850s, and in 1860 the Jardin became the site of the first large aquarium on the Continent, a building displaying fourteen separate thousand-liter tanks (Osborne 1994). Later in the century, to enhance gate receipts and recognizing the inextricable links between the study of exotic peoples and animals, the Jardin hosted a series of ethnographic exhibitions of African, Latin American, and North American peoples (see, e.g., Coutancier and Barthe 2002; Hale 2002).

The initial plan for the largest zoo in modern Paris, the Parc zoologique de Paris in the Bois de Vincennes, stemmed from Isidore Geoffroy Saint-Hilaire's concern for the welfare of animals at the cramped museum menagerie. He recognized that throngs of visitors, small cages, and other factors compromised the animals' health and reproductive behavior. He and other Museum zoologists discussed rotating the animals away from permanent display, and about 1860 he proposed an annex to the menagerie at the Bois de Vincennes, an urban forest southeast of Paris. Much later, France celebrated its colonization of Algeria with the 1931 International Colonial Exposition at the Bois de Vincennes. Some thirty-three million people visited the exposition, which included animals furnished by the Hagenbeck firm of Hamburg. The museum's Society of Friends purchased the giraffes, ostriches, and other exotic beasts and transferred them to the Parc zoologique de Paris, opened in 1934. An additional collection of tropical aquatic animals, including fish from the French colonies and later a Nile crocodile, was retained in the basement aquarium at the Palais de la Porte Dorée, constructed as part of the 1931 exposition (Osborne 1996).

Aside from its role in zoo history and the exotic animal trade, the society embraced nature conservation at the turn of the twentieth century. Its president, zoologist and director of the museum Edmond Perrier, continued to view animal nature as a resource. The society's move toward conservation, parallel with Progressive Era movements in the United States, led among other things to the 1906 creation of a society of friends of the elephant. Later, the society created a system of nature reserves in France including an ornithological park in Brittany (1912), the vast zoological and botanical Camargue reserve in the delta of the Rhône River (1927) and other reserves in the Pyrénées (1935). The society also co-organized the first International Congress for the Protection of Nature (1923) and cofounded the International Union for the Conservation of Nature (1948).[3] Although its official name still includes acclimatization, the society is now known as Société nationale de protection de la nature, completing the evolution begun by Perrier. Many other scientists have rallied to Perrier's call, and the society has now become a font of modern preservationist ideology, and histories moderate its colonialist heritage (Luglia 2015). The Jardin d'acclimatation is still a zoo but is focused on children's attractions.

CONCLUSION

This brief overview, focused on France, shows that an instrumentalist concept of animals need not exclude concerns about their conservation, preservation, and well-being. Going back to Aristotle, humans valued animals in multiple ways. Exotic animals held particular social and cultural value, as their display in elite menageries demonstrated. But the animals at Versailles also held scientific value and, to some extent, gained recognition as individuals. The development of selective breeding practices gave additional instrumental value to certain animals in the nineteenth century but also drew closer attention to their environment, both in and outside zoos, and to their welfare. Modern techniques of captive breeding originated in nineteenth-century zoos, although modern motivations differ considerably from those of nineteenth-century acclimatizers. Wildness and domesticity were viewed as constantly fluctuating categories (see also Rothfels, this volume).

The multiple uses and meanings of animals in zoos also play a historical role. Much that has been written under the rubric of the modern discipline of animal studies views the premodern zoo in particular as a benighted time and place of cruelty and oppression. Cruelty there certainly was, but the premodern zoo was also a site of new knowledge about animals and new appreciation of their qualities. As we contemplate the future of zoos, the zoo as a historical and cultural artifact, with a deep past and a complex context, requires acknowledgment rather than dismissal.

CHAPTER TWO

The World as Zoo: Acclimatization in the Nineteenth Century

Harriet Ritvo

In 1826, "Zoological, or Noah's Ark Society" was the *Literary Gazette*'s headline for a brief report on plans for the menagerie soon to open in Regent's Park (quoted in Scherren 1826 [1905], 22). At least in the early years of the Zoological Society of London, the second element of that headline turned out to have a double meaning. Its primary reference (and doubtless the only one intended) was to what has mostly remained the standard "stamp collection" principle for selecting and acquiring zoo animals: at least one of each, or two if additional specimens were anticipated or desired. But as it turned out, the headline contained an allusion to Noah's mission as well as to his passenger list: As God says to Noah after the flood, "Bring forth with thee every living thing that is with thee, of all flesh, both of fowl, and of cattle, and of every creeping thing that creepeth upon the earth; that they may breed abundantly in the earth, and be fruitful, and multiply upon the earth" (Gen. 8:17 [KJV]).

On its website, the Zoological Society of London characterizes itself, along with similar institutions that opened in Paris and Dublin at the same period, as among the first "modern" and "scientific" zoos, although a visitor from the twenty-first century would be unlikely to describe them in those terms. But greatly though they might differ in mode of display and practice of husbandry, they embodied implicit (if provisional) answers to many of the questions that continue to engage the managers of contemporary zoos and aquariums. The most obvious one—which animals to include—implied prior decisions about several others. Is it rarity or charisma or foreignness or some other quality that makes an animal worthy of or appropriate for display? How many kinds should be included, and how many of each kind? Should zoos display only wild animals, and, if so, how could wild animals be distinguished from their domesticated relatives? Should interbreeding between closely related individual animals or hybridization between closely related kinds be prevented or encouraged? What, besides the entertainment and enlightenment of visitors, was the proper goal of a collection of living animals? Of course, the overlap between the strategic challenges faced by early nineteenth-century zookeepers and those faced by their contemporary equivalents is not complete. The original menagerie of the Zoological Society of London was assembled and maintained to serve the purposes of public edification and scientific research, not with any idea of species preservation or large-scale propagation—at least not of propagation in remote natural habitats. But the animals on display in Regent's Park were not the only chattels of the fledgling zoo. Even before they possessed either creatures or a place to put them, the society's founders had outlined sweeping and ambitious goals, of which one was unconstrained by the limits of their prospective acreage: "the introduction of new and curious subjects of the Animal Kingdom" to enhance and diversify existing British livestock.

THE ZOO AS FARM

In accordance with this agenda, when the zoological garden at Regent's Park opened to the public in 1828, a small area was reserved for what were termed breeding experiments (Mitchell 1929, 93). These quarters almost immediately proved inadequate and unsuitable, and in 1829 the inhabitants were moved about ten miles west to a farm in Kingston, abutting Richmond Park, which offered more space and fewer distractions while still being easily accessible from the center of London. The animals at Kingston were not different in kind (or kinds) from the animals in the menagerie, but the range of species was significantly smaller, and there was room for more individuals of each. They included various species of deer, sheep, goats, zebras, kangaroos, zebu cattle, rabbits, ostriches, emus, gallinaceous fowls (wild relatives of chickens and turkeys), ducks, and geese (Scherren [1826] 1905, 66–73). That is, allowing for a few imaginative flights of fantasy, they were close allies of the domesticated animals that occupied most British barns and fields.

They had also been selected to accommodate a much narrower human constituency than the one drawn to Regent's Park. The fellows of the Zoological Society of London were all, of course, interested in wild animals from the perspective of natural history. But many of them were also elite agriculturalists who shared more practical concerns with regard to animal breeding and husbandry, as aquarists later in the nineteenth century would share the concerns of commercial fishermen (see Muka, this volume). Indeed, the original governing council of the society was fairly evenly split between anatomists and naturalists based in the capital and gentlemanly nonspecialists of much grander social position. Its twenty-one members included a duke, a marquis, two earls (and two earls-to-be), a viscount, and several baronets, most of whom owned significant rural estates (Mitchell 1929, 26). In its 1829 Report the council defined the objectives of the farm expansively to include providing a retreat for animals that found the menagerie environment stressful and "conducting experiments in all matters relating to breeding and points of animal physiology connected therewith." But its core aim was much more focused: "effecting improvements in the quality or properties of [domesticated Quadrupeds and Birds] used for the table; and likewise in domesticating subjects from our own or foreign countries, which have not hitherto been inmates of our poultry or farm yards" (Scherren [1826] 1905, 43). This was why the zoo farm circulated a questionnaire asking poultry breeders to rate types of domesticated fowls and game birds on such qualities as beauty, productivity, and courage; why it mated zebras with donkeys to see if the resulting hybrids were better in some way; and why it responded positively to requests like this: "The Duke of Bedford presents his compliments [to the ZSL secretary], and begs to know whether he can spare him a Cereopsis male goose" (Scherren [1826] 1905, 66–69).

As it turned out, the farm's first years were its only years. Maintaining live animals was, as it continues to be, expensive, and the young society's finances were fragile. It was not feasible to sink scarce funds into an enterprise that engaged only the wealthier fellows of the Zoological Society (and not all of them) and none of the visitors who paid at the gate. (The farm was not the only casualty of the vaulting aspirations of the society's founders—another ambitious early project, a zoological museum to accommodate the remains of especially interesting or distinguished animals who died at the zoo, also came to an untimely end, although not quite so untimely.) By 1832 the society had begun to auction off the farm animals, along with a few menagerie animals considered to be surplus; they had all been sold by 1834. They did not fetch impressive prices, especially compared with the stud fees the farm had been charging—for example, a pair of Chinese geese sold for ten shillings and a wapiti doe for four guineas (Scherren [1826] 1905,

73–74). (Stud fees had ranged from five shillings for the services of a zebu to two pounds for those of a zebra (Ritvo 1987, 237), and even these were trifling compared with the fees fetched by elite bulls, rams, and stallions of ordinary breeds.)

PRIVATIZING IMPROVEMENT

Although this failure or set of failures seemed definitive, it was not catastrophic. That is, it primarily showed that the Zoological Society was not the appropriate base for such endeavors and that the market for exotic domesticates or potential domesticates was limited. Differently situated and institutionalized, both interest in the farm's agenda and attempts to realize it persisted through the nineteenth century. There were, however, a few intrinsic oddities about this agenda. It was a distinctively nineteenth-century product, but it was not widely shared by nineteenth-century practitioners of animal husbandry. From the middle of the previous century, elite agricultural improvers had been obsessed with breeding—with the development of superior breeds of livestock (among other things, of course, such as soil chemistry and farm machinery and crop rotation). This was the period when the apparatus that continues to certify excellence in domesticated animals (pedigrees, shows, breed societies, and so forth) emerged. Overwhelmingly this obsession increasingly led breeders to restrict the range of genetic contribution to prized strains, often choosing to pair only close relatives. The farm of the Zoological Society of London offered a diametrically opposed mode of enhancement. Nor were wild animals, whether exotic or indigenous, generally considered to enhance the countryside. On the contrary, Britons often pointed to the medieval extinction of such creatures (in particular, the wolf the bear, the boar, and the beaver) as an indication of the superiority of their culture; imperial officials suggested that a similar process of elimination was among the many benefits they offered their colonial subjects.

So it is worth thinking about why, at a period when purity was highly prized as both a means and an end in breeding (human as well), and when "civilization," which can be considered the human version of domestication, was widely understood as an index of national and racial superiority, the possessors of valuable strains of domesticated animals should have desired to mix their rarefied blood with that of wild animals of different species or foreign animals of different breeds (similar issues preoccupied the equine aficionados that Nigel Rothfels discusses in his contribution to this volume). And why should they have attempted to supplement the indigenous fauna of Britain (and ultimately of British settler colonies) with alien species? Why, that is, should such mixture and admixture have been regarded, at least in some circumstances, as embellishment rather than contamination?

There were many indications of continuing interest in such genetic infusions. Although the farm of the London zoo existed only briefly, it would not have existed at all had not some of its influential patrons desired to avail themselves of the stud services and surplus animals it was established to offer. And after the institutionalized support of the Zoological Society was withdrawn (as before), they continued to pursue their interests in domestication and hybridization on their own estates. Edward Smith-Stanley, thirteenth Earl of Derby, who served as the president of the Zoological Society from 1831 until his death in 1851, owned an assemblage of living animals that rivaled that of Regent's Park. But the selection and distribution of species at Knowsley Park was very different, more like that of the defunct zoo farm than that of the menagerie. Indeed, the earl's massive collection of exotics represented a hypertrophy of his early interest in game birds and dog breeding; it was intended to serve utilitarian purposes as well as scientific ones. Thus he produced hybrids between several Asian pheasant species and the local pheasant population (also ultimately derived from Asia, but at a much earlier period), and among all four species (two domesticated and two wild) of South American camelids. He maintained a breeding herd of elands, the largest and meatiest of the antelopes, and therefore a persistent object of gustatory desire. Unfortunately, at least for the inhabitants of the menagerie and their admirers, his son's interests ran more to politics than to zoology (he was soon to become prime minister), and he auctioned off the animals as soon as possible after his father's death. The coverage of the auction makes it clear that the menagerie prioritized creatures that people (at least some people) would like to eat (Fisher 2002, 87, 102, 114, 116).

At the end of the century, another president of the Zoological Society, Herbrand Russell, eleventh Duke of Bedford (who served even longer, from 1899 to 1936), made this inclination more explicit. He also accumulated an impressive collection of exotic wild animals, almost exclusively ungulates (and a few other grazers, like kangaroos and wallabies), including various deer, goats, cattle, gazelles, antelopes, tapirs, giraffes, sheep, zebras, llamas, and asses. A summary census printed in 1905 made it clear that the duke collected with a view to acclimatization. It explained that animals that appeared temperamentally unsuitable for domestication were summarily dispatched, while nature was allowed to take care of those that turned out not to thrive in the English climate ("A record of foreign animals . . .").

ORGANIZING ACCLIMATIZATION

As was the case in many spheres of endeavor, during the nineteenth century such heroic individual efforts were increasingly subordinated to public or institutional ones. Impressive as his menagerie was, for example, the Duke of Bedford did not deal with the Zoological Society of London as an equal partner as his predecessor the Earl of Derby had done six decades earlier. The institutions that emerged in response to this persistent desire to integrate attractive exotic attributes into domestic barns and bloodlines were not zoos per se; at least, that is, they labeled themselves as acclimatization societies, not as zoological gardens. (Of course, like many distinctions, this one becomes fuzzier the more it is contemplated.) In their heyday—the second half of the nineteenth century—there were a lot of them, scattered across Europe as far to the east as Russia, across North America, and throughout the colonized world.

The British version, the Society for the Acclimatisation of Animals, Birds, Fishes, Insects and Vegetables within the United Kingdom, was established in 1860. It had a tenuous connection to the Earl of Derby, having been inspired by a zoological dinner held at a London tavern in 1859, at which the gathered naturalists and menagerists enjoyed the haunch of an eland descended from the his herd. The declared objects of the society were grandiose and diffuse: to introduce, acclimatize, and domesticate "all innocuous animals, birds, fishes, insects, and vegetables, whether useful or ornamental"; to perfect, propagate, and hybridize these introductions; to spread "indigenous animals, etc." within the United Kingdom; to procure "animals etc., from British Colonies and foreign countries"; and to transmit "animals, etc. from England to her colonies and foreign parts." (Very similar language had been included in the prospectus of the Zoological Society of London—another indication of the permeable boundaries between these institutions.) Unsurprisingly, the Acclimatisation Society made little progress toward achieving these noble goals. Despite their ambitious founding proclamation, its members mostly confined their attention to birds and sheep, to very little effect. The society itself survived only through 1866, when it was absorbed by the Ornithological Society of London and vanished without a trace (Lever 1977, 29–35; Ritvo 1987, 239; Lever 1992). The American Acclimatization Society, founded in 1871, was similarly unsystematic and capricious in its activities and similarly evanescent. It did, however, leave a more persistent historical footprint, since its release of English starlings proved to have massive and unforeseen environmental impact.

It is likely that these two societies were weak because they were virtual. That is, they consisted only of human members, who often felt free to fantasize unconstrained by concrete considerations. More robust acclimatization societies controlled both animals and real estate. The French Société d'acclimatation possessed both of these advantages, and it also enjoyed government support and more distinguished scientific sponsorship (for example, the Geoffroy St. Hilaires had a family interest in improving domesticated animals through naturalizing and interbreeding—they far outweighed the eccentric Englishman Francis Buckland in gravitas). Founded in 1854, it immediately began holding monthly meetings and publishing the proceedings (Anita Guerrini and Michael A. Osborne offer a more detailed discussion of French animal collecting in their chapter in this volume).

Along with more wide-ranging introductions, hybridization was an early preoccupation of the members. The first issue of the society's *Bulletin* included several articles about the virtues of angora goats and other long-haired varieties that lived in Central and Southwest Asia; the author of one of them earnestly recommended that the society "procure . . . animals of the most desirable races . . . in order to try to acclimatize them and study the results of crossing" with French varieties (Sagra 1854, 30). This was a particular theme to which members returned repeatedly over the months and years; a typical article was titled "On angora goats and cashmere goats, and the advantage of crossing those races" (Bourgeois 1854, 268). And more generally (that is, focusing on other mammals and birds), many members pursued experiments in hybridization. At the society's annual prize giving in 1911, the president Edmond Perrier (a zoologist best known for his study of invertebrates) praised the recipients for having "created the most beautiful and productive races, the best adapted to our needs, which have become for us a source of profound joy" (Perrier 1911).

Before the society was a decade old, its "Jardin zoologique d'acclimatation" opened its gates in the Bois de Boulogne. As its name suggests, it was conceived as a hybrid institution. It included the attractions that had become standard for a zoological garden and were consequently required by the general public—big cats, elephants, and other iconic animals. But these constituted only a part of its collection, and not the most important part, at least in theory or in principle. Its ostensible core mission was much more pragmatic, distilled in the term "applied zoology." It was intended as a laboratory for the study of acclimatization, and its priorities were distinctively utilitarian. This characteristic emphasis helped determine both its acquisition policy and its exhibition policy. Its initial displays emphasized the economic potential of animals from French colonies (Algeria turned out to be a special preoccupation, as both a source and a target of acclimatizable animals), and it housed the largest collection of exotic agricultural animals in Europe (Osborne 1994, 98–129). Unlike its English and American analogues, it lasted into the twentieth century. But although its animals and its data were available to all French agriculturalists, it did not effect any noticeable transformation or enhancement in either the wild or the domesticated fauna of their nation.

The most persistent and effective acclimatization societies were established on the other side of the world, in the British colonies that became Australia and New Zealand. There was no discernible time lag in the transplantation of these metropolitan institutions to the antipodes, nor was there much modification of the acclimatization agenda in acknowledgment of environmental differences. In 1861 the Acclimatisation Society of Victoria described its objectives in much the same language that had been used in Britain: "the introduction, acclimatisation, and domestication of all innoxious animals, whether useful or ornamental; the perfection, propagation, and hybridisation of races newly introduced or already domesticated, [etc.]"; and this language was subsequently borrowed by similar societies elsewhere in Australia and New Zealand (Acclimatisation Society of Victoria 1861). These colonial acclimatizers felt strongly aware of their predecessors, and they were determined to surmount the obstacles that had impeded earlier attempts. Thus a hortatory lecture delivered in Sydney regretted that the farm of the Zoological Society of London was "either from the expense or some mismanagement . . . given up," after introducing only three species of geese and a few other minor birds, and warned that the dispersal of the Earl of Derby's stock "must be a sufficient lesson of the uncertainty of private collections and the necessity of founding societies based on a general and public interest" (Bennett 1846, 4, 40). (They shrewdly took as their model the French acclimatization society rather than the English one.)

The range of mooted targets was, however, as ambitious and fanciful as it was in Britain—including babirusas (an Indonesian relative of the pig) and giraffes, as well as the more plausible antelope, deer, sheep, and goats. As in France, angora and cashmere goats were objects of special fascination. And as in France, at least some of these societies had their own premises. Within a few years of its foundation, the Acclimatisation Society of Victoria owned camels, llamas, alpacas, several species of goats, sheep, and deer, hares, and various kinds of birds and fish ("Acclimatisation Society of Victoria" [1864]). The focus on eating was equally strong. As in Europe, it was hoped that antelopes would offer a change from the monotony of beef and mutton and that the appealingly large South American curassow could supplement more pedestrian fowl (Bennett 1862, 13–16, 24).

Even the Australian fauna, which were routinely derogated by acclimatizers on both utilitarian and aesthetic grounds, could be drafted to serve in this campaign. Adventurous eaters claimed that "the flesh [of the wombat] is always . . . a great treat"; "the opossum is good . . . especially when curried or stewed"; and "the monitor lizard . . . , if one could overcome the repugnance of its appearance, is delicate and excellent food" (Bennett 1862, 19). A celebratory dinner held in a Melbourne hotel in 1864 was described as a "fitting tribute to the cause of acclimatisation . . . [and] also a complete triumph of gastronomy. . . . Among the *entrees* were curries, *pates* and *salmis* in which wombat, bandicoot, and parroquet figured conspicuously" ("Acclimatisation Society's Dinner" 1864).

Despite the eccentric tone of much acclimatization-related discourse (and behavior, for that matter), in both Australia and New Zealand acclimatization societies received public support, albeit inspired by varying degrees of official enthusiasm and inspiring varying degrees of taxpayer outrage. At least in some places, they were endorsed by God as well as by Caesar; thus parish clergymen in Victoria were encouraged to persuade their flocks that "a society which multiplies . . . the gifts of an All Bountiful Creator . . . is worthy of the support not only of the Philosopher but also of the Christian" (Wilson 1864).

In both places the acclimatizers also had to deal with mounting suspicion that some of their introductions were doing more harm than good, so that, for example, the Australian societies repeatedly swore they were not to blame for what had quickly become a plague of rabbits. Even the journal *Nature* chimed in from twelve thousand miles away, stating that "the English Acclimatisation Society fortunately came to an end before it could do any harm here; but its example has been mischievous in our dependencies" (*Nature* 1872, quoted in Wellwood 1968, 24). Of course, that was putting it mildly. The faunas of both Australia and New Zealand have been radically altered over the past two centuries. This was not completely the fault of acclimatization societies—much of this transformation resulted from the operation of forces that Alfred Crosby has labeled "ecological imperialism" and (with a different geographic focus) "the Columbian exchange"—that is, the effect of transplanting an integrated European system of agriculture. But the deer and stoats of New Zealand, and the rabbits and foxes of Australia, among many other kinds, constitute a continuing memorial to the aspirations and inclinations of their nineteenth-century introducers.

THE CALL OF THE "WILD"

Exploring nineteenth-century attempts to revivify extant faunas by introducing new species or by enhancing existing ones with exotic blood, whether sponsored by zoos or in other contexts, reveals a lot of inconsistency and confusion. But two general points do emerge from this story. One is the persistent appeal of the alien and the wild, even when the dominant consensus suggests otherwise. And the other is the difficulty of deciding what is wild and what is alien. Britain offers many examples of these complicated or confused understandings: for example, the allegedly wild white cattle of Chillingham Castle in Northumberland (really just a herd that continued to experience a medieval mode of husbandry) and the allegedly indigenous Herdwick sheep of the Lake District (which resemble many other British upland varieties and, like them, descend from a species indigenous to the eastern Mediterranean region). (The British were not alone in having difficulties establishing such boundaries, as Ben A. Minteer shows in his discussion of the "zoo" that existed briefly in Yosemite National Park [this volume].)

Of course the somewhat (but not too much) subversive call of the wild (or of other threatening attributes) has been frequently recognized, in the nineteenth century and in other periods. It accounts for much of the appeal of the gothic novel and its successors, and of men who are "mad, bad, and dangerous to know." It is the reason nineteenth-century travelers, like their twenty-first-century successors, admired stark mountain landscapes and stormy seas. Of course they, like us, were more inclined to admire and enjoy such thrills if they occupied a position of safety, such as has traditionally been provided by the vicarious experience of reading, or such as has been increasingly provided by technological defenses against natural threats. The conventional zoo or menagerie provided a low-tech example of such defenses—that is, dangerous creatures were confined behind metal bars. The activities of acclimatization societies, especially those that put their theories into practice, but even those that mostly just talked, offered a more nuanced version. The exotic and wild was converted into the domestic or domesticable, at least notionally.

This transmogrification may help make sense of one of the most immediately striking or puzzling features of the lists of desiderata repeatedly promulgated by acclimatizers. They are notably miscellaneous—it would be difficult to derive a coherent sense of the faunal assemblage that was the ultimate goal from analyzing them. Perhaps another way of saying the same thing is that there is often a radical disjunction between the pragmatic goals espoused by the acclimatizers and the species they suggest as the means of realizing those goals. To give an example, an Australian society suggested the following targets for acclimatization and domestication: kangaroos, kangaroo rats, the bandicoot, the echidna, the emu, the bustard, hares of several species, the chinchilla, antelopes of several species, wild sheep of several species, the hedgehog, the robin, and the hyrax (a diminutive relative of the elephant) (Acclimatisation Society of Victoria 1864, 7–16). (And this example shows relative restraint—some acclimatization advocates dreamed of enhancing their homelands not only with purely ornamental animals, but with large exotic carnivores.)

Putting it this way suggests that acclimatizers were at least unrealistic and possibly foolish. But their wish lists gave concrete expression to an ambiguity that also bedeviled sober naturalists. Identifying species—that is, the limits of species—had always been both essential and problematic. The conventional definition of the boundary between similar organisms—the ability to produce fertile offspring—was clearly disregarded by many animals (and even more plants). Nineteenth-century zookeepers enjoyed experimenting with interspecies and intergenus hybrids, and zoogoers admired the resulting hybrids between, among others, donkeys and zebras, domestic cattle and bison, and dogs and wolves (Ritvo 1997, 92–25). This ambiguity has been distilled in the classification of domesticated animals. That is, none of them has become sufficiently different from its wild ancestor to preclude the production of fertile offspring, and some mate happily (or at least effectively) with more distant relatives. Despite these persuasive demonstrations of kinship, however, from the eighteenth-century emergence of modern taxonomy, classifiers have ordinarily allotted most types of domestic animal their own species name (rabbits and pigs are exceptions). Domestic sheep are still classified as *Ovis aries* while the mouflon is *Ovis orientalis*, and dogs are still classified as *Canis familiaris* while the wolf is *Canis lupus*. This practice has much to recommend it in terms of convenience, but it also constitutes a simultaneous acknowledgment both of the difficulty of distinguishing between wild animals and domesticated ones and, nevertheless, of the felt importance of doing so.

The implications of making or not making such distinctions extend beyond the intellectual realm. As the activities of nineteenth-century acclimatizers show, they construct the physical world at the same time that they describe it. And they still do so. The advent of DNA analysis in recent decades has made it both easier to distinguish between domesticated animals and wild ones and more difficult. For example, the Scottish Wildcat Association was established in 2007 to protect the small remaining British subpopulation of the very widely distributed species ancestral to domestic cats. (Of course, that such creatures are considered worthy of protection signals a distinctively modern valuation of wild animals; Victorian gamekeepers hunted down the ancestors of these animals and nailed their skins to their barn doors.) The targeted felines strongly resemble domesticated tabbies, although they tend to be larger and more irascible. Perhaps for this reason, the distinction between pure wild animals and those contaminated by miscegenation features prominently on its website: "The latest surveys suggest less than 100 individuals survive in the wild. Numbers originally decreased due to deforestation and human persecution, but today the primary threat is cross-mating with feral domestic cats, a process called hybridisation. This gradually waters down the true wildcat genes" (Scottish Wildcat Association website, accessed June 26, 2016). The association advocated "improving legal protection, launching a public awareness campaign, supporting the captive breeding program and creating special reserves for wildcats which would in turn benefit many other species." The Scottish wildcat was declared a "priority species" (in Scotland). In an ironic gloss on the efforts of nineteenth-century acclimatizers, it therefore became eligible to benefit from the establishment of a studbook, a captive breeding program, and other measures that blur the cultural boundary between wild and domesticated even as they attempt to reinforce the genetic boundary that separates them.

PART II

The Rise of US Zoo and Aquarium Conservation in the Nineteenth and Twentieth Centuries

CHAPTER THREE

Historic and Cultural Foundations of Zoo Conservation: A Narrative Timeline

Vernon N. Kisling Jr.

Keeping wild animals in captivity has a long history, extending back to the ancient civilizations of Mesopotamia, Egypt, China, and India as well as those of Central and South America. However, the zoo as we know it today emerged out of the nineteenth-century menagerie. The time of these menageries was also a time of exploration and discovery of new species, the maturation of natural history into professional science, an increase in the public's knowledge about animals and nature, and an increased use of natural resources, resulting in the decline of wildlife habitats. The social context also led to menageries' becoming cultural institutions with a focus on recreation and education. These menageries operated as living museums with a level of design and husbandry commensurate with the level of knowledge at the time. Displaying specimens of each species was appropriate and important as the public and naturalists alike learned about the abundant diversity of life in the nineteenth-century world (Bell 2001; Kisling 2000a).

As modern zoos (here including zoological parks/gardens, safari parks, aquariums, oceanariums, aviaries, and related facilities) emerged in the twentieth century, conservation and science became functions that supplemented the traditional focus on recreation and education [see the contributions by Barrow and by Henson, this volume]. This chapter therefore emphasizes twentieth-century zoo accomplishments. These conservation and scientific achievements had modest beginnings at the start of the twentieth century because they were still not significant concerns at the time, but they intensified as conservation became an important social issue and wildlife habitats rapidly disappeared, especially at midcentury, after World War II.

Though several particular zoo conservation milestones are presented here, many other institutional and personal efforts also existed.

The first serious effort to explain captive wildlife husbandry came with the publication of *A Handbook of the Management of Animals in Captivity in Lower Bengal* in 1892, written by Ram B. Sanyal, director of the Calcutta Zoo. It covered housing, diets, breeding, health, general care, and natural history of the zoo's 241 mammal and 402 bird species. However, because an Indian wrote it about a non-European zoo it was largely ignored.

Pere David's deer was saved from extinction in the 1890s to early 1900s when the eleventh Duke of Bedford at Woburn Abbey brought together eighteen animals from various European collections for a captive breeding program.

The Lacey Act of 1900 provided a national effort in the United States to enforce many existing state laws protecting wildlife—primarily game species. Although the law was a modest effort that controlled interstate commerce of these protected species, future expansions of the act provided substantial wildlife protections, including regulating the acquisition, trade, and transport of wildlife among zoos (see Bean 1983 for more information on all the wildlife laws).

The American Bison Society was established in 1905 to protect the bison, which had become extinct in the wild except for a small number of plains bison in Yellowstone National Park and a few wood bison in Canada. William Hornaday, director of the New York Zoological Park, led the effort to create this society and its bison reserves (as discussed by Barrow, this volume, and Henson, this volume). Bison from his zoo, other zoos, and private herds provided the stock for these breeding efforts, which eventually succeeded in reestablishing herds, though whether they are the same as the original wild herds is a valid question, as is the question whether that matters and why (see figure 3.1; also Rothfels, this volume; Barrow, this volume).

The Philadelphia Zoo established the Penrose Research Laboratory in 1904 to study wildlife health, nutrition, and husbandry. It produced some of the first zoo feeds that incorporated the necessary nutrients for each species and led the way in veterinary care for tuberculosis, a very important advance because of the loss of animals to TB in early zoos (as discussed in several chapters in this volume).

Although several zoo magazines had been published since the mid-nineteenth century, in 1907 the New York Zoological Society began publishing a scientific journal, *Zoologica*. It presented research conducted at the zoo as well as the society's field studies. It was discontinued in 1973, but in

FIGURE 3.1. "The Herd, 1860." Former herd of bison as drawn by M. S. Garretson based on an eyewitness account describing a valley one mile wide and many miles long completely filled with bison. DeGolyer Library, Southern Methodist University.

1982 it was replaced with the commercially published *Zoo Biology*, which covers research from all zoos.

William Hornaday's *Our Vanishing Wild Life* appeared in 1913 (fig. 3.2). This book assessed wildlife conservation at the time and is considered by some as a precursor to the more modern Red Data Books (see also Barrow, this volume; Henson, this volume). Although Hornaday served as director of the New York Zoological Park, his book did not discuss the role of zoos in conservation. He mentions only three zoos: the National Zoo, the Philadelphia Zoo, and his own zoo. He concludes that only his zoo was doing conservation work, but he does not mention what those conservation efforts were. Hornaday discusses efforts to improve native wildlife living freely on the zoo grounds but says nothing about propagating the species kept in the zoo collection, how the zoo would conserve these species in the wild, or how the zoo would help save the endangered species listed in his book.

The last passenger pigeon died in captivity at the Cincinnati Zoo in 1914, and the last Carolina parakeet died there in 1918 (see Barrow, this volume). These birds did not attract the same attention from zoos or the public as the bison did (fig. 3.3). Conservation was not yet important enough to include all species; it covered just the more impressive megafauna.

The New York Zoological Society established a field research station in British Guiana in 1916. While not specifically meant for conservation pur-

FIGURE 3.2. William Temple Hornaday (1854–1937). Hornaday was the first director of the New York Zoological Park, and he also established the National Zoo and founded the American Bison Society, which reestablished the wild bison herds in the United States. Smithsonian Institution Archives, record unit 95, box 13, folder 39. Public domain image.

poses, it was an early effort by a major zoo to embrace science and in situ field studies.

Compilation of wisent (European bison) studbook data began in 1923 as this species became extinct in the wild, the first such effort for a wild species. It was published in 1932 by a consortium of European zoos to track heredity and to coordinate breeding efforts by the participating zoos. The second studbook was published in 1959 for the Przewalski's horse, another species that was also extinct in the wild (see Rothfels, this volume).

The American Association of Zoological Parks and Aquariums (AAZPA, now AZA) began in 1924 as an affiliate of the American Institute of Park

Executives (later the NRPA). AAZPA had a Conservation of Wild Life Committee from the start. Hornaday was appointed chair of this committee, but he declined the position and felt that such a zoo organization was unnecessary. This lack of cooperation among some zoos impeded the progress and good intentions of the AAZPA for many years.

The Fish and Wildlife Coordination Act of 1934, with major revisions in 1946 and 1958, was an early effort to protect and conserve nongame wildlife species. It was one of the first efforts to recognize that many wildlife species were important and needed protection, not just the game species.

The International Union of Directors of Zoological Gardens (IUDZG—now the World Association of Zoos and Aquariums, WAZA) was established in 1935. The International Union for the Conservation of Nature (IUCN—now the World Conservation Union) was established in 1948. This organization created the Species Survival Commission (SSC) in 1949 and the Conservation Breeding Specialist Groups (CBSG) in 1980. The World Wildlife Fund (WWF) was established in 1961.

Red Data Book is a generic term used for publications that assess the conservation or endangered status of species. The term was first used in the

FIGURE 3.3. Former flock of passenger pigeons. Illustrated Sporting and Dramatic News, July 3, 1875. Public domain image.

1960s when the IUCN published books under this title, with red bindings. However, precursors existed: Hornaday's *Our Vanishing Wild Life* in 1913, three publications of the American Committee for International Wildlife Protection (two for mammals in 1942 and 1945 and one for birds in 1958), and the International Council for Bird Preservation's publication on the status of birds in 1958. Many Red Books/Lists have since been published at the international, national, and regional levels.

An exhibition on the need for wildlife conservation was held at the Muséum national d'histoire naturelle in Paris in 1955 and an international symposium on habitat destruction in 1956 resulted in the publication of *Man's Role in Changing the Face of the Earth*. Appropriately, this symposium recognized the work of George P. Marsh, who in 1864 published *Man and Nature, or Physical Geography as Modified by Human Action*, widely considered one of the foundational texts of the American conservation movement (fig. 3.4).

FIGURE 3.4. George Perkins Marsh (1801–82). Marsh wrote Man and Nature (1864) and was also the US ambassador to Italy, US minister resident to the Ottoman Empire, and a congressman from Vermont. Public domain image.

FIGURE 3.5. Rachel Louise Carson (1907–64). Carson, a marine biologist, wrote *Silent Spring* (1962) as well as other books about the environment. US Fish and Wildlife Service National Digital Library. Public domain image.

None of this, however, had the impact of Rachel Carson's *Silent Spring*. Carson's book appeared in 1962 (fig. 3.5). It assessed the effect of pesticides and pollution on the loss of wildlife species and had a significant influence on the growing environmental movement. It was part of many social changes that occurred because of World War II and that culminated in popular movements during the 1960s. These developments included the widespread public awareness of environmental degradation, loss of wildlife habitat, and the plight of endangered species, and they also opened the way in the United States for new environmental legislation, important wildlife conservation efforts, and major changes in zoo husbandry and management in the 1970s.

Arabian oryx were exterminated in the wild in 1972; however, the beginning of a new world herd was established at the Phoenix Zoo in 1962 (see Allard and Wells, this volume) with the contributions of animals from private collections in Saudi Arabia and Kuwait.

Heini Hediger published *Man and Animal in the Zoo* in 1963 (in German, with a US edition in 1969). This book established the scientific perspective of captive husbandry known as zoo biology.

The US Environmental Protection Agency was established in 1970. The first Earth Day occurred on April 22, 1970, and it has since been an annual event. The Laboratory Animal Welfare Act was passed in 1966 and expanded in 1970 to include animals exhibited in zoos. This revised Animal Welfare Act has had the greatest effect on zoos, since it requires their licensing and determines the management criteria for species kept in zoos.

AAZPA separated from NRPA to become an independent organization in 1972. This increased the professionalization and cooperation of zoos. Most important, it developed the accreditation of zoos with standards that surpassed those of the Animal Welfare Act, and it developed more effective conservation programs.

The US Marine Mammal Protection Act of 1972 passed, as did the US Endangered Species Act of 1973, which expanded the powers of the Endangered Species Preservation Act of 1966 and the Endangered Species Conservation Act of 1969. The Convention on International Trade in Endangered Species (CITES) was created in 1973. These laws and regulations covered animals in captivity as well as in the wild.

The Chengdu Giant Panda Breeding Research Station, begun in 1973 at the Chengdu Zoo, was the first of several now found in China.

An International Species Information System (ISIS), created in 1973, maintained zoo animal collection records. This greatly improved zoo animal record keeping, established accurate heredity information on captive animals, and promoted more coordinated breeding programs among the zoos.

The National Zoo established its Zoological Research Division in 1965 and its Conservation and Research Center (now the Smithsonian Conservation Biology Institute) in 1975. The San Diego Zoo established its Center for Reproduction of Endangered Species (CRES—now the Institute for Conservation Research) in 1975 (Tubbs, this volume) and began its Frozen Zoo program in 1976 (Ryder, this volume). The Cincinnati Zoo, the Kings Island Wild Animal Habitat, and the University of Cincinnati formed the Center for Research of Endangered Wildlife (CREW) in 1981.

Species Survival Plans (SSP) began in 1981 as AZA cooperative population management and conservation programs for specific taxa, starting with the Siberian tiger, golden lion tamarin, Asian wild horse, gaur, barasingha, and Bali mynah.

The AZA Field Conservation Committee was established in 1991 to coordinate in situ (in the wild) with ex situ (in captivity) conservation efforts within the zoo community. It also recognized that, to be effective, zoos need to cooperate with other types of conservation efforts.

Zoos as modern Noah's arks involve a concept that gained popularity in the 1990s. This idea viewed zoo collections as resources for captive breeding and reintroduction to the wild, therefore preserving species for the future even if they become extinct in the wild. Previously this term was used to describe zoos as repositories for individual specimens of as many species as possible for display purposes only: the stamp collection concept that resembled the biblical Noah's ark. The modern concept is a reflection of the changes that have taken place in zoos over the years (Norton et al. 1995; Wemmer 1995). These changes, in turn, are a reflection of public expectations as the public has become more knowledgeable about wildlife and the need to conserve endangered species (see the chapters in this volume by Norton and by Cerezo and Kapsar).

The *World Zoo Conservation Strategy* published in 1993, with revised editions in 2005 and 2015, states that wildlife conservation "can only be achieved through the awareness of all nations, including all strata of their societies, their governments, other institutions and organizations. This also includes zoos and aquaria found in nearly every country of the world. . . . Until now, however, the magnitude of the collective potential of the global zoo and aquarium community has never realistically been expressed" (World Zoo Organization 1993, x). Such is the status of zoo conservation at the end of the twentieth century.

CONCLUSION

As the evolution of zoo conservation continues into the twenty-first century, we need to determine the effectiveness of these efforts, how to better integrate them into the overall global efforts, and how to realistically evaluate where we go from here. The *World Zoo Conservation Strategy* is a guide for what needs to be done (Barongi, this volume). The potential is there, but only implementation will bring effective action. While we can be critical of past zoo conservation efforts, we also need to be critical of all conservation efforts. The real concern is whether any of these attempts have lasting value, or whether they have merely slowed down the inevitable. As the human population continues to increase and wildlife habitat continues to decrease, we will have limited success in saving wildlife in the wild. If we can successfully conserve wildlife and preserve its genetic diversity in captivity, then our zoos may actually become a modern Noah's ark, a wilderness within the zoo rather than the other way around.

The problem is how to do this realistically when the limitations of zoos are well known. Biblical scholars who studied Noah's ark and its capacity to save the animals had the same problem as the number of new species discovered during the age of exploration continued to increase. Our Earth is the Noah's ark of last resort, but soon its capacity will also be limited, and the last of its wildlife will be managed in exhibits we call national parks. Earth will always be accepted as an ark, but it cannot prevent the loss of its wildlife species. For zoos to be arks that both house and conserve wildlife, they must do more.

Historically zoos, like the ark, have been physical structures. Instead, the twenty-first-century zoo should be a natural space that emphasizes its wildlife, not the institution that houses the wildlife. Visitors should be able to appreciate the wildlife without disliking the institution managing it. Zoos need better captive breeding strategies for conserving genetic diversity, both in living animals and in the laboratory. And zoos need to be more than a space tied to the local community. They must support natural habitats worldwide through cooperation with those managing existing habitats and through the purchase of additional habitats, in order to better integrate their ex situ and in situ efforts.

Menageries transformed into zoological parks and then into conservation centers. These concepts are simply incremental improvements on man-made structures with modest advances in conservation, research, and education: cultural institutions that have been living museums. Moving beyond these traditional boundaries will not be achieved easily, but doing so is important. Wildlife in these ex situ places and also their native in situ places must be conserved both for its own sake and so that the quality of human life will continue.

CHAPTER FOUR

Teetering on the Brink of Extinction: The Passenger Pigeon, the Bison, and American Zoo Culture in the Late Nineteenth and Early Twentieth Centuries

Mark V. Barrow Jr.

INTRODUCTION

The emergence of a broad wildlife conservation movement and the establishment of the first modern zoological gardens were among the many profound transformations in human-animal relations that occurred in the United States at the end of the nineteenth century and beginning of the twentieth (see, e.g., Kete 2007; Grier 2006; McShane and Tarr 2007; Turner 1980).[1] As the nation became increasingly urbanized and industrialized, a coalition of sportsmen, naturalists, and nature lovers banded together not only to raise public consciousness about the increasingly desperate plight of America's native fauna but also to lobby for protective legislation. Sportsmen were among the first to embrace the conservation cause when they pursued a long series of state laws aimed at safeguarding their continued access to declining game birds and mammals through hunting licenses, closed seasons, and bag limits (Tober 1981; Reiger 2000; Herman 2001).[2] Ornithologists and nature lovers focused on the problem of shrinking bird populations and, working through newly organized Audubon societies, secured state laws protecting nongame species across much of the nation (Barrow 1998; Graham 1990; Doughty 1975). These groups joined forces to oppose market hunting and to gain passage of the first federal wildlife legislation: the Lacey Act of 1900, which prohibited interstate transportation of birds and mammals taken in violation of state laws, and the Migratory Bird

Treaty Act of 1918, which codified a treaty between the United States and Canada (then a part of Great Britain) to protect both game and nongame birds that moved seasonally (Cioc 2009; Dorsey 1998).

This flurry of wildlife protection occurred just as the first modern zoos were opening in the United States. The inspiration for these publicly accessible collections of wild animals was the zoos that European cities had begun establishing with increasing frequency over the course of the nineteenth century with an eye toward projecting imperial power, promoting civic pride, importing economically valuable species, producing scientific knowledge, and entertaining and educating visitors (Ritvo, this volume; Ritvo 1987; Rothfels 2002; Baratay and Hardouin-Fugier 2002; Hoage and Deiss 1996; Kisling 2000a). Across the Atlantic, menageries and traveling animal shows had long been popular with American audiences. It was not until 1874, however, that a group of naturalists and civic-minded citizens opened the first large, permanent public zoo in the United States, the Philadelphia Zoological Gardens, a forty-two-acre facility in Fairmount Park that initially housed just under a thousand animals, including 282 mammals, 674 birds, and 8 reptiles (Kisling 2000b, 150–52; Howard 1879). Other American cities quickly followed suit, creating sixty more zoos over the next six decades (Henson, this volume; Kisling 2000a; Hanson 2002; Bender 2016).

By the end of the twentieth century, the ex situ conservation of endangered species would become a prominent part of the mission for many American zoos. The California condor, Arabian oryx, black-footed ferret, and red wolf are several widely touted conservation success stories where zoos hosted captive breeding programs that helped revive species once teetering on the brink of extinction (Tudge 1992; DeBlieu 1991; Rabb 1994; Koebner 1994; Allard and Wells, this volume). But a century or so earlier, when public zoos were just beginning to get off the ground in the United States, they remained more focused on educating and entertaining the public than on addressing wildlife decline. With only a couple of exceptions explored in this chapter, nineteenth- and early twentieth-century American zoos remained largely indifferent to threatened fauna. Ironically, natural history museums, which rose to prominence in the United States about the same time as zoos but exhibited dead specimens rather than living animals, generally proved more responsive to wildlife decline during this period. In building the collections that supported the creation of an authoritative published inventory of the world's fauna, museum curators developed both detailed knowledge about the fate of wildlife and a strong desire to counter the threat of extinction.[3]

This chapter explores why the first modern American zoos failed to play a larger role in the turn-of-the-century wildlife conservation movement by focusing on the stories of two iconic animals that proved central to that movement: the passenger pigeon and the bison. Both were once superabundant species with geographic ranges that spanned much of the North American continent, and both experienced nearly simultaneous population crashes in the second half of the nineteenth century. Although many factors contributed to these unprecedented declines, voracious market hunting, which was made possible by the rapid growth of telegraph and railroad networks linking the nation's urban markets to its vast rural hinterlands, proved especially critical. The nearly coincident collapses of America's passenger pigeon and bison populations sounded the alarm about the myriad threats facing wildlife more generally, offered powerful warnings about the dangers of insufficiently regulated market hunting, and provided fodder for those arguing that the nation needed to reexamine its deeply entrenched ideas about the inexhaustibility of nature (Barrow 2009, 78–134). As these two species became increasingly rare in the wild, America's newly established zoos sought to acquire them both, but only the bison would become successfully incorporated into a pioneering zoo-based captive breeding program that helped pull the species back from annihilation. In the absence of the same level of concern, support, and commitment, the passenger pigeon ultimately fell victim to extinction.

THE PLIGHT OF THE PASSENGER PIGEON

Known to science as *Ectopistes migratorius,* the passenger pigeon was a migratory species that numbered as many as three to five billion birds when Europeans first arrived on North American shores (Schorger 1955; Greenberg 2014; Blockstein 2002; Cokinos 2000, 195–278). Arguably, it was one of the most abundant birds in the world. In addition to its rapid flight, a characteristic feature of the species was its tendency to gather in massive flocks that could blacken the skies for days at a time, provoking awe and wonder in those who witnessed them. Native Americans and European settlers both held celebrations to mark the periodic arrival of "pigeon years," when enormous flocks descended on an area with a large enough crop of beech, oak, or chestnut mast, the passenger pigeon's primary food (Price 1996, 1–56). In addition to massing in vast flocks, the birds were highly synchronous in their reproductive behavior, with nesting, hatching, and fledging occurring at roughly the same time across a given flock. This survival strategy, known as "predator satiation," served the species well before the advent of intensive market hunting in the second half of the nineteenth century, when it proved fatal (Blockstein and Tordoff 1985).

Commercial exploitation of the passenger pigeon, which began in earnest in the 1830s and 1840s, increased dramatically with the advent of the railroad. In the second half of the nineteenth century, large urban dealers recruited expansive networks of buyers, hunters, and trappers, who were alerted to the movements of the species by railroad express agents equipped with telegraphs (Schorger 1955, 144–57; Greenberg 2014, 78–90). Local residents hoping to cash in on the birds' periodic visits joined professionals who descended on promising sites. Although many birds were shot or knocked from their nests, the most efficient method was using large baited nets that captured a thousand or more each time they were sprung (Schorger 1955, 167–99; Greenberg 2014, 91–108). As late as the 1870s, market hunters were shipping millions of pigeons from nesting sites in Michigan and Wisconsin. Most were sold for food, but sportsmen also used many live pigeons as targets for trapshooting competitions (Schorger 1955, 157–66; Greenberg 2014, 89–90, 109–19).

During the 1880s, sightings of the pigeon became sporadic, and the number shipped to market declined dramatically. Estimates of individual flock sizes that once ranged in the hundreds of thousands or millions had now been reduced to thousands and hundreds. As early as 1862, New York passed a state law that prohibited disturbing pigeons at their nesting grounds or discharging a gun within one mile of those sites, although it appears these and other early state wildlife laws were rarely enforced (Schorger 1955, 226–27). In 1897 the Michigan legislature passed the only measure giving complete protection to the passenger pigeon, but by then it was too late. Claims of pigeon sightings continued well into the twentieth century, but the last specimen taken from the wild was probably shot in Indiana in 1902 (Greenberg 2014, 173). In 1909, Clark University biologist C. F. Hodge launched a nationwide search for the species, offering a reward of more than $1,000 for anyone who could find even one undisturbed nest ("Last Attempt" 1910; "Notes and News" 1910, 1914; Hodge 1910, 1911, 1912). When none of the many claimed sightings panned out, he declared an end to the project in 1912. By that point the only passenger pigeon known to still exist was a single aging female at the Cincinnati Zoo.

The passenger pigeon might have been rescued through a systematic captive breeding program, but there were no sustained attempts to establish one, even as its predicament became more widely publicized. A handful of individuals and institutions dabbled with raising the species, but as a hobby, for scientific research, or for exhibition rather than as part of an effort to rescue the bird from extinction. One example was the Milwaukee bird fancier and swimming-school owner David Whittaker, who in 1887 obtained four passenger pigeons from northeastern Wisconsin (Deane 1896, 1908). Although one of the original birds soon escaped and another died after injuring itself on the wire netting used to enclose it, by 1895 Whittaker's flock grew to fifteen birds. He sold all his pigeons to the famous University of Chicago biologist Charles O. Whitman in 1896 and 1897, but Whitman returned seven of them a year later. By 1909, all of Whittaker's birds had perished.

Whitman's flock met a similar fate at roughly the same time. A gifted scientist, teacher, and administrator who had begun keeping domesticated pigeons as a boy, Whitman maintained an extensive colony of about 550 pigeons and doves that he used to study avian evolution, behavior, and genetics (Pauly 2000, 162–64; Deane 1908). In what now seems like cavalier treatment of a species known to have been facing extinction, each summer he shipped his flock back and forth between Chicago and Woods Hole, Massachusetts, where he served as founding director of the Marine Biological Laboratory. The Whitman passenger pigeon population peaked at sixteen birds in 1902 before beginning to decline; that year two of his female birds escaped while they were at Woods Hole, and he donated another female to the Cincinnati Zoo. By 1907 the last of Whitman's passenger pigeons died of tuberculosis, leaving only two infertile male hybrids, the progeny of a cross between a male passenger pigeon and a female common ring dove. At that point the nation's attention turned to the Cincinnati Zoo, home of the last living passenger pigeons.

Sometime around the time of its opening in 1875, the Cincinnati Zoo acquired several passenger pigeons, although the exact numbers and dates remain hopelessly confused in the meager, conflicting sources that survive (Schorger 1955, 28–30; Cokinos 2000, 258–78; Greenberg 2014, 185–89).[4] What is known is that although the zoo boasted twenty passenger pigeons by 1881, it does not seem to have made a concerted effort to save the species, which where displayed in an eighteen-by-twenty-foot aviary. By 1907 the flock had declined to three birds: two older males and a single female that Whitman had donated five years earlier. One male died in 1909, leaving an elderly pair of passenger pigeons that zoo officials named after America's first president and first lady, George and Martha Washington. When George died a year later, Martha was alone, the final passenger pigeon known to survive anywhere in the world. Sometime on the afternoon of September 1, 1914, she drew her final breath, and officials shipped her ice-encased body to the Smithsonian Institution, where it was photographed, skinned, dissected, and mounted for exhibition in the National Museum of Natural History (Shufeldt 1915).[5]

Almost a century after Martha's death, passenger pigeon aficionados launched an ambitious, controversial plan to bring back the species that had become a prominent icon of modern extinction (Rich 2014; Minteer 2015). In 2011 the environmentalist and technoenthusiast Stewart Brand, best known for his countercultural *Whole Earth Catalog*, experienced an epiphany after hearing zoologist Tim Flannery lecture on mass extinction as part of the Long Now Foundation's Seminar about Long-Term Thinking. In discussions following that talk, Brand realized that the genomic technologies Harvard biologist George Church had been developing to try to resurrect the woolly mammoth might also be used to restore life to the passenger pigeon. Passenger pigeon DNA could probably be extracted from the hundreds of specimens in museum collections around the world and sequenced. Using the resulting DNA code as a blueprint, researchers might then deploy genetic engineering techniques to modify parts of the genome of a closely related species, the band-tailed pigeon, until it resembled the living passenger pigeon. The resulting cells could then be introduced into a band-tailed pigeon's embryo, which, if everything worked as planned, would hatch into a band-tailed pigeon with either passenger pigeon sperm or passenger pigeon eggs.

With backing from Brand and his wife, Ryan Phelan, in 2012 the Long Now Foundation launched Revive & Restore, which seeks to coordinate, promote, and support efforts to bring back the passenger pigeon and several other iconic lost species (Revive & Restore, n.d.). Critics worry that de-extinction programs divert money and attention from more tried-and-true conservation efforts, fail to address habitat loss, and promote a hubristic belief in the power of technofixes to solve the growing problem of mass extinction (Minteer 2015; Pimm 2013). Unmoved by these concerns, Revive & Restore continues to aggressively promote experiments aimed at resurrecting the passenger pigeon, while projecting that it will be in a position to begin breeding the species in captivity by 2022 and releasing it into the wild a decade later. Whether the majestic passenger pigeon will ever again grace North America's skies remains to be seen.

THE FALL AND RISE OF THE BISON

The story of the decline of the bison paralleled that of the passenger pigeon in several key ways, but the response proved remarkably different. Although the American bison (*Bison bison,* more popularly known as the buffalo) once occasionally ranged as far east as the Atlantic Coast, its population center was the Great Plains, especially the short-grass prairie that dominated the western portion of the region (Isenberg 2000; Dary 1974; Barsness 1985; Barrow 2009, 92–96, 108–24). As many as twenty to thirty million bison once lived there, helping to shape the grassland habitat through nomadic browsing. For much of the year the species roamed in relatively small bands, seeking forage and shelter from extreme weather. During the summer rutting season, however, bison gathered into vast herds that frequently awed the Euroamerican visitors who ventured into the region. As one of countless examples, in 1834 American ornithologist John Kirk Townsend reported finding the species during an expedition to the Columbia River: "The whole plain, as far as the eye could discern, was covered by one enormous mass of buffalo" (Danz 1997, 17).

Several factors contributed to the near-extinction of this once superabundant species. Native Americans working in the fur trade began the destruction in the first half of the nineteenth century. By the middle of the century, California-bound immigrants regularly traveled through bison territory, a decade-long drought struck, and bovine diseases introduced by European livestock all exerted additional downward pressure on the bison population (Isenberg 2000, 110). The commercial bison trade received a significant boost in the 1860s and 1870s with completion of the transcontinental railroad, manufacture of the first accurate long-bore rifles, and development of techniques for tanning buffalo hides into leather that was in high demand for industrial belting (Isenberg 2000, 130–32). After the Civil War, thousands of hunters descended on the Plains hoping to cash in on the newly expanded market for bison leather, and the total slaughter rose to two million or more animals a year (Isenberg 2000, 136–37). In addition to overhunting, which was the primary cause of bison loss, additional casualties resulted from the introduction of Texas cattle fever in the southern part of its range, a series of abnormally severe winters, occasional grass fires, periodic draught, and, perhaps most important, competition from introduced cattle. By the 1870s the southern Plains bison had been reduced to a few hundred stragglers; a decade later, the northern Plains herd faced a similar fate. A species that had once blackened the landscape as far as the eye could see was barely hanging on.

As with the case of the passenger pigeon, the precipitous decline of the bison did not pass entirely unnoticed. As early as 1876, while much of the nation celebrated the technological and economic progress the United States had enjoyed during the previous century, Harvard naturalist and museum curator Joel Asaph Allen wrote a series of publications lamenting the precipitous decline of American wildlife that had accompanied the "transformation of hundreds of thousands of square miles of wilderness into 'fruited fields,' dotted with towns and cities, and intersected by a series of railways and telegraph lines" (Barrow 2009, 78–84; Allen 1876b, 794). In a lengthy monograph, *The American Bison, Living and Extinct,* and a popular article, "The North American Bison and Its Extermination," Allen chronicled the systematic destruction of this iconic species, predicted that "the period of his *extinction* will soon be reached," and proclaimed its destruction to be "one of the most remarkable instances of extermination recorded, or ever to be recorded in the annals of zoology" (Allen 1876a, 180, 1876c, 216).

Ten years after Allen's distressing report, William Temple Hornaday, chief taxidermist at the US National Museum, estimated that there were only about three hundred bison left in the wild.[6] Greatly alarmed at his discovery, Hornaday rushed to Montana to collect specimens of the species for the Smithsonian Institution before it was too late.[7] After trips in May and September 1886, he returned with twenty-five bison skins, a series of fresh and dry skeletons, two bison fetuses, and a young calf named Sandy that he exhibited on the mall in Washington, DC, for two months before it perished. The expedition resulted in the creation of a habitat group, featuring Sandy and five other Montana specimens that Hornaday painstakingly mounted to serve as a proper "monument to the American bison" (Wonders 1993, 122; Shell 2002). It also led to Hornaday's conversion from ambitious taxidermist to ardent conservationist. Three years after his return to Washington he published a monograph, *The Extermination of the American Bison,* that featured the hyperbolic language found in many of his later conservation writings. There he blamed the extermination of the bison on "the descent of civilization, with all its elements of destructiveness, upon the whole of the country inhabited by that animal. From the Great Slave Lake to the Rio Grande, the home of the buffalo was everywhere overrun by the man with a gun; and, as has ever been the case, the wild creatures were gradually swept away" (Hornaday [1889] 2002, 464).

The Montana bison expedition and the capture of Sandy also led Hornaday to call for a new Smithsonian-affiliated zoo that would not only serve as home to a breeding stock of endangered species but also help raise public awareness about the plight of America's wildlife (Dolph 1975, 566–648; Horowitz 1973–74; Dehler 2013, 66–70). At Hornaday's urging, in October 1887 George Brown Goode organized a Department of Living Animals as part of the US National Museum, and he soon named Hornaday as its curator. He gathered the first specimens for the experimental venture during a monthlong western trip in the fall of that year and housed them in a makeshift wooden structure just south of the original Smithsonian building. By January he had fifty-eight animals under his care. The popularity of the living animal exhibit helped persuade Smithsonian secretary Samuel P. Langley to support Hornaday's proposal for a new National Zoo in Washington, DC. Hornaday played a leading role in nearly every aspect of bringing this idea to fruition, from selecting the 166-acre site at Rock Creek to negotiating the purchase of the land, and from drafting authorizing legislation to preparing detailed plans for the grounds and animal accommodations. Throughout the two-year process, he repeatedly stressed that one of the proposed zoo's main purposes would be to provide "a suitable place in which to preserve representatives of our great game animals before they are all exterminated" (Dolph 1975, 582, 572, 575). A little more than a year after legislation authorizing establishment of the National Zoological Park became law on April 22, 1889, Hornaday abandoned the project, resigning from the Smithsonian after differences with Langley over his role in the fledgling institution. Without Hornaday's leadership the new zoo quickly devolved into a popular recreational space for Washington citizens and tourists rather than a place where threatened North American species might be rescued from extinction.

Hornaday soon secured another opportunity to implement his vision for a new kind of conservation-focused zoo. After leaving the Smithsonian, he spent several years working for a real estate firm in Buffalo, New York, before receiving an invitation to direct an institution that officials of the New York Zoological Society were trying to get off the ground.[8] The president of the society was paleontologist Henry Fairfield Osborn, who was also professor of zoology and dean at Columbia University, head of the Department of Vertebrate Paleontology at the American Museum of Natural History, and later longtime president of that institution. When Osborn invited Hornaday to head the effort, he jumped at the chance.

Soon after becoming director in April 1896, Hornaday recommended the South Bronx site, one of several locations under consideration for the proposed zoo. In a telling metaphor, he wrote that if Noah were to arrive there "with his arkful of animals and turn them loose . . . each species would promptly find its own suitable place." In addition to gathering a miscellaneous collection of animals with the aim of entertaining and educating "the general public, the zoologist, the sportsman, and every lover of nature," the new zoo would serve as a modern-day Noah's ark that would attempt to rescue the many "native animals of North America" that were struggling to survive. He went on to declare that the institution's first priority would be to collect "a liberal number" of the continent's notable animals threatened with extinction, for "nearly every wild quadruped, bird, reptile, and fish is marked for destruction" (Bridges 1974, 31–32). As Hornaday and his colleagues envisioned it then, the New York Zoological Park was to be a hybrid institution, a mixture of a more conventional zoo stocked with a variety of exotic creatures and a reserve for threatened North American wildlife.

Obtaining bison to stock the twenty-acre range he planned for the park's southeastern corner ranked high among Hornaday's priorities for what became popularly known as the Bronx Zoo soon after its opening in November 1899.[9] After several inquiries, he negotiated the purchase of four bison from Texas rancher Charles Goodnight and three from Oklahoma rancher and businessman Ed Hewins (Hagan 2007).[10] Hornaday also hoped to establish breeding populations of several other large western mammals that were rapidly declining—antelope, caribou, mule deer, and others—but he faced numerous difficulties in achieving this goal. During its early years, for example, the newly opened zoo received sixteen pronghorn antelope in two shipments, but they all soon fell victim to disease or attacks from roaming dogs. Indeed, as had long been (and would long remain) the case for zoos, successfully capturing wild animals, safely transporting them great distances, and determining how to properly feed, care for, and breed them proved very challenging. As a result the mortality rates of most wild species kept in zoos initially remained high, undoubtedly contributing to their failure to play a larger role in the late nineteenth- and early twentieth-century wildlife conservation movement.[11]

As with the antelope, Hornaday initially struggled to maintain a self-perpetuating bison herd at the Bronx Zoo, one of the institution's most popular early attractions. He initially blamed their recurrent deaths on gastroenteritis, which he thought was caused by eating rank grass, but the animals continued to die off in large numbers even after the grass was burned over, the topsoil removed, and the animals prevented from grazing freely (Bridges 1974, 263–67). Although forced to relinquish plans to raise large numbers of other North American game animals, Hornaday persevered with the bison. After repeated infusions of purchased or donated specimens, the herd eventually stabilized and then began to increase through the birth of new calves. The experiment proved successful enough to inspire an ambitious project to begin restocking the West using animals from the New York Zoological Society collection.

In March 1905, not long after President Theodore Roosevelt began establishing the first federal wildlife reserves, Hornaday received authorization to offer the federal government twelve bison from the Bronx Zoo (Bridges 1974, 258–61).[12] The hope was to transplant these zoo-raised animals to the Wichita National Game Reserve in Oklahoma Territory, which Roosevelt had established on a former Indian reservation recently opened to settlement. Hornaday and his colleagues now admitted that zoo-confined animals, "even where the enclosures were as large as the New York Zoological Gardens," would fail to perpetuate the species over the long haul. They also feared that existing privately established herds of bison could be sold at any time or crossed with cattle. The only way to maintain the bison as a purebred species "in full vigor for the next two hundred years, or more" would be to establish herds on public lands "in ranges so large and diversified that the animals will be wild and free" (Hornaday 1908, 55–56, 1910, 17–18; "National Buffalo Herd" [1907]; "Zoo's Bison Herd Accepted" 1906; Hornaday 1907).

To rally support for this and other protective measures, later that year Hornaday joined forces with nature writer, naturalist, and conservationist Ernest Harold Baynes to found the American Bison Society, an organization that would soon grow to four hundred members (Barrow 2009, 118–24; Coder 1975, 118–70). The first order of business for the fledgling society was to complete the Wichita bison reintroduction project. After visiting the site in February 1906, J. Alden Loring, a former Bronx zoo curator, issued a report recommending twelve square miles on the western portion of the reserve as the most suitable range for the species (Loring 1906). Congress then unanimously appropriated the funds to erect a fence around this section of the reserve, a project completed in 1907. In October of that year, Hornaday supervised the challenging process of loading fifteen bison into crates that were then shipped cross-country from New York by rail and hauled twelve miles by wagon to the newly established reserve.[13] Ever mindful of the value of publicity, Hornaday made sure photographers were on hand to document the move, which appears to have been the first time an endangered mammal had been bred in a zoo and then reintroduced to its natural habitat.

Hornaday and the American Bison Society also lobbied for creation of the National Bison Range in western Montana, on the grounds of the Flathead Reservation that the federal government had opened to white settlement

(Hornaday 1910). In 1908, with the backing of Roosevelt, Congress authorized this refuge and provided the funds needed to fence its thirteen thousand acres. With more than $10,000 raised by private subscription, mostly from wealthy easterners, Hornaday purchased thirty-four bison from Alicia Conrad in Montana. The society also received several bison donations from Charles Goodnight and the New Hampshire entrepreneur Austin Corbin that were released at the new refuge.[14]

By 1910, after the creation of the federal National Bison Range in western Montana, Hornaday declared that "the future of the American bison, as a species" was "now secure." He offered to resign as president of the American Bison Society but was persuaded to stay on for a final year (Hornaday 1910, 17–18). Most of his colleagues felt more remained to be done to ensure that the species was truly safe, however, and in 1913 the organization successfully lobbied for establishment of another federal herd at Wind Cave National Park in South Dakota (Palmer 1916). Once again the New York Zoological Society provided a nucleus stock for the refuge, shipping fourteen animals westward. By that point the society's annual census claimed that more than three thousand full-blooded bison resided in North America, including five hundred under federal protection.

Almost a century later, the survival of this iconic species now seems certain, with as many as 500,000 bison in private hands and 30,000 on public lands (Defenders of Wildlife, n.d.). But numbers alone do not necessarily tell the full story. Over the past decade genetic studies have shown that almost all bison today carry DNA from domestic cattle. Apparently most American bison, even those on public lands, are the descendants of five herds kept by ranchers who engaged in bison-cow hybridizing experiments that continue to affect the bison gene pool today. Some biologists, conservationists, and wildlife managers worry that cattle gene introgression threatens the bison's long-term viability by subtly changing its behavior, fertility, and disease resistance. Others dismiss these concerns, pointing to the impressive growth of bison populations over the past century as evidence that the small levels of cattle introgression discovered thus far have not negatively affected the species.[15]

CONCLUSION

What accounts for the profound differences between the ultimate fates of the passenger pigeon, which fell victim to extinction, and the bison, which did not? Both species gained widespread publicity as their numbers became thinned, but the passenger pigeon failed to capture the public imagination like the bison, which was not only the largest living land mammal in the United States but also widely seen as central to America's identity as a nation. As a charismatic creature that once blackened the Plains, the bison had sustained the lives of native peoples and aided the western migration of white settlers. The shaggy, lumbering beast's continuing place in the nation's collective imagination has been commemorated in paintings, museum exhibits, coins, paper currency, and stamps, and its near demise prompted the creation of an organization devoted to rescuing it from extinction.[16] Biology also played a role in the divergent experiences of the two species. Like many ungulates, the bison proved relatively easy to breed in captivity.[17] Several ranchers had established self-sustaining bison herds before the founding of the American Bison Society, and indeed, these privately held animals became the starting point for the bison that the New York Zoological Society bred and shipped back West to stock newly created federal reserves. Although there were scattered, short-term successes with raising the passenger pigeon, its long-term needs proved more challenging for human captors to fulfill. Moreover, initial efforts to maintain the species in captivity did not come until its numbers had already been dramatically reduced.

While all these factors played important roles in how the story of these two threatened creatures unfolded, it remains that no one appears to have made a concerted effort to breed the passenger pigeon while it remained possible to do so. Well into the 1870s and 1880s, professional hunters were still capturing hundreds, even thousands of live passenger pigeons with a single netting, but only a handful of these birds ended up in facilities where they might have been properly cared for and methodically bred. At the same time, while the dozens of American zoos established in the second half of the nineteenth century sought to acquire rare animals to exhibit, they failed to do so with an eye to perpetuating threatened species. Even if they had, early zookeepers lacked the knowledge of how to properly care for and successfully breed most animals, especially over the long haul.

The exception that proves the rule, William T. Hornaday, was among the first Americans to recognize that in an era of rapid wildlife decline, zoos might play an important role in rescuing endangered species, and he worked to translate this realization into action, first at the US National Zoo and then at the New York Zoological Society. Not until the era of the modern environmental movement in the 1960s and 1970s did a large number of American zoos begin to follow Hornaday's lead, adopting conservation as one of their core missions (Kisling 2000b, 173; Bender 2016; Kisling, this volume).

Indeed, in the face of the modern biodiversity crisis, mounting ethical qualms about confining wild animals, and heartfelt commitments from their patrons, leaders, and staff members, zoos have begun to redefine themselves as wildlife conservation centers, sites where not just captive breeding, but also conservation education aimed at behavior change among visitors, support for in situ protection, modeling of environmentally sustainable practices, and scientific research on endangered species routinely take place (Barongi, this volume; Knapp, this volume; Grow, Luke, and Ogden, this volume; Traylor-Holzer, Leus, and Byers, this volume; Zimmermann et al. 2007).

Even as they have begun to expand the scope and scale of their wildlife protection efforts, zoos and other conservation-oriented institutions will need to remember the lesson learned from the stories of the passenger pigeon and the bison: the importance of timely action in the face of impending extinction. Zoo officials, conservationists, and naturalists responded to the decline of the bison before its numbers dropped below a critical threshold from which it could not recover, something they failed to do so with the endangered passenger pigeon. New genomic techniques, like those that Revive & Restore promotes to bring back the passenger pigeon, might eventually allow us to reanimate some lost species. But such techniques, if they ever prove successful, will never be fast or cheap enough to save all the species threatened by what is now often referred to as the sixth extinction, the first mass extinction to be caused entirely by humans. Nor can those techniques restore the social and behavioral traits or critical habitat that species need to survive in the wild. As the threats to wildlife in the twenty-first century continue to proliferate, aggressive, thoughtful, and timely action to prevent species extinction—like the New York Zoological Society's pioneering efforts not only to use captive breeding to increase the number of bison but also to return zoo-bred bison to protected habitat—remain crucial approaches to addressing the biodiversity crisis.

CHAPTER FIVE

American Zoos: A Shifting Balance between Recreation and Conservation

Pamela M. Henson

INTRODUCTION

American zoos have changed significantly over the past two centuries. What were the goals and practices of nineteenth-century zoos, and how did they evolve into the conservation centers of the twenty-first century? This historical overview of American zoos will highlight the major factors propelling changes in purpose and zoo husbandry and trace the threads that continued throughout these changes. Late nineteenth-century American zoos had their roots in European menageries and acclimatization gardens that displayed animals from other regions, but zoos in the United States were more influenced by the urban parks movement. Several of these zoos included an early emphasis on conservation of native species, such as bison, and conservation ideals affected both their design and their operation. Natural-looking exhibits in large enclosures sought to recreate the environments where hoofed stock roamed or beavers built dams. However, the focus on conservation proved hard to sustain as zoos competed to attract audiences and income as visitors looked for exotic creatures and dramatic displays. In addition, native species proved hard to maintain in urban environments, leading zoos to discontinue those species or to alter exhibits for greater habitat control. After the turn of the century, rapid loss of such species as the passenger pigeon and the Carolina parakeet transformed zoos from Noah's arks to refuges for the last specimens of these species (Barrow, this volume). Conservation efforts moved to associations such as the Audubon Society and Sierra Club and focused on limiting hunting and preserving natural habitats.

In the early twentieth century zoos rapidly increased in number, often led by charismatic directors who competed to secure and display exotic species such as pandas and gorillas. Animal health research focused on keeping these exotics alive in zoo enclosures. Two world wars and a global economic depression kept zoos struggling to stay afloat. In the mid-twentieth century, a renewed emphasis on conservation was stimulated by works such as Rachel Carson's *Silent Spring* (1962) and by awareness of an increasing number of endangered species. At the same time, the animal rights movement questioned the basic concept of keeping animals ex situ in zoos, citing abnormal behavior and short life spans in caged animals. Postcolonial nations placed limits on animal collecting, and new laws limited animal trafficking. Zoos focused on keeping and breeding the exotic animals they did have, since they could not be replaced from the wild. Then, as the science of animal care improved, zoos began to play a major role in breeding critically endangered native and exotic species such as the California condor and the cheetah. Competition and secrecy were slowly replaced by cooperative Species Survival Plans and an emphasis on sharing knowledge to preserve species (Kisling, this volume). But zoo administrators still had to balance the books, bringing in enough income to care for facilities, animals, visitors, and, increasingly, specialized equipment and staff members who spent time doing research.

Today zoos cultivate cooperative relationships worldwide to maintain species that no longer exist in the wild. As natural environments shrink or disappear, zoos play a larger role as reservoirs for critically endangered species, and the line between in situ nature and ex situ zoos has become blurred. Conservation has waxed and waned during the history of American zoos, but it is at the core of virtually all twenty-first-century zoos. Despite earlier efforts, more flora and fauna than ever are at risk from development, disease, climate change, and hunting, ensuring that zoos will face significant conservation challenges in the coming decades.

AMERICAN ZOOS IN THE NINETEENTH CENTURY

The first zoos in the United States were created as part of the urban parks movement in mid-nineteenth-century America to create restorative retreats from the stresses of urban life. They appeared in the same decades as the first urban and national parks and reflected a similar belief in the health benefits of contact with nature. As Guerrini and Osborne (this volume) and Ritvo (this volume) discuss, they were also influenced by European traditions of maintaining acclimatization gardens to adapt exotic species to the local environment as well as menageries and pleasure gardens to entertain the public (Osborne 2000; Ritvo 2012; Stott 1981, 54–55). New York's Central Park had animals on display in a small menagerie since 1864, but in 1899 the New York Zoological Society supplemented it with a large rural park, the Bronx Zoo. In 1874 the Philadelphia Zoo opened in Fairmount Park to the public to provide education and entertainment, with over eight hundred animals in a carefully landscaped environment. The Cincinnati Zoo opened modestly in 1875 with a small collection of both native and exotic species such as buffalo and monkeys. Founded by German immigrants to copy the zoological gardens of their homeland, its goals were adventure, conservation, and education (Hanson 2002, 11–16; Kisling 1996, 115–21). The Smithsonian's US National Museum had a small menagerie of live animals used as models for taxidermists that soon proved popular with the public. In addition, the chief taxidermist became aware of the rapid decline in native American species such as the bison and committed himself to conservation of American animals. In 1891 the live animal collection was moved to a large urban park in Washington, DC, where it could grow and inspire interest in saving North American species (Hanson 2002, 46, 134).

As Barrow's chapter in this volume demonstrates, rapid environmental changes in the nineteenth-century United States led to the sudden decline of species and the possibility of dramatic extinctions. Where once millions upon millions of bison roamed the western Plains, by the mid-1880s the plowing of the prairie and uncontrolled slaughter left only a few remnant herds. Americans became more aware that extinction could occur through human action, once an unimaginable idea. The National Museum taxidermist, William Temple Hornaday, traveled to the West in 1886 to collect bison for display but found only a few small groups hiding in remote ravines. Formerly an avid hunter, Hornaday returned East committed to keeping the American bison alive. To encourage Americans to conserve native species, he created a magnificent display for the National Museum that presented a realistic bison life group (Barrow 2009; Coffman 2013, 10–30; Horowitz 1996, 126–32; Kisling 2013, 129–30; Hornaday 1886, 1887a).

Hornaday also brought back live animals and broached the idea of creating a zoological park that would teach the public about native American species (Kisling 2013, 129–30). Hornaday's conservation goals met with a positive reception from his supervisors, Smithsonian secretary Spencer Fullerton Baird and National Museum director G. Brown Goode. Both were naturalists who were becoming aware of the decline of animals, especially fish. Baird served simultaneously as US commissioner of fish and fisheries, with Goode as his assistant. They had become increasingly concerned about the numbers of fish in the Greater Gulf of Maine fishery region, as well as salmon stocks across the United States.

Baird and Goode embarked on a massive study of fish, fishing techniques, and the role of fish in human societies, and they published an eight-volume treatise titled *The Fisheries and Fishery Industries of the United States* (Goode 1884–87). They created the US Fish Commission Station at Woods Hole, Massachusetts, set up carp ponds on the National Mall in Washington, DC, prepared fisheries displays at international exhibitions, and urged the passage of legislation to manage fish stocks. Their goals were practical as well as scientific: fish were an important food source, and their decline or extinction would have dire economic consequences (Allard 1978, 262–79). As Muka (this volume) discusses, aquariums to study and breed fish stocks drew a popular audience to view and learn about aquatic species. Baird and Goode also responded sympathetically to Hornaday's concerns about the conservation of other animals.

Unfortunately, Spencer Baird died as Hornaday returned from the West in 1887, and the next Smithsonian secretary, astrophysicist Samuel P. Langley, did not share Hornaday's vision. Hornaday nonetheless found a suitable and beautiful site in Rock Creek Park, "a pleasant carriage ride from the city," with a varied landscape and water sources. The noted landscape architect Frederick Law Olmsted was hired to design the new park, and Hornaday garnered strong public and political support. The zoo opened in 1891 with the small menagerie from behind the Smithsonian Castle and animals on loan from a traveling circus. A bison exhibit was an important part of the new park. Yet Hornaday's independence irked Secretary Langley, who forced him out of the Smithsonian, leaving behind Hornaday's dreams of museum exhibits and zoo displays that would nurture a conservation ethic in the United States. Langley hired Frank Baker, a rather timid man who failed to inspire legislators or donors to provide much funding, and Hornaday's great dream dwindled into a small, struggling menagerie for many decades (Hanson 2002, 24–26; Horowitz 1996).

Hornaday's idea was rekindled in a different way by the New York Zoological Society less than a decade later, when they called on him to plan the Bronx Zoo. Led by a private group of New York City elite, mostly members of the exclusive Boone and Crockett Club, they also were concerned about the demise of the wilderness and native species. Hornaday was thus able to create several naturalistic displays in the southern part of the park—notably for the American bison, jackrabbits, pronghorn antelope, beavers, and prairie dogs. Hornaday also displayed less charismatic animals, arguing that "our country possesses the greatest variety of squirrels and marmots to be found in any one country," so these smaller animals needed to be preserved as well (Bridges 1974, 4–20, 24–31; Hanson 2002, 46–47).

Despite some successes, the northern part of the park was a great disappointment to Hornaday, with enclosures more typical of the zoological gardens of Europe and exotic animals caged in elaborate buildings. Over time, Hornaday grew more frustrated by the direction of the zoo—its visitors and its neighbors as well as its board of governors, as nearby residents cut down trees and shot songbirds and visitors teased exotic species. Most disturbing, however, were problems with native species. The hoofed stock, including his beloved bison, developed gastroenteritis and grass poisoning from the local grasses. A display, Hornaday learned the hard way, had to do more than look natural—it had to function naturally or the animals would not thrive or even survive. Over time, many of the Hornaday's favored native hoofed stock were eliminated from display or not replaced as they died prematurely (Hanson 2002, 134–35; Horowitz 1975, 444).

Hornaday could take consolation, however, that the fruits of his other labors had been more successful. His writings, including *The Extermination of the American Bison* ([1889] 2002) and *Our Vanishing Wild Life: Its Extermination and Conservation* (1913), and his efforts to organize conservation advocacy did help galvanize the American conservation movement. Yellowstone National Park provided a safe environment for a small but growing bison herd. The International Migratory Bird Treaty was ratified in 1918. But the conservation movement did not see zoological parks as important resources. As Kisling has pointed out, even Hornaday did not write about zoos as sites of conservation (Kisling 2013, 127–29). Conservationists such as Aldo Leopold now led the movement with a new set of environmental ethics (Leopold 1933; Worster 1985, 271–90). They advocated for wilderness preservation and for managing native species in their natural habitat rather than removing them from their normal environment. Instead of looking forward, parks like the Cincinnati Zoo became melancholy arks housing the last members of species on the verge of extinction—the last passenger pigeon in 1914 and the last Carolina parakeet in 1918 (Stott 1981, 59–61; Barrow, this volume).

The loss of birds as numerous as the passenger pigeon, which once darkened the skies for days during migrations, shocked many Americans into action. The conservation movement made steady progress on many fronts.

Opposition to bird predation by the millinery industry and by egg and nest collectors made life safer for many birds (Barrow 1998, 107–20). The nature study movement taught children to value and respect the natural world, as did children's scouting movements. Families now had more leisure time, which was often spent at the new national and local parks, increasing awareness of the importance of conserving these regions. And Progressive movement technocrats argued that they could manage the natural world for the benefit of native animals and plants as well as humans. Conservation societies grew rapidly, including the Audubon Society founded in the 1890s by advocates for birds, the Sierra Club started in 1892 by outdoor enthusiasts influenced by John Muir (Cohen 1988, 1–27), the American Bison Society founded by Hornaday and Theodore Roosevelt in 1905, the National Wildlife Federation begun in 1926 by hunters and outdoor enthusiasts, and the Wilderness Society established in 1935 to protect natural areas (Cohen 1988, 109–19). Local organizations sprang up across the country to preserve natural areas and local species and to encourage people to visit and support natural parks (Worster 1985, 185, 261–78, 283–84).

A NEW CENTURY: 1900–1950

By the late nineteenth century, zoos were well established as sites for healthy recreation. Although conservation concerns had spurred the creation and design of some zoos, as the American conservation movement developed zoos did not play an important role. By 1900 zoos were established across the country as the population grew and leisure time increased, with more than one hundred American zoos founded between 1880 and 1930. In the early twentieth century, zoos solidified their importance as recreation venues offering a glimpse of exotic worlds. Subscribers to *National Geographic* visited local zoos to see the unusual animals they read about. At the same time, zoos concentrated more on attracting audiences through displays of exotic animals and performances by trained animals such as chimpanzees and seals. Taking a page from amusement parks, children rode ponies and Galápagos tortoises as buses and trains transported visitors around the parks (Hanson 2002, 79–87). Zoo directors became well-known public figures who competed to secure the most exotic animals (Hanson 2002, 109–16). These directors did secure rare species—a baby gorilla at the National Zoo in 1928, a Komodo dragon at the Bronx Zoo in 1934, and a panda at the Brookfield Zoo in 1936—but these creatures rarely lived long since little was known about their needs, diet, and habits (Bridges 1974, 442; Hanson

2002, 90; Kisling 2000b, 172). When directors did succeed in caring for animals, they kept their approach, such as food recipes, secret to maintain a competitive advantage.

Despite challenges, some zoos did move forward with new areas of research in the first two decades of the century. Notable among them was the Philadelphia Zoo, whose Penrose Research Laboratory conducted studies on veterinary pathology, animal nutrition, acclimatization of tropical animals to temperate climates, use of glass walls to prevent disease transmission, and treatment of tuberculosis—a scourge of many zoos. The New York Zoological Society established the first veterinary clinic in a zoo in 1916, as well as a field research program, the Department of Tropical Medicine, with a station in British Guiana. But as interest in conservation grew, zoos were not considered part of the solution. Zoo research focused primarily on keeping zoo residents alive and healthy longer, with an occasional breeding success (Kisling 2000b, 164–72).

Zoos hosted school tours and special programs focused on celebrity animal birthdays and animal shows. Public engagement emphasized recreational activities that would bring income, such as sale of train tickets, animal crackers, and stuffed animals. They offered little or no real conservation education, and little education at all (Hanson 2002, 34–40). The Bronx Zoo planned a twelve-acre conservation exhibit and broke ground in 1949, but it was never completed owing to lack of funds. In 1945 a Question House opened at the center of the zoo where visitors could ask all sorts of questions—but most were for requests for directions and other logistics (Bridges 1974, 414–17, 425–26, 470).

The first half of the twentieth century was a challenging time for many zoos. During the world wars, zoos lost staff members, had limited budgets, and lost access to foreign animals. During the Great Depression, they struggled to feed their denizens as crops failed in the Dust Bowl and breadlines took precedence over zoos. Several zoos did manage to use the Works Progress Administration and other work programs to build or upgrade zoo buildings and facilities (Kisling 2000b, 169; Mann 1977, 36–39). And a few new zoos were also established, notably the San Diego Zoo, from the small menagerie left over from the 1915–16 Panama-California International Exposition at Balboa Park. In the next decades, the San Diego Zoo rapidly expanded into a major nature park with innovative displays and animal shows, enjoying strong support from the local community (Kisling 2000b, 165).

CAN'T WE ALL GET ALONG?

By the mid-twentieth century, zoos increasingly came under fire as places that merely held exotic animals captive in bad conditions for public amusement (Minteer and Collins 2013, 44; Lindburg 1999). Charismatic zoo directors still competed to secure exotic animals that would draw larger audiences. Pandas, chimpanzees dressed as children, and oddities such as white tigers were covered in news reports and treated as celebrities. Perhaps the most charismatic of these directors was Marlin Perkins, director of the St. Louis Zoo and host of the television program *Wild Kingdom* that brought zoos into suburban American homes in the 1960s through 1980s, increasing the popularity of zoos in general (Mitman 1999, 85–91, 134–56). One notable public education campaign was the US Forest Service's Smokey Bear project, which advocated fire safety in natural areas. Its mascot, a singed cub found clinging to a tree, lived out his life at the National Zoo (Morrison 1976, 1–44). But with the rise of the animal rights movement and the activities of organizations such as People for the Ethical Treatment of Animals (PETA), a negative spotlight was cast on zoos as prisons rather than arks and preserves (PETA 2008).

Exotic animals also became harder to secure. In a postcolonial world, governments in Africa, Asia, and South America restricted the export of animals. US laws and regulations also limited acquisition of foreign and rare animals. The US Animal Welfare Act of 1966, Marine Mammal Protection Act of 1972, and Endangered Species Acts of 1966, 1969 and 1971, as well as the Convention on International Trade in Endangered Species (1975), increasingly regulated actions related to endangered flora and fauna (Hanson 2002, 166–69). As more species approached extinction in their natural environments, zoos began to develop new techniques for caring for the animals they had, trying to increase life spans and produce offspring. These efforts would allow zoos to play a new role in conservation (Barrow 2009; Kisling 2000b, 173–74; Kisling, this volume).

Zoos developed programs in veterinary medicine, nutrition, behavior studies, and breeding, with research and breakthroughs published in scientific journals. Several zoos created nonpublic animal care facilities to study and breed endangered species. These larger enclosures allowed animals more normal behavior and privacy. Primate labs demonstrated learning by orangutans rather than tricks by costumed chimps. Zoo animal care became an accepted field of scientific research, hiring PhDs rather than circus trainers (Reed 1989, April 14, 10–16, and October 13, 31, 36–48; Shumaker et al. 2001).

The hallmark of zoos in the second half of the twentieth century was a slow but steady movement from competition to cooperation. A major change in interactions between zoos came with the adoption of Species Survival Plans in which all organizations with individuals of a critically endangered species agree to work cooperatively to ensure the species' survival. The plans determine which zoos get to keep, display, and breed animals and which animals breed with other animals to ensure genetic diversity and avoid inbreeding problems (Rabb 1994; Reed 1989, August 3, 103–4). Earlier zoo directors could not have imagined giving up a popular rare animal so it could be bred elsewhere, but in the 1990s that became the norm. Today the International Species Information System serves as a cooperative resource for zoos and nature preserves worldwide (Flesness 2003). Advanced breeding studies included behavior, hormone levels, genetic diversity, artificial insemination, in vitro fertilization, surrogacy, and pre- and postnatal care. There have been significant successes with animals such as the black-footed ferret (Dobson and Lyles 2000; Allard and Wells, this volume) and California condor (Ralls and Ballou 2004). But there were also heated disputes over whether endangered animals like the California condor should be kept in a lead-free natural environment or confined to a zoo's captive breeding program (Balmford, Mace, and Leader-Williams 1996; Snyder and Snyder 2000). Despite these efforts, the fates of such animals as the white rhino and clouded leopard remain precarious—and for the white rhino, seemingly hopeless (Stack 2015).

Several global conservation organizations have played critical roles in negotiating and ensuring this cooperation, as discussed by several of the contributors to this volume. The International Union for the Conservation of Nature and Natural Resources was established in 1948 under the aegis of UNESCO to encourage international cooperation across the entire spectrum of nature conservation (Gärdenfors 2001). IUCN began publishing the Red Data Books on the conservation status of species in 1964, and later the Red List of Ecosystems (Burton 2003; Lamoreux et al. 2003; Rodrigues et al. 2006). Its Species Survival Commission provides information on biodiversity conservation. Its Conservation Breeding Specialist Group (CBSG) is dedicated to increasing the effectiveness of conservation efforts worldwide (Traylor-Holzer, Leus, and Byers, this volume).

CBSG employs the scientifically managed breeding of threatened wildlife to create and maintain populations that enable, support, or enhance the conservation of wild populations. CBSG links conservation breeding institutions (such as zoos, aquariums, and botanical gardens) with other stakeholders, helping each to contribute more effectively to the conservation of species in wild habitats. Within North America, Species Survival Plans (SSP) and Population Management Plans are managed through the Association of Zoos and Aquariums' Species Survival Program. The SSP determines which animals are to be conserved and how. Breeding programs census animals available in zoos and in the wild, maintain a genetic information bank, determine which animals should be bred with which other animals, and decide where offspring are to be reared. Currently, AZA members are involved in over three hundred SSPs working on behalf of almost six hundred species (Balmford, Leader-Williams, and Green 1995; Hutchins and Conway 1995; Reed 1989, August 13, 103–4, and October 13, 56–57; Ryder and Feistner 1995).

But zoos still need to attract large audiences. They have developed sophisticated public engagement strategies with membership societies, magazines, and special events like holiday lights and Boo-at-the-Zoo. Stuffed toys representing charismatic megafauna such as elephants and pandas bring income into zoo gift shops, supplementing admission fees and restaurant profits. As the conservation message has grown stronger, endangered species "adoptions," modeled on Save the Children appeals, have provided additional income while offering detailed information on the species being conserved. Other conservation messages have been harder-hitting. The Bronx Zoo even set up an annual "graveyard of extinct species" in Baird Court to remind the public of the rapid decline in bird populations (Bridges 1974, 500–504).

Thus, by the late twentieth century zoos addressed the growing conservation challenges in several ways: by renewed efforts at conservation education, by serving as genetic reservoirs for endangered species, by sharing information and resources rather than competing, and by developing new techniques for species preservation. The modern zoo contains many of the developments noted earlier: advanced veterinary, nutrition, and behavior programs; field research on natural behavior; and enriched enclosures that functionally mimic the natural environment. Enclosures crowded with a variety of species were converted to more adequate facilities for fewer species. In situ field research was supplemented by education of wildlife managers in third and fourth world countries—sharing knowledge beyond zoos to natural environments (Morgan et al. 2011; Wemmer et al. 1993). Nonpublic zoo preserves, sophisticated medical treatments, advanced breeding research and techniques, and restoration of some species to their original homes also began. Zoo Atlanta, Zoo Oregon, and the National Zoo bred the golden lion tamarin and returned it to its natural range in Brazil in the

1980s and 1990s (Kleiman and Mallinson 1998; Stoinski and Beck 2001). The San Diego Zoo's Maui Bird Center is working to return the Hawaiian crow to nature. The Arizona-Sonora Desert Museum has focused on local species and those rarely addressed at zoos—plants, insects, small mammals that live below ground, and reptiles. Of note is their research on the pollinators that ensure the survival and diversity of plants in the desert environment and therefore the survival of many other species (Meffe 1998). But as "natural environments" become more stressed through development and climate change, the line has become blurred between ex situ, or zoo- and aquarium-based, research and conservation and in situ, or field-based, biological research and conservation practice. And as Rothfels discusses in this volume, there are questions about whether ex situ animals are really the same as in situ species. Even these new efforts have raised complex ethical issues as organizations such as PETA have questioned the ethics of disturbing these animals' natural environments. Mendelson (this volume) asks us to consider whether keeping a frog ex situ in a glass box or storing frozen genetic material really constitutes saving that species. Thorny questions about even the most advanced attempts to save endangered species are still being debated (Minteer and Collins 2013, 44).

TODAY AND TOMORROW

In many ways twenty-first-century zoos fulfill the dreams of early idealists such as Hornaday: they serve as arks for endangered species whose native habitats are gone or severely threatened. As exotic animals became more difficult to secure, zoos focused on health and reproduction, backing into a conservation role. Zoos now provide sophisticated care in all areas: nutrition, environment, behavior, medicine, and reproduction. Despite intense pressures to distinguish themselves so as to attract audiences and income, they cooperate with other zoos, research institutions, nature preserves, and conservation organizations in the shared task of saving natural environments and endangered species (Conway 2011). The twenty-first-century norm is that zoos share knowledge, animals, and credit for success, although as Cerezo and Kapsar (this volume) and Norton (this volume) discuss, they are also being pushed beyond this ark model.

Outreach focuses on the conservation role of zoos, engaging the public in person and online through webcams and social media and through structured educational programs with schools and at zoo "camps" (Patrick et al. 2007). Escaped animals capture public attention with blogs and tweets about their adventures: entire cities were put on watch for "Rusty," the fugitive red panda (Ruane and Patel 2013). Charismatic megafauna, such as giant pandas, still receive a disproportionate amount of attention and publicity, but this is justified as a way to channel public interest to other animals and issues. Successful preservation and reintroductions are rightly celebrated. American bison still roam the West, California condors patrol the skies of the Southwest, and black-footed ferrets are keeping prairie dog populations in check. Golden lion tamarins roam the Brazilian rain forest, and panda populations are increasing at nature reserves and zoos. But animals such as the clouded leopard remain difficult if not impossible to breed because of behavioral issues, despite the best possible medical equipment, staff, and techniques (Wielebnowski 2002). The survival of existing animals at zoos and in natural environments has had major setbacks, such as recent northern white rhino deaths that move the species to the brink of extinction (Parks 2015; Stack 2015). Despite the mixed record of conservation success, zoos today have succeeded in changing public perceptions of these institutions. Recent surveys showed that most people believe the primary function of twenty-first-century zoos is conservation (Reade and Waran 1996; Tribe and Booth 2003).

CHAPTER SIX

(Re)Introducing the Przewalski's Horse

Nigel Rothfels

In an October 1904 article about a new Asiatic deer house in the New York Zoological Society's *Bulletin*, there is a photograph of the building with two horses in the foreground (fig. 6.1). Over a century later, this photograph is less remarkable as a record of a newly constructed building at the Bronx Zoo than as a rare photograph of two horses that have absolutely nothing to do with the article. It turns out that these two unlikely-looking creatures, called "Prjevalsky Horses" in the caption, today stand as key figures in what is generally seen as an exemplary story of how zoological gardens can play a central role in conservation.

To begin to understand why these horses are important, we can jump ahead exactly ninety years—to October 1994—and an article in the *New York Times*. With the headline "Rare Przewalski's Horse Returns to the Harsh Mongolian Steppe," the article declares, "Captive breeding saves a species from extinction." According to the article, by 1970 the Przewalski's horse, described as "one of only two surviving species of horse, the other being the domestic horse," had become extinct in the wild and survived only in captivity. Through international collaborations between scientists, conservationists, and zoological gardens, however, the horse was being brought back from the brink of total extinction. The article describes a stout tan horse with a short, upright mane, a horse that should not be confused with "wild mustangs, which are really feral domestic horses" (Possehl 1994), and makes it clear that, were it not for the sustained, international efforts of zoos over decades, this animal would simply have gone the way of the quagga, a subspecies of the plains zebra that was extinct by the end

FIGURE 6.1. "Asiatic Deer House. Photographed previous to occupancy by deer. The Prjevalsky Horses, in the foreground, temporarily occupy one corral." Reproduced by permission of the Wildlife Conservation Society Archives.

of the nineteenth century. Facing over a century of criticism that they are little more than a form of lowbrow entertainment justified by overreaching claims to science, education, and conservation, zoos today consistently deploy the story of the return of the Przewalski's horse to Asia along with other reintroduction narratives as proof that the modern zoo can be a force for good in conservation efforts. As Harriet Ritvo points out in this volume, the nineteenth century zoo was an "ark in the park" only insofar as the institution hoped to exhibit at least two of every species. If so, reintroduction stories like that for the Przewalski's horse suggest that today's ark in the park can be a refuge for endangered species until the animals can be returned to their former ranges in "the wild."

Despite the enthusiasm for reintroduction efforts among leading zoological gardens, the story of the Przewalski's horse leads to important questions at the heart of all reintroduction efforts: Does the very act of human involvement in reintroduction call into question the premise that what is being restored is a "wild" creature? How can a species that is being closely monitored and controlled by human observers ever be considered successfully reintroduced? Leaving aside significant questions about the desirability of reintroduction, is true restoration to some sort of pristine state, as Ben Minteer also explores in this volume, even theoretically possible in the Anthropocene? It is undeniable that the reintroduction of the Przewalski's horse points to the significant potential of coordinated, or "One Plan," ex situ and in situ efforts to preserve biodiversity in the short term. It is also the case, though, that the modern history of this horse confuses any simple account of what these horses even really are, let alone what they might represent for twenty-first-century conservation.

A CONSENSUS HISTORY OF THE PRZEWALSKI'S HORSE

In 1959 Dr. Erna Mohr, a taxonomist at the Zoological Museum in Hamburg, published *Das Urwildpferd* (*The Asiatic Wild Horse*), a small book on the Przewalski's horse that included details on the animals' history, conformation, biology, behavior, and husbandry. The book also included a registry recording all Przewalski's horses known to have been brought into captivity and all their descendants. This published international studbook was constructed from responses to questionnaires sent out to all institutions and individuals known to have (or have had) the horses in their collections and represents a milestone not just in the preservation of the horses but in the development of conservation strategies for endangered species. Mohr based her work on personal familiarity with many of the key figures in the breeding of the horses, her own records and memories, and the technical expertise she gained as the studbook keeper for the European bison. The resulting book provided a modern history of the Przewalski's horse and helped lead to coordinated efforts to save the animal (Volf 1994). Over recent decades that history has been expanded and refined by geneticists, conservation biologists, zoo curators and scientists, and Przewalski's horse enthusiasts. From their combined efforts, a general consensus has emerged about the history of the horses in captivity (see especially Bouman and Bouman 1994; King et al. 2015). It is largely this history that has informed the reintroduction efforts.

In 1878 Colonel Nikolai Przewalski brought to the Zoological Museum of the Academy of Sciences in St. Petersburg a skull and hide he had been given during his second expedition exploring Central Asia. After returning from his third expedition in 1880, Przewalski announced his discovery of wild horses, creatures he claimed to have observed in two groups. Then in 1881 I. S. Poliakof, the taxonomist at the Zoological Museum, published his assessment of the materials in the *Proceedings of the Imperial Russian Geographic Society*. He named the animal *Equus przewalskii* and argued that it was an intermediate form between asses and horses. For Poliakof, what distinguished the Przewalski's horse was primarily the "erect mane, absence of forelock, and tail only partly furnished with hair" (Poliakof 1881, 26): unlike domestic horses, Przewalski's horses have only short guard hairs at the root of the tail, with longer hairs farther down.

With the reports of Przewalski, Poliakof, and others, news about small, stout tan wild horses in Mongolia that had erect manes, no forelocks, and other "primitive markings"—including, in various accounts, dorsal stripes and horizontal striping on their legs—began to circulate among horse aficionados. This was a time when there was an intense focus from taxonomic, evolutionary, and economic perspectives on how equids might be related to each other and how they might be useful for agriculture, exploration, and colonial efforts. This was the time, too, when Walter Rothschild had photographs taken of zebras pulling his carriage; when Carl Akeley assembled an expedition to Somalia to collect specimens of another tan, striped equid, the Somali wild ass; when figures like James Ewart in Edinburgh were crossing zebras with ponies and horses to study heritable characteristics; and when major international voices in taxonomy continued to maintain that phenotypic differences between individuals were essentially sufficient to distinguish species. Beyond all this, it was also a time when there was great interest in prehistory, spurred by a variety of remarkable paleontological finds, including the discovery in the German Neandertal of prehistoric human remains and the discoveries in France and Spain of prehistoric cave drawings. And in those caves were images of horses—stout creatures with erect manes, some without forelocks. As more and more caves were discovered, images of horses with tan bodies and dark legs and manes completed a thought that seems to have occurred to some almost immediately after the discovery of the Przewalski's horse—these horses might be the remnants, the last survivors of a virtually unchanged primeval horse that once ranged across Europe and Asia. In the intellectual conjuncture of a central preoccupation with horses and an intense interest in prehistory, it is perhaps not too surprising that there was an almost immediate, powerful desire to bring some of these living horses—living relics of the past as they were often (and are still) described—back to Europe for study.

At this point two prominent figures promoting the acclimatization of exotic species into Europe took the lead: the German Friedrich von Falz-Fein, who wanted to acquire some horses for Askania Nova, his estate in Crimea, and Herbrand Russell, the eleventh Duke of Bedford, president of the Zoological Society of London (1899–1936), who maintained a large collection of animals at his estate, Woburn Abbey. Falz-Fein made the first attempt to capture the horses in 1897–98. The plan was to capture foals in spring 1897 and bring them west in fall and winter 1897–98. They captured five horses, but all soon died. A second attempt was launched in spring 1898, with seven horses (six fillies and one colt) captured; four of the fillies arrived at Askania Nova early the next year, and the stallion arrived in 1904. A small catch of three horses occurred in 1900, and then the Duke of Bedford got involved. He contacted the Hamburg animal dealer Carl Hagenbeck, who had already successfully imported a variety of animals from Russia and Asia. Hagenbeck agreed to try to acquire some horses. His people were in Mongolia in May 1901. In October, in what would be the largest shipment of these animals ever, Hagenbeck received in Hamburg twenty-eight foals (fifteen colts and thirteen fillies, apparently of fifty-two that had been captured). Five colts and seven fillies in this shipment went to Woburn, and the others went to zoos and research institutions in London, Manchester, Berlin, Halle, Paris, and New York. In 1902 at least eleven more horses arrived in Hamburg (five soon died, two went to New York, two to Edinburgh, and two stayed in Hamburg), and another four arrived in Askania Nova. In 1903 two more mares arrived in Askania Nova. In 1938 one mare was captured and stayed in Mongolia, and in 1947 another mare was captured and eventually ended up in Askania Nova. Many of these animals did not live long, and of those that did survive many were never successfully bred, at least to other Przewalski's horses. Of all these horses, there is agreement that only twelve are represented in the current genetics of about two thousand living Przewalski's horses in the world. Two of these "founder" specimens had gone to Askania Nova early on, nine came from Hagenbeck's two shipments, and the twelfth was the mare caught in 1947. A thirteenth founder was a Mongolian domestic mare, brought with the first Hagenbeck shipment of foals as a surrogate dam, that was bred to the young Przewalski's stallion sent to the Agricultural Institute at the University of Halle in Germany.

In 1908 Hagenbeck published an account of his firm, devoting chapters both to thematic issues—catching, training, caring for, and breeding wild animals—and to animals of particular interest like apes, reptiles, elephants, and ostriches. One of the longer stories in the chapter on catching animals centers on the Przewalski's horses. According to Hagenbeck, Wilhelm Grieger traveled to Mongolia for Hagenbeck in winter 1900–1901, and the hunt started in mid-May. There are reasons to be skeptical about the technique Hagenbeck describes, but he claimed that a large group of local people mounted on fast horses would startle a herd of Przewalski's horses into running and chase after the dust cloud. Eventually the foals would not be able to keep up. He writes, "When at last the foals are quite worn out, they stand still, their nostrils swelling and their flanks heaving with exhaustion and terror" (Hagenbeck 1910, 86). At that point the animals were gathered up, brought back to camp, and put with domestic mares whose own foals had been slaughtered. According to Hagenbeck, despite there being very few herds, within a few weeks they had obtained fifty-two foals. "With these," he writes, "the long journey home was commenced, the party consisting not only of the wild foals, but also their foster-dams" (Hagenbeck 1910, 87).

Hagenbeck also claims "there were no less than three varieties of the wild horse in the neighbourhood closely resembling one another in form but showing differences of color. They all had wavy hair over the body and legs and blackish eyes while in the foals the colour is variable" (Hagenbeck 1910, 86). On this point, if on few others, Mohr agreed. Reviewing photographs of the original horses, she noted striking color variations, differences in the extent of the whiteness of the nose, mulelike qualities in the dark filly that went to Halle, the long, convex face of the stallion in Edinburgh, the presence of a forelock on several horses, and the far from optimally erect manes of the horses in their winter coats. Part of these variations she credits to the stresses of travel, the absence of adult horses to groom the younger ones, and the animals' poor physical condition. Pointing to the "mixed up contingents of yearlings and foals that were imported from Mongolia" (Mohr [1959] 1971, 33), however, she reminded her readers that "amongst the imported horses, foals that were born in freedom, there were some that were not liked" (Mohr [1959] 1971, 13) and that many people believed some of the imported horses were not "pure" wild horses. In the end, and clearly responding to controversial efforts by Heinz Heck to eliminate "Mongolian domestic" blood through euthanasia, Mohr insisted that no one could really know the "exact wild components of those that were captured" (Mohr [1959] 1971, 14), that there was no reason to favor one line of the horses over another, and that it was "beyond doubt that over the years of perseverance and clever mating by selection and elimination of extreme components, a limited 'local breeding line' e.g. 'breed' [had been] created"

(Mohr [1959] 1971, 40). In the end, with the work of Mohr and others, a great deal became known about the critical early history of the Przewalski's horses in captivity. Within this history, though, especially in Mohr's concern about the reduction in phenotypic variability of the horses through selective breeding and in Hagenbeck's claim that there were several varieties of the horse in the wild, there were already evident concerns and confusions about the distinctiveness—the peculiar nature—of these horses.

"PREJEVALSKI HORSES" ARRIVE IN NEW YORK

This brings us back to the photograph of the two horses in the New York Zoological Society's *Bulletin* and to relevant surviving correspondence between Hagenbeck and William Temple Hornaday, then director of the Bronx Zoo, an institution that, as other contributors to this volume make clear, has played an important role in many discussions about the connections between zoos and conservation (see esp. the chapters by Barrow, by Henson, and by Kisling, this volume). On December 31, 1902, Hornaday wrote to Hagenbeck, "The two Prejevalski horses arrived last night at nine o'clock, in good condition" (Hornaday 1902). With this quick notice, Hornaday acknowledged receipt of two animals that that had been excitedly anticipated by members of the New York Zoological Society. According to Mohr, though, the horses "were not favorably regarded" (Mohr [1959] 1971, 114) and were soon sent on to Cincinnati. Inge and Jan Bouman provide more detail, writing that "the stallion was in good condition, but the mare was not, she had difficulty standing" and that although she "had improved by the spring of 1903, the management of the Bronx Zoo wanted to have another pair" (Bouman and Bouman 1994, 25). Indeed, it is clear from the correspondence between Hornaday and Hagenbeck that the filly did have some difficulty after arriving in New York, but there was another issue beyond her physical health. On January 20, 1903, Hornaday wrote to Hagenbeck:

> I have not had time to write you since the horses arrived until now; but I greatly fear we are going to lose the female of the pair. From the first she was very stiff in her legs; but we hoped this was due only to confinement on board ship, and that she could soon be better of this trouble; but she does not improve, and seems really to be growing weak on her legs,—so much so that she sometimes leans against the fence for support. We have both animals in the open air, and with no artificial heat. The male is in good health, and is a fine animal every way, and fit to represent the species. The female does not seem to me to be very good stock, as her legs are very thick in the lower joints. I am sorry to say that

> Professor Osborn is not very well pleased with her; but of course we will give her every chance to come around all right, and will do everything that we know of to help her along. (Hornaday 1903)

The news that Henry Fairfield Osborn, an expert on living and fossil equids, was "not very well pleased" with the young female points to an occasional theme in the correspondence between Hornaday and Hagenbeck: Hornaday was often skeptical about Hagenbeck's claims and worried that the German dealer might be giving his best animals to presumably preferred customers in England and Germany. In a letter from March 26, 1904, complaining about a number of recent shipments from Hagenbeck, for example, Hornaday listed an "Indian Sambar," represented by the dealer as the "finest buck," that turned out to be a "hornless fawn one year old" and a "Giant Bear" from Japan that turned out to be nothing more than "a small dwarfish animal that has not grown a particle since arrival" (Hornaday 1904).

Mohr originally brought up the story of the two horses' being not "favorably regarded" to point out the selectiveness of the buyers, something she found continued in efforts to breed the animals. Although she does not speculate about what caused the selectiveness, it seems likely to have arisen because original buyers (like Hornaday) had learned to question the claims of animal dealers and because the phenotypic variability of the original horses, along with rumors of "impurity," meant the buyers found themselves choosing one horse (as being a better specimen) over another. Of course, none of these people had ever seen a Przewalski's horse, but they felt they knew what the horses should look like because of the published description and because the animals had been described as "primitive." For these buyers, such a creature would naturally show characteristics of both the modern domestic horse and what was imagined as a more primitive "stage" of horse evolution: the asses and zebras. Przewalski's horses, then, should have horselike ears, but their erect manes, lack of forelock, tan color, dorsal stripe, and general stoutness should recall the asses and zebras. In fact, Przewalski's horses very quickly became that much sought after object of taxonomic science in the period—a link between species, and as such, the more "linkish" the specimen was, the better. Beyond this, it seems likely that the cave drawings themselves were used to help establish the horses' approved conformation. In the end, facing phenotypic variability, concerns over intermixture, and quickly established standards for the horses, buyers and breeders sought, through careful purchases and selective breeding, to minimize reputed domestic "blood" in the population while cleaving to increasingly exacting conformation standards. In short, even fifty years after the horses first arrived in the West, they were something very different from what was captured in Mongolia at the beginning of the century, and this situation has not changed in the past sixty years.

"PERFECT" REPLACEMENTS FOR NEW YORK'S ORIGINAL HORSES

In his *Popular Official Guide to the New York Zoological Park*, Hornaday (1909) describes the replacement Przewalski's horses in the Bronx garden. First he discusses the species, noting the mane, the tail, the lack of forelock, and the dorsal stripe, and explains that Przewalski's horses should be understood as the "connecting link" between horses and the other equids. Hornaday then turns to the two animals being exhibited. He notes that the "parents of these horses were captured in 1900 by an expedition sent out by Carl Hagenbeck" and that they "are very perfect and typical representatives of their species" (Hornaday 1909, 48–49). In establishing a believable (if inaccurate) pedigree for the animals and insisting that they are "perfect and typical representatives," Hornaday is clearly speaking to an audience expecting the zoological garden to exhibit "the real thing." He may also have been speaking to those who might have remembered the first two specimens of the horse to have been shown at the zoo.

In comparing the photograph of the two horses in this *Guide* (fig. 6.2) with the zoo's earlier horses photographed in front of the Asian deer house (fig. 6.1), it is clear why the first two animals were not liked. The problem is largely with the original mare, known in the studbook as 18 Bijsk 8 (at the left in fig. 6.1): to put it simply, she does not look like what a Przewalski's horse was supposed to look like.

It turns out, though, that her case is a bit more complicated. First of all, she is not even the horse listed in the studbook. In a letter of December 13, 1902, Hagenbeck wrote to Hornaday that although he had shipped the horses the day before, he had "had to take a mare from this year instead of the mare from last year, as the animal got sick two days before we intended to ship her and she looks rather queer that I did not risk to send her." He continued: "The young female I did send is a very strong animal for its age, and I think, will turn out an excellent beast" (Hagenbeck 1902). A follow-up letter from Hagenbeck on December 31 shared the news that the older filly (the actual Bijsk 8) had died on December 27, probably to quell any concerns Hornaday may have had about a last minute bait-and-switch. Per-

FIGURE 6.2. "Prjevalsky Horses" from Popular Official Guide to the New York Zoological Park (1909). Reproduced by permission of the Wildlife Conservation Society Archives, Bronx, New York.

haps it is not surprising that the less than one-year-old filly actually sent to New York (she should perhaps be recorded as Bijsk M in the studbook) was weak when she arrived; however, despite Hornaday's apprehension (and Hagenbeck's assurance that he would replace her if she died), she rallied.

She didn't die, but Hornaday didn't find her suitable for the zoo either. Two years after the animals arrived at the Bronx Zoo, Hornaday insisted to Hagenbeck that *both* of the first two horses be replaced because the male did not meet his expectations for the size of a Przewalski's horse and the mare, he argued, was actually a Mongolian domestic horse, not a Przewalski's horse. On January 7, 1905, he wrote to Hagenbeck that he wanted to put the "unfortunate matter" of the Przewalski's horses behind him. Noting that the animals had cost $1,800 and that Hagenbeck was currently selling young stallions for $600, he did not mince words: "Regarding this Mongolian pony mare,—she is of no value to us, and we would rather not keep her. The stallion is, as you saw, very much under-size, and he is now old enough that we know he never will be any larger. He does not properly represent the size of the Prjevalsky horse, and we wish that you would take him back again with the mare, and send us a new pair, that will be satisfactory" (Hornaday 1905). In short, the stallion was too small and the mare was simply not a Przewalski's horse. Hagenbeck complied, and the replacement horses thrived, had thirteen offspring, were frequently written about and photographed, and are two of the founders of today's herd. The original horses, though, also thrived at their new home, the Cincinnati Zoo, where they had one offspring, a colt designated 113 Cinc 1 in the studbook, and through him they too became important founders.

LEGACIES AND LESSONS

It is important to know more about the original horses sent to the Bronx Zoo in 1902 *not* to add another volley in the long battle about the "purity" of Przewalski's horses. That debate has more to do with late nineteenth- and early twentieth-century ideas about race, eugenics, and prehistoric life than with the current and future status of these animals. Indeed, Mohr's efforts in the 1950s should have put to rest the question of "purity." Recent studies of the genetics of living, founder, and ancient Przewalski's horses make it clear *again* that over the thousands of years since the lines leading to modern domestic horses and Przewalski's horses split off from each other, there has been ongoing genetic sharing among them (Der Sarkissian et al. 2015; see also Bowling et al. 2003; Geyer and Thompson 1988). In the case of the Przewalski's horse, "purity" is simply not as biologically, ecologically, or historically useful a concept as has generally been assumed. With that said, the correspondence between Hagenbeck and Hornaday should serve first as a reminder that documents like studbooks must be regarded as historical artifacts, not collections of objective facts. The correspondence is also important for those interested in the genetic history of the horses, because if the original New York mare is an F1 hybrid (as she has been generally seen since the mid-1980s), then she is potentially even more important in reducing inbreeding depression because she is less likely to have been sired by the same stallions as the other foals in Hagenbeck's first shipment, many of whom presumably were closely related, and because if judgments of the time are correct and the original New York mare was actually a Mongolian domestic horse, she becomes a more substantial source (among many others) of domestic horse genes in the current world herd.

In the end, the reason to know more about the first Bronx horses is that they highlight how ex situ conservation and reintroduction efforts are rarely (if ever) the simple win-win stories that the marketing departments (and some directors) of zoos seem to love. If zoos wish to be seen as organizations committed to science and knowledge, then telling the vexed stories of trying to help animals should be a more important task than advancing the often extreme claims that get made about reintroduction. The story of the mare should make it clear that the "Przewalski's horse," the "Mongolian wild horse," and the "takhi" (the name used today for these horses in Mongolia) are creatures that have at some significant level been *made* by Western and now international expectations. I am not claiming that the horses being "introduced" or "reintroduced" into Mongolia and China today are not the descendants of horses that were brought into captivity beginning at the end of the nineteenth century; I am arguing that they are much more than that, and much more interesting.

As Oliver Ryder noted in 1993, just before reintroductions of the Przewalski's horse began, "After prolonged and expensive efforts to minimize genetic changes accompanying the captive propagation of wild horses, human interference with wild patterns of behavior of released animals would be antithetical to the goals of reintroduction." Thinking about the prospects of successful reintroduction, he hoped that "free-living wild horse populations should integrate with and not be separate from the other biotic components of their habitat" (Ryder 1993, 15). Part of that habitat for thousands of years, he noted, has been the human population and its domestic livestock. That Przewalski's horses have been interbreeding with domestic horses over millennia points to perhaps the most controversial question raised by the story of the first Bronx horses: How justified can we be in "protecting" reintroduced Przewalski's horses from introgression from domestic horse genes (seen as a threat in IUCN and other documents [Wakefield et al. 2002]) when we know that interbreeding was a part of the long history of the horses before they became known in the West; that two of the thirteen founder horses were understood at the time to be Mongolian domestic horses; and that efforts to preserve the "purity" of the horses over the past century have been driven—often disastrously—more by ideology than anything else?

Answering this question will require that we look beyond creating positive messages in zoos as a way to foster the public's affective relationships with animals and commitments to conservation. Those who advocate for almost exclusively feel-good messages over scientific and historical facts cannot but undermine the credibility of the whole zoo industry. Similarly, those who cherry-pick zoo "success stories," like that of the Przewalski's horse, to make claims about the significant role zoos can play in efforts at international conservation only make it more apparent that most zoos are only indirectly and minimally engaged in sustaining wild populations of endangered species. If there is a credible role for zoological gardens in international conservation, it will not be to tout self-aggrandizing stories of "saving" this or that species from extinction; rather, the role, as many other contributors to this volume make clear (see esp. Palmer, Kasperbauer, and Sandøe; Ivanyi and Colodner; Monfort and Christen; Norton; Cerezo and Kapsar; and Mendelson), must be to chart new directions for zoo conservation that focus on efforts to teach and foster sustainability both locally and in distant settings.

ACKNOWLEDGMENTS

I am grateful for comments on earlier versions of this chapter from the editors of this volume and the many participants in this unique project. I particularly thank Madeleine Thompson of the Wildlife Conservation Society Library and Archives for her knowledgeable and thoughtful engagement.

CHAPTER SEVEN

Conservation Constellations: Aquariums in Aquatic Conservation Networks

Samantha Muka

Aquariums[1] are an important node in the aquatic conservation network. Similarly to zoos, aquariums work to conserve species both in situ and ex situ; aquatic conservation measures are diverse and diffuse, ranging from education and fund-raising to captive breeding and stocking. However, unlike zoos, aquariums often work closely with local wildlife agencies and commercial industries to achieve a wide range of goals. Conservation measures are enacted by a large network of participants, including public aquariums, universities, private conservation groups, fisheries biologists, aquatic organism dealers, anglers, hobbyists, and volunteers. Most conservation efforts include several of these groups working together, and many larger initiatives include all of them. It is uncommon to see the same aquariums involved in every conservation effort; large aquariums often operate at local, national, and international levels, while smaller institutions participate primarily within local or regional networks.

Although zoos and aquariums have expressed a desire to strengthen these networks (Barongi, this volume; Grow, Luke and Ogden, this volume; Knapp, this volume; Lukas and Stoinski, this volume; Grajal, Luebke, and Kelly, this volume; Cerezo and Kapsar, this volume), little work has been done to highlight the actual history and function of these networks. In addition, while discussing networks can be useful to illuminate the requirement of diverse communities in realizing conservation goals, the expansiveness of these networks can be daunting to trace and can actually obscure the avenues and collaborations by which conservation is enacted (Boyes and Elliot 2014). For these two reasons I refer to each community working toward a single conservation goal as a *constellation* to signify that, while they operate in a large network of marine conservation professionals, they are distinct configurations where that effort is concerned. Using the term constellation instead of network helps us clearly identify the conservation effort that members of the network focus on, the members of the network, and the role of individual members. These three variables are obscured when using only the term network—when analyzing a large network the term aquarium stands in for all aquariums regardless of size or function—but it can be revealed when looking at a smaller formation.

The third variable, the role of the individual aquarium in a conservation constellation, is the hardest to examine with a wider lens. Aquariums often perform a variety of conservation functions depending on total resources and on those required for each endeavor (Grow, Luke, and Ogden, this volume). These institutions have a variety of resources that make them integral to a wide range of conservation initiatives. Permanent tanks offer space to maintain organisms for breeding and rehabilitation. This means they might play an important role in ex situ conservation such as captive breeding programs, studying life cycles of endangered species, and rehabilitating injured specimens. Aquariums also contain advanced laboratory equipment, allowing them to necropsy and process a wide range of samples. These necropsies can help in understanding life cycles, aquatic infectious diseases, and other causes of death that inform in situ conservation efforts. Finally, public aquariums reach wide audiences by displaying animals, allowing them to educate visitors about environmental issues and consequently raise funds directly for conservation initiatives (Spring, this volume). While some larger aquariums perform all these functions in different constellations, smaller aquariums might participate in only a few constellations and perform the same function in all of them (the most common being education).

This chapter highlights the role of aquariums in aquatic conservation constellations.[2] The work is divided into three sections: history and current work on conservation of game fishes; recent conservation efforts for ornamental tropical species; and the role of aquariums in marine mammal conservation. This division is important because the issues and avenues of conservation for game, ornamental, and mammal species are distinct, and aquariums have different roles in each of these constellations. I emphasize one major conservation program in each section. These examples are not meant to stand for "best practice" programs but instead exemplify how a constellation is constructed and the role of aquariums in these programs. In addition, I examine conservation only of aquatic species. While conservation programs focusing on avian or amphibian species exist at aquariums, these programs more closely resemble those operated through zoological parks and are often run in conjunction with them (Mendelson, this volume).

GAME AND COMMERCIAL FISH

The first aquariums grew out of the earliest recognition that fisheries were being depleted and the subsequent growth of fisheries programs throughout the United States. Americans considered their aquatic resources robust and inexhaustible (this myth of superabundance occurred in the terrestrial realm as well; see Barrow, this volume; Henson, this volume), but in the 1860s those thriving on coastal production began to notice a change in catches. Spencer Fullerton Baird, assistant secretary of the Smithsonian (Henson, this volume), was asked to investigate claims by fishermen in both Massachusetts and Rhode Island that using certain types of nets caused a decrease in fish stocks. Baird was given limited time and resources to investigate these claims. He spoke with local fishermen in both states to gauge stock depletion based on local knowledge and presented these findings to both states. Each state ruled differently (Rhode Island banning certain nets, Massachusetts seeing no evidence for doing so), and the outcomes convinced Baird of the need for a more systematic investigation. According to Baird, "This remarkable contradiction in the results of the two commissions showed the necessity of a special scientific investigation on this subject, to be prosecuted in the way of direct experimentation on the fishes themselves, their feeding and their breeding grounds" (Baird 1873, viii; Allard 1978). The US Fish Commission (USFC) was founded in 1871. The federal structure of the USFC, combined with state agencies, increased scientific interest and research into managing and strengthening fisheries in the United States (Smith 1994).

Participation of the USFC and its state partners in fisheries expositions led to the first public aquarium exhibits. Beginning in 1880 in Berlin, a series of fisheries exhibitions were held, moving to Edinburgh in 1881 and London in 1883. In addition to cooking demonstrations and live music, countries were invited to display their fishing economies and cultures. The Americans exhibited regional fishing gear including native Alaskan nets and boats, commercial goods such as Tiffany's pearl brooches and isinglass, and tanks filled with native fishes and reptiles. The exhibitions proved extremely popular with visitors and resulted in temporary displays during the Chicago and St. Louis world's fairs and the American fisheries exhibitions in Chicago (1893) and Tampa (1898), as well as a permanent space at the fisheries laboratory at Woods Hole (Hillard 1995, 88–90). These exhibitions provided the first glimpse of underwater scenes for the general public and whetted their appetite for larger permanent displays (Hubbard 2014; LaChapelle and Mistry 2014).

Many states eventually converted their traveling exhibitions into permanent aquariums. For example, the Pennsylvania exhibit at the St. Louis World's Fair consisted of thirty-five tanks of live fish lit from above. Visitors walked past an indoor waterfall and pool into a grotto structure where a darkened passageway made the fishes more visible. The Pennsylvania Commission sent three train car loads of fish (the first two loads died from heat and high aluminum levels in the local water). However, interest was so high that people were touring the exhibit even when tanks were empty (Lambert 1905). The city of Philadelphia, seeking to capitalize on this interest, suggested creating a permanent aquarium out of the St. Louis tanks, but public interest in aquatic exhibits was not enough to secure funding. It was proposed that the aquarium also work with the state fisheries to function as a hatchery. The relationship would be symbiotic: the aquarium would receive fry from the fisheries department, rear them to adulthood, displaying them to the public, then return mature fishes to the state for release into local waters. This system proved mutually beneficial, and the Fairmount Aquarium in Philadelphia opened in Fairmount Park in 1911.

This relationship between fisheries and aquariums was strengthened as personnel from fisheries departments moved to aquariums. At Philadelphia, William Meehan, former director of the Pennsylvania Fish Commission, was the first director, and many staff members also came from that community. Charles Haskins Townsend, second director of the New York Aquarium (the first after it was purchased by the city) was a scientific investigator from the US Bureau of Fisheries, and many of his staff, including his successor, came from that group (Burnett 2012, 297–99). Other major aquariums, including the Belle Isle Aquarium at Detroit and the Washington, DC, Aquarium, were headed by fisheries personnel and worked closely with the federal and state fisheries departments to further fisheries research and initiatives through those institutions.

The strong symbiotic relationship between aquariums and fisheries departments established conservation concerns at these institutions for the first half of the twentieth century. In March 1914 the New York Aquarium (NYA) received fifteen brook trout from the New York Conservation Commission. On May 16, 1914, they sent fifteen hundred brook trout fry to the New York Conservation Commission at the Cold Spring Harbor hatchery for release into local waterways (Charles Townsend Papers, director's log,

Wildlife Conservation Society Archives). These trout were reared in a special exhibit room of the NYA where visitors could view them. Docents explained the development process as well as how to catch, clean, and cook the fish. Signage throughout the aquarium pointed to important game fish displays and informed visitors of the aquarium's efforts to enhance their numbers (fig. 7.1). The aquarium bred common game fishes for introduction into local waters and educated visitors on their importance to the economy. In this way the aquarium operated as both a hatchery and an educational unit in this conservation constellation.[3]

Over time, the methods by which aquariums engage in fisheries conservation has shifted, but their integral role in the constellations dedicated to conserving overfished species remains. Public aquariums functioned as auxiliary hatcheries for fisheries services until the mid-1930s. First-generation aquariums in New York, Philadelphia, Boston, and Detroit, and even the Shedd Aquarium in Chicago, had dedicated hatchery rooms and explicit agreements with their local fisheries bureaus. However, as more aquarists were trained in academic laboratories or in the aquarium setting (as opposed to being trained in the field with fisheries experts), breeding for eventual introduction decreased. This shift changed the form fisheries conservation took, but it did not end it.

FIGURE 7.1. Sign from the New York Aquarium ca. 1905. "New York Aquarium Fish Exhibit Labels Battery Park." © Wildlife Conservation Society. Reproduced by permission of the Wildlife Conservation Society Archives, Bronx, New York.

One example of a robust fisheries-based conservation constellation is the "Saving the Sturgeon" (STS) program. Lake sturgeon, an important commercial species, were overfished and eventually extirpated from the Tennessee River by the 1970s. After initial research in the 1980s by the Tennessee Valley Authority (TVA), the Tennessee River Lake Sturgeon Working Group (TRLSWG) formed specifically to reintroduce lake sturgeon to the area. This constellation included the Tennessee Aquarium, Tennessee Aquarium Conservation Initiative (research arm of the Aquarium TNACI), Tennessee Tech University, University of Tennessee, TVA, US Geological Survey, US Fish and Wildlife Service, World Wildlife Fund, Conservation Fisheries, Inc. (a nonprofit geared toward restoring and maintaining biodiversity in the southeastern United States), Tennessee Clean Water Network, and Wisconsin Department of Natural Resources. In addition to formal partnerships and funding, TRLSWG works with both commercial and recreational anglers to monitor the health of released fish, providing incentives for citizen scientists to participate as well. All together, TRLSWG members released 7,000 to 14,000 sturgeon yearly; as of 2012, over 115,000 have been released into the Tennessee River and surrounding locations (George, Hamilton, and Alford 2013).

The Tennessee Aquarium performs several roles within this constellation. TNACI, their research arm, is one of several facilities that hatch, release, and record. The facility receives fry that they rear for one year. They release roughly 90 percent of these, retaining 10 percent for tagging programs to gather data on the range and life span of released specimens. In 2015 TNACI received a $17,000 grant from the AZA to expand their sonic tagging program and to participate in the Saving Animals from Extinction (SAFE) program (Grow, Luke, and Ogden, this volume). The aquarium uses its role as educator to raise awareness and money for this conservation initiative. Visitors at the sturgeon touch tank can feel the bony fish's spines while a docent tells them about natural history and the importance of the reintroduction program. In addition, the aquarium promotes conservation learning through classroom projects. Each year, students in Gap Creek Elementary School's fifth-grade class rear a sturgeon, taking daily measurements. Alongside their lessons on aquatic vertebrate biology, the students learn about conservation and water quality issues. They attend the annual release of the fishes. In this way the aquarium functions as the public face of a much larger constellation of groups involved in this successful initiative (George, Hamilton, and Alford 2013).

The work done by the Tennessee Aquarium and TNACI relies on a similar constellation of work done by the New York Aquarium at the turn of the twentieth century. The major differences are reflected in the understanding of conservation and the science involved, but it is clear that these institutions have been integral to reintroduction of game and commercial fishes since their earliest years.

TROPICAL AND ORNAMENTAL FISH

The earliest aquariums did not have conservation initiatives regarding tropical and ornamental fishes. The first US aquariums were in temperate areas. These institutions, including New York, Philadelphia, Detroit, and Boston, sent a collector to Key West twice a year to acquire specimens for display. The colorful species created a jewel box effect and were widely advertised to bring in visitors (Bonner 1926). However, many of these displays were short-lived. Temperature and mineral fluctuations in the water supply caused massive mortality, and the distance to collecting grounds meant many tanks were filled temporarily with local commercial or game fishes loaned by the fisheries services until the next collecting trip. Unlike local specimens, which continued to be brought to the aquarium by fisheries personnel, local fisherman, and aquarists throughout the twentieth century, tropical and subtropical fishes became increasingly commercialized. Most tropical fishes were bought from private collectors specializing in procuring them for the professional aquarium and hobbyist (pet store) trade (Grier 2006, 332–34).

Throughout the twentieth century, advances in technology and husbandry made it possible to keep ornamental tropical fishes in captivity longer. In addition to fishes, beginning in the 1970s advances in the husbandry of coral created a surge in tropical displays (Lewis and Janse 2008). Both of these advances made it possible for aquariums to maintain intricate reef scenes complete with species never before viewed. As the popularity of these displays increased, professional collection for hobbyist and public aquarium displays eventually contributed to the decline in fishes and damage to the world's coral reefs (one study estimates an 80 percent mortality rate for ornamental specimens during transit) (Wadnitz et al. 2003; Sadovy and Vincent 2002). This decline led to concerns about the availability and health of tropical species, and eventually to conservation initiatives.

The Syngnathidae, including seahorses, sea dragons, sea moths, and pipefishes, is one family around which a large conservation initiative has arisen in the past twenty-five years. The earliest aquariums displayed seahorses to great public delight. The New York Aquarium displayed over three hundred local seahorses in Battery Park in the early twentieth century and also displayed Australian and tropical species when they could. The animals became so popular that the aquarium used an illustration of them on the cover of the official tour guide (fig. 7.2) (Townsend 1919). The seahorse continues to be a staple display at public aquariums. While many aquariums lack charismatic megafauna, crowds seem drawn to these creatures and, especially in the 1990s, whole exhibits were dedicated to the entire family Syngnathidae. It was about this time that a variety of communities began to notice the decline in natural populations and began concentrated efforts to decrease stress on them.

Project Seahorse is a conservation initiative aimed at preserving ocean ecosystems by raising awareness of the human impact on a single species. In 1998 thirty-five participants from seventeen countries attended a workshop

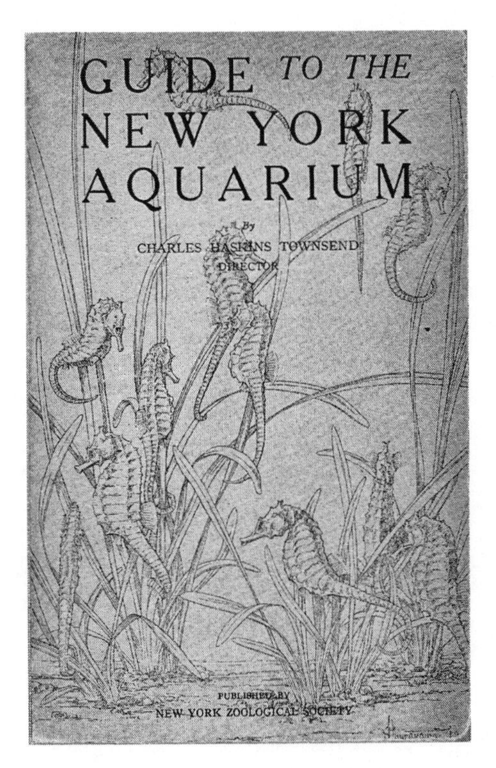

FIGURE 7.2. The front cover of the New York Aquarium's tour guide. The image was drawn by H. Muriyama from life and depicts the seahorse tank at the aquarium ca. 1918. Courtesy of the Wildlife Conservation Society Archives.

in Cebu City, the Philippines, on the management and culture of marine species used in traditional medicines. The participants came from a variety of fields, including the traditional medicine trade, fishing, aquaculture, conservation, aquariums (specifically from the John G. Shedd Aquarium in Chicago), medical anthropology, and gender research. The meeting ended with several "outputs" including "creating a priority list of species that should (and could?) be produced by aquaculture, with input from both the TM (traditional medicine) and aquaculture communities; developing an educational exhibit of live seahorses, to tour areas with high concentrations of TM use to promote interest in conservation, with a reciprocal exhibit on the use of marine species in TM, to take to Western zoos and aquaria; promoting research into seahorse biology, ecology and behaviour, and the marine medicinals trade" (Moreau, Hall, and Vincent 2000, 8). The workshop led to a conservation constellation focused on decreasing pressure on native populations from the traditional medicine trade and from the aquarium/hobbyist fish collection community.

Aquariums were integral to achieving the outcomes outlined in this meeting. In particular, aquariums focused on three roles: studying fish biology, distributing captive-bred specimens, and education.

Biology

Aquariums work with each other and with dealers to understand the biology of the Syngnathidae. Correctly taxonomically identifying species in captivity is important to maintaining pure breeding lines (see Ryder, this volume; Rothfels, this volume). Larger programs, such as the Shedd's, sponsor direct field research to identify species in situ. Ex situ, the Association of Zoos and Aquariums (AZA) runs a taxon advisory group (TAG) for several species of seahorses (AZA 2014e; Vincent and Koldewey 2007).

Captive Breeding

Since the 1990s, aquariums have advanced husbandry techniques to produce seahorses in captivity. Closing the cycle for a variety of species has allowed aquariums to maintain displays without continually collecting. It also allows them to trade these organisms throughout the network of aquariums, thereby operating as specimen dealers and alleviating collecting stress on wild populations from the larger aquarist community. Husbandry techniques and breeding centers operate at aquariums throughout the United

States (Koldewey and Martin-Smith 2010; Koldewey 2005; Foster and Vincent 2004).

Education

Finally, aquariums have shifted their displays to integrate and disseminate conservation information from Project Seahorse. The Birch Aquarium, the public aquarium for the Scripps Institution of Oceanography, combines its husbandry research with displays for public education. In displays reminiscent of the early twentieth-century hatcheries, Birch's "There's Something about Seahorses" exhibit displays Syngnathidae in various growth stages. Throughout the display (which takes up an entire room) visitors learn about the threat to these species in the wild and the process of breeding them in captivity. Similar to its predecessors, this display is the public face of a variety of conservation initiatives.

Project Seahorse is a large constellation. Each aquarium involved plays a separate role in achieving the program's goals. The difficulty of describing this constellation is in teasing apart the varied roles played by the aquarium community. The Shedd functions in all three areas (see Knapp, this volume), while a small aquarium might participate only by displaying captive-bred specimens and including conservation-oriented language in display information for a tank.

MARINE MAMMALS

The history of successfully keeping marine mammals in captivity is brief. As with charismatic megafauna for zoos, aquariums coveted marine mammals for their ability to draw crowds (Hanson 2002). The New York Aquarium developed specialized shipping tanks for transporting dolphins from the only American dolphin fishery, in Cape Hatteras, North Carolina, to their Battery Park location. The aquarium also purchased manatees from Florida and seals from California and Maine.[4]

However, mammals proved difficult to maintain in captivity.[5] Most of the dolphins and manatees died en route, and those that survived lived only a short time (under six months in most cases). Aquarists had limited success at maintaining cetaceans in captivity; seals and manatees proved more amenable but still required large enclosures and resources. In addition, these aquariums were in the North, and seasonal complications, such as finding indoor space during the winter, prohibited long-term displays.

The Fairmount Aquarium's solution was a sort of seal time-share: during the summer, seals lived in a space next to the aquarium where they could frolic on the muddy banks of the Schuylkill River, and in the fall they were sent by train to the Million Dollar Pier in Atlantic City, New Jersey, to serve as a tourist attraction in a large indoor atrium. These complicated schemes eventually broke down, and after initial enthusiasm they became increasingly sporadic. Aquariums accepted bycaught or stranded mammals from fishermen and citizens, but by the 1950s they were more judicious in their mammal displays, choosing to purchase and display species with which they had known success.[6]

No public aquarium or marine park has ever bred marine mammals for conservation purposes.[7] A few early aquariums sought to establish breeding colonies of threatened mammals. The Fairmount Aquarium in Philadelphia requested several seals from the US government to build a breeding program for the threatened but valuable Pribilof fur seals. This colony was meant to be in line with the fisheries-based conservation initiatives of these institutions. William Meehan, aquarium director, was former commissioner of fishes in Pennsylvania, and he felt that establishing a breeding colony in Philadelphia would provide enough revenue to eventually run the whole aquarium and provide a new industry for the textile-rich region. This scheme failed because the US government would not provide the animals, but it demonstrates that fisheries principles were applied to mammals as well as fishes (Townsend 1893). Other aquarists took note of the low numbers of certain mammal species. The New York Zoological Society ran a story in 1910 highlighting their exhibit of three Caribbean monk seals, which they stated was "the most noteworthy exhibit in the building" because they were "three flourishing specimens of a large species near the verge of extinction" (fig. 7.3). Although the article discussed documenting the biology and habits of these species for posterity, there was no robust conservation conversation regarding mammals at these institutions until the last quarter of the twentieth century (*Zoological Society Bulletin* 1910, 644).

Although public aquariums do not breed for conservation, they have found other avenues to contribute to marine mammal conservation. For instance, both public aquariums and marine parks have become integral nodes in regional Marine Mammal Stranding Networks (MMSN) as part of conservation efforts for threatened and endangered species. A marine mammal stranding is defined (by the US National Marine Fisheries Service) as "any dead marine mammal on a beach or floating near shore; any live

FIGURE 7.3. Two of the last Caribbean monk seals seen before their extinction. The New York Aquarium received these from a dealer in the Yucatán Peninsula in 1909. Zoological Society Bulletin, 1910, 644. Courtesy of the Wildlife Conservation Society Archives.

cetacean on a beach or in water so shallow that it is unable to free itself and resume normal activity; or any live pinniped which is unable or unwilling to leave the shore because of injury or poor health" (Wilkinson 1991). The passing of the Marine Mammal Protection Act prompted the formation of regional stranding networks, formally established along the US coast in 1981. While regional networks originally operated independently (administered by the National Marine Fisheries Service), by 1991 the stranding network became nationally centralized in order to collect information and regulate stranding policies. While there is currently a coordinator for each regional group, the network is made up of a large community of interested parties, including universities, aquariums, marine parks, citizen scientists, and fisheries (Wilkinson and Worthy 1999).

One such program is the Southern Sea Otter Research Alliance (SSORA). It includes the University of California at Santa Cruz, the University of California at Davis, the Monterey Bay Aquarium (MBA), the US Geological Survey, the US Fish and Wildlife Service, and the California Fish and Wildlife Marine Wildlife Veterinary Care and Research Center. Within the constellation, the MBA runs the Sea Otter Research and Conservation

(SORAC) program that has treated more than seventy stranded newborn sea otter pups. The program receives stranded pups from local waters and, with the help of the constellation, performs research on the species, rears and reintroduces stranded pups, and educates the public about a declining population (Stevens 2009; Kieckhefer et al. 2007).

Biology

Stranded sea otter pups are transferred to the MBA. These animals are nursed back to health in hopes of reintroduction, but they are also studied during captivity to understand the basic physiology of the species. For instance, between 2009 and 2011, seven stranded pups (one male, six females) were observed for a study to measure the energy demands of nursing on female sea otters. This study found that energy shortfalls contribute heavily to the mortality rate of sea otters in the Monterey area (Thometz et al. 2014).

Rear and Release

SORAC works to rear and release stranded otters. They have refined their methods, and in 2007 they reported that an effective technique includes pairing stranded pups with older surrogate captive females. While reintroduction is the ultimate goal, the process of rearing and releasing also provides basic information about the species and contributes to in situ conservation goals of the SSORA (Nicholson et al. 2007).

Education

Many stranded otters are not suitable for release. These specimens play a huge role in the MBA's conservation public outreach and education. Not only are otters held in-house at the MBA, they are distributed throughout the United States.

Marine mammal conservation is one of the newest forms of conservation at public aquariums. While many institutions take part in these initiatives, they play significantly different roles depending on their place in the constellation in which they are situated. Some are at the center of that constellation (such as MBA and SORAC), while others have only minimal roles. Regardless of the actions aquariums take for conservation, they always work within a constellation of actors.

CONCLUSION

Public aquariums perform marine conservation within a series of constellations composed of a constantly shifting contingent of organizations, institutions, and volunteer groups. The history and current status of these conservation initiatives is difficult to trace because each aquarium operates within a variety of conservation constellations, performing a wide variety of functions within them. An aquarium could belong to an MMSN and a TAG and display and disseminate information regarding game and commercial fisheries. That would be an aquarium with three separate conservation constellations. The same institution might perform multiple roles in each constellation depending on resources, conservation goals, equipment, and staff. This picture of conservation at aquariums is dynamic and constantly shifting.

Public aquariums are integral to marine conservation initiatives, yet we lack even a basic understanding of the history of these institutions and their environmental and ecological work. Consistently, researchers studying zoo conservation have extrapolated findings and applied theories to aquariums. For instance, one zoologist discussing conservation at zoos stated that their "adopted *raison d'être*—[was to] become arks for the sorely pressed wild forms of the day. (I include aquariums when I refer to zoos in this discussion)" (Lindburg 2007, 437). In fact, most aquariums do not engage and never have engaged in ark conservation initiatives.

Zoos and aquariums share many conservation goals and sometimes work in tandem, especially on projects concerning reptiles and amphibians, but clear differences can be seen. One major difference is that, unlike zoos, aquariums acknowledge, and in many instances encourage, visitors' recognizing displays as pets or food sources. Conservation constellations involve hobbyists, specimen dealers, anglers, and fisheries experts not as antithetical to their goals, but as integral to the successful implementation of programs. Whereas zoo conservation is dedicated to ending the exploitation and consumption of their animals for entertainment, food, or medicine, aquariums must acknowledge and work within constellations that consider their displays acceptable in all these categories.

The casual acceptance within the zoo community, from scientists, zoologists, and historians alike, that aquariums are similar to zoos in their conservation activities has greatly hindered our understanding of their actual work in the marine conservation community. It is important that we study them on their own terms to understand the development and continued growth of marine conservation initiatives.

PART III

Zoo and Aquarium Conservation Today: Visions and Programs

CHAPTER EIGHT

Committing to Conservation: Can Zoos and Aquariums Deliver on Their Promise?

Rick Barongi

Despite significant contributions by some zoological institutions to saving animals in the wild, zoos and aquariums are still perceived by some as exploiters rather than champions of wildlife conservation. The newly released WAZA Conservation Strategy, *Committing to Conservation: The World Zoo and Aquarium Conservation Strategy* (Barongi et al. 2015; fig. 8.1), should provide further evidence of the increasingly vital role modern zoological facilities play in biodiversity conservation. Maximizing this potential is the focus of the WAZA Conservation Strategy and of this chapter. WAZA, the World Association of Zoos and Aquariums, represents a global community of zoological facilities united for the care and conservation of living fauna and flora.

As defined in the WAZA Conservation Strategy, conservation is "securing populations of species in natural habitats for the long term." This proactive publication provides a wealth of best practices and business models that will assist the leaders of all institutions with live animals in their care to realize a core mission of their organizations—the conservation of species in the wild. If they can deliver on this promise, zoos and aquariums can rank as one of the most powerful forces for conservation on the planet.

Zoological institutions are uniquely positioned to be species champions, with their broad-based community support and over seven hundred million annual visits worldwide (waza.org). As stated by Inger Anderson, director general of the International Union for the Conservation of Nature (IUCN), in the foreword to the WAZA Conservation Strategy, "Zoological facilities have an unrivaled platform to engage the general public in conservation." Zoos and aquariums must back up inspiring messages about animal care

FIGURE 8.1. The newly released WAZA Conservation Strategy.

and conservation with direct support for saving animals in the wild. Supporting field conservation with monetary donations is certainly not the only way zoos and aquariums contribute to saving species in the wild. Raising awareness and teaching pro-environmental behavior are also critical to the behavior-change campaigns being promoted by many organizations (Barongi et al. 2015, 44–49). However, it is the direct support of field conservation and the positive influence of that support that have the most immediate effect on wildlife.

One key fact WAZA organizations have discovered since the last conservation strategy, published ten years earlier (WAZA 2005), is that when visitors understand that zoos and aquariums are working to save animals in the wild, their support for these institutions increases dramatically (Barongi et al. 2015, 23). However, zoos and aquariums are rarely perceived as conservation organizations because of historical practices and the inconsistent commitment within the zoological profession toward conservation initiatives.

The shared vision for all WAZA members is to create a stronger connection between their resident animals and their counterparts in nature. This is the basis of the One Plan Approach to field conservation (Byers et al. 2013; fig. 8.2; see also Traylor-Holzer, Leus, and Byers, this volume). The One Plan Approach is an integrated species conservation plan that considers all populations of species (inside and outside the natural range), under all conditions of management, and engages all responsible parties and resources. This planning strategy encourages a more direct connection between the exhibits and the wild habitats they showcase, so that every zoological institution can demonstrate a measurable impact in saving species in the wild.

Although money is not the only way to address conservation, it is a critical element for supporting long-term programs and is one measure of an institution's commitment. The three hundred WAZA member organizations contribute approximately $300 million to wildlife conservation each year (Gusset and Dick 2011). However, most of these funds come from 10 percent of the WAZA membership, while the other members designate less than 2 percent of their annual operating budgets to field conservation. If all WAZA institutions designated 5 percent of their operating budgets to conservation, over $1 billion would be generated annually.

These estimates are aspirational goals, but they are not unrealistic. Still, zoological facilities will require a new business model and branding

ONE PLAN APPROACH

Definition: Integrated species conservation planning that considers all populations of the species (inside and outside the natural range), under all conditions of management, and engages all responsible parties and resources from the start of the conservation-planning initiative.

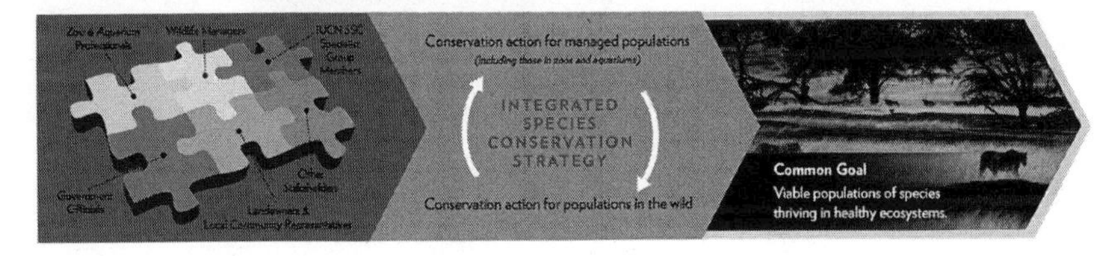

FIGURE 8.2. The One Plan Approach.

7 STEPS TO CONSERVATION LEADERSHIP

Step 1: Inform

Educate your governing authorities and staff about the status of wild populations of animals on a regular and ongoing basis, and demonstrate how everyone can play a noteworthy role in reversing the declines.

Step 2: Mission

Update the mission statement and strategic plan of your zoo or aquarium to include: a *declaration* that your institution exists for a higher purpose—wildlife conservation; a *pledge* that your institution will commit resources to this effort; a *plan* for creating a culture of conservation in your staff, communities, governing authorities and donors that gives everyone the opportunity to make a measurable difference.

Step 3: Budget

Assess how much your institution currently spends on field conservation according to the WAZA definition of conservation, and benchmark that with similar regional institutions.

Step 4: Revenue

Work with staff to identify dedicated streams of revenue that can be used for field-conservation programmes. Ideally, these are streams generated both internally (from operating budgets and events) and externally (visitor, donor or government funded).

Step 5: Partnerships

Leverage resources by collaborating and partnering with other zoological institutions, conservation organisations, centres of learning, government agencies and high-net-worth individuals that share our passion for animals and conservation.

Step 6: Priorities

Identify and prioritise species which allow you to deliver conservation victories that clearly demonstrate the impact the animals in zoos and aquariums have on our ability to save their wild counterparts. Connect your animals to field conservation with personal stories of organisational commitment, both financially and with staff expertise.

Step 7: Communication

Develop a communications plan that is positive and proactive about your commitments and actions. Cultivate respected, independent spokespersons to deliver conservation stories to visitors, the greater community and society.

FIGURE 8.3. Seven Steps to Conservation Leadership.

campaign to achieve them. The Seven Steps to Conservation Leadership in the "Appeal to Zoo and Aquarium Directors" (fig. 8.3) encourages information sharing, partnerships, and creating new revenue streams. The first steps must come from within the organization. An entire section of this 2015 conservation strategy is devoted to creating a culture of conservation within your organization, from the frontline employee to the CEO to the board chair. This internal commitment to conservation integration at all levels of operation is essential for developing a strong brand of conservation with your visitors, members, and donors.

The WAZA Conservation Strategy provides guidelines for leaders to develop a stronger case for supporting the conservation and animal welfare missions. Resident animal welfare is distinct from conservation, but the two missions are dependent on each other. Modern zoos and aquariums must actively address both the welfare of their resident animals and the "welfare" and survival of their counterparts in the wild (Palmer, Kasperbauer, and Sandøe, this volume).

No single document can be a recipe for success. It must be accompanied by an unwavering commitment and a sustainable and innovative business plan. While setting aside funds from the operating budget for conservation may be part of the plan, these funds can also be leveraged to raise additional support from external sources such as donation kiosks, cause-related donor events, corporate sponsorships, and bequests. A strong conservation brand should attract a previously untapped pool of donors who have been identified and cultivated by impassioned staff members and field conservation partners.

EASIER SAID THAN DONE?

How do zoos and aquariums commit more resources to saving animals when they are faced with expanding operational costs and unpredictable economies and with inclement weather affecting attendance at their outdoor venues?

The best practices of WAZA member zoos and aquariums do more than raise awareness–they also result in direct support of field conservation programs and in measurable results. Zoological facilities are making a difference in fighting extinction with both behavior-changing programs and innovative revenue streams to enable frontline conservationists to save animals in the wild (Ahmad and Grow 2015).

The 2015 Conservation Strategy has a new format that uses compelling images to reinforce the text, making it more user-friendly for a wider audience. The focus is on implementing leadership guidelines to drive conservation action and results. The critical issues of sustainability of the animal populations in human care and of how zoological institutions can play a more significant role in mitigating climate change are also addressed (Barongi et al. 2015, 55). Notably, this document aligns with the United Nations *Strategic Plan for Biodiversity* 2011–2020, including the Aichi Biodiversity Targets (CBD 2010).

This 2015 Conservation Strategy is primarily directed at the leaders of our zoos and aquariums, making a stronger case for being proactive in preparing for future challenges and driving conservation action and results. In a world that depends on technology to service a growing human population and its inherent resource consumption pressures, we continue to lose wildlife and wild places at an alarming rate. Population sizes of vertebrate species, for example, have declined 52 percent since 1970 (WWF 2015). Technological advances need to work with nature—not against it—because human beings are totally dependent on nature. To borrow a phrase from Conservation International, "Nature doesn't need people. People need nature" (Conservation International 2016).

APPLIED RESEARCH: ZOO TO FIELD

Significant advances have been made in conservation-relevant scientific research in the past ten years (Barongi et al. 2015, 36–41). This form of applied research has proved of enormous benefit to managing wild populations, when laboratory and zoo- and aquarium-based scientists collaborate with field researchers and wildlife managers. Examples include the assessment and treatment of an individual's health by experts in their field, the development and improvement of contraception and assisted reproductive techniques (sometimes referred to as conservation breeding), and pretesting satellite tracking devices on animals in zoological facilities to ensure safe and effective monitoring of wild animals. There is still enormous untapped potential in applying the knowledge gained from our zoo and aquarium animals to saving conspecifics in their natural ranges. Research carried out on elephants (elephant endotheliotropic herpes virus [EEHV] and tuberculosis) and on frogs' chytridiomycosis in zoo-operated facilities is proving to have significant benefits for testing and protecting wild populations. There are several excellent chapters in this volume specifically devoted to the advances and effects of facility-based research on wildlife populations (Knapp, this volume; Mendelson, this volume).

CREATING A CULTURE OF CONSERVATION

Before you can cause change on the outside you must have a shared vision and philosophical buy-in at all levels within your organization. As stated in the WAZA Conservation Strategy, "Creating a conservation culture requires clear lines of communication to all personnel about the conservation work being undertaken, and celebrations of success when conservation objectives are achieved." Everyone—staff, board members, donors, and volunteers—must feel they are playing a meaningful role and making a difference.

Once an internal culture of conservation is integrated throughout the organization, then visitors and local communities must be engaged. This is accomplished by explaining how a visit to their institution, or their actions in daily life, directly helps to save animals in the wild. The Monterey Bay Aquarium Seafood Watch program empowers people to make more responsible choices that support a healthier ocean and preserve diverse marine ecosystems (Spring, this volume). Supply-chain activism, such as Don't Palm Us Off, the sustainable palm-oil buying campaign created by Zoos Victoria, is another effective program that has involved the entire community as well as the corporate world (Zoos Victoria 2016).

Developing a strong conservation brand at a zoo or aquarium not only is the right thing to do, it is what the educated public expects of the zoological community (Barongi et al. 2015, 23). It also makes good business sense. Zoos Victoria has determined that their conservation work is the second most important reason the Melbourne community visits their zoos. This is predicated on some very compelling marketing campaigns that convey the exciting conservation stories of Zoos Victoria all over Melbourne and surrounding areas. Whether being a recognized leader in conservation motivates more people to visit zoological institutions might be debatable in some regions. What is certain is that conservation provides some fantastic stories that can be communicated to guests during a visit. These conservation victories will more than likely influence visitors' decisions about making return visits, purchasing memberships, and donating to conservation programs.

To illustrate this vital connection and the potential effect on wildlife conservation, I'll share a recent story from my former employer. In May 2015 the Houston Zoo opened a new $28 million Gorilla Forest habitat. This was the culmination of a five-year project that included planning, design, fundraising, and construction. When this project was approved in 2010, the conservation team began to research gorilla conservation programs that would be a good fit for the ethos at the Houston Zoo. The zoo wanted to become a true partner in every aspect of the program, not just a source of financial support. Two major projects were selected—Gorilla Doctors in Rwanda and GRACE (Gorilla Rehabilitation and Conservation Education) Center in the Democratic Republic of the Congo (see Lukas and Stoinski, this volume). Not only was a long-term funding commitment developed, but the Houston Zoo also assisted with graphic design, community-education programs, website upgrades, staff expertise, and active participation on the board of directors for each organization. Such wide-ranging involvement ensured that the Houston Zoo was an integral part of the conservation stories delivered to our 2.5 million annual guests and not just doing "checkbook conservation."

One month before the public unveiling of Gorilla Forest, the marketing team at the Houston Zoo arranged for a local newscaster to accompany a member of the conservation team to Rwanda to film a one-hour TV special on the zoo's role in helping to save gorillas in the wild. In Rwanda the reporter was amazed at the impact the Houston Zoo had on the local community and their mountain gorilla "neighbors." He witnessed firsthand the health and education programs supported by the Houston Zoo to improve the quality of life in the region. The reporter conveyed this holistic approach to conservation in his broadcast, which aired during prime time back home in Houston one week before the much anticipated opening. While the TV special helped to increase visitation rates to the new exhibit, its primary purpose was to demonstrate how a visit to Gorilla Forest helps to save endangered wild gorillas. Having a popular TV spokesperson confirming this direct connection between captive animals and those in the wild is worth more in positive public relations than money can buy. This is a practical example of how the One Plan Approach should work for every species that lives in a zoological institution.

CONNECTING EXHIBITS TO FIELD CONSERVATION

Zoos and aquariums must always remain places for enjoyable experiences and social outings, but these recreational excursions can be enriched with compelling stories of animal welfare and conservation. Caring for live animals in zoological institutions carries an enormous responsibility—not only must they receive the best possible care and enrichment, but they should be conservation ambassadors that connect to their wild counterparts (Barongi et al. 2015; Mellor, Hunt, and Gusset 2015). While tremendous progress has been made in the development of welfare protocols for captive wildlife (Maple and Segura, this volume) there remains tremendous inconsistency in the interpretation and execution of these established guidelines (Mellor, Hunt, and Gusset 2015a).

Exhibiting live animals in artificial habitats is challenging, but if done "right" it can be a win-win for all concerned—the resident animal, the visitors, and the endangered wild animal. The "right way" is to display captive-born animals, or animals that have been properly acclimated, so that all their physiological and psychological needs are addressed.

For these zoo animals to be accepted as true "conservation ambassadors" for their wild counterparts, they must thrive in human care. Creating the most naturalistic space is imperative, and expensive, but it serves to evoke compassion and understanding of the real world.

Our conservation actions must correlate with our conservation messages, as we are being held more and more accountable in the world of instantaneous information sharing. While there is no substitute for seeing animals in the wild, most of the general public will never have an opportunity to see these species in their natural range. Creating a unique live-animal experience that inspires care and action should be the goal of every zoo and aquarium exhibit. Despite the increasingly popular computer- generated virtual reality events, there remains a strong human need for real-life experiences (Louv 2008). The unscripted and unpredictable adventures of a well-planned live-animal exhibit can be a powerful tool to engage visitors and elicit a positive change in environmental behavior. With more than seven billion people consuming more natural resources every day, a paradigm shift in the way live-animal habitats are designed is essential. All new exhibits must have a direct connection to the animals and habitats they represent, clearly demonstrating the effect a visit has on saving wild animals and supporting the people who share their natural habitats. Ultimately we all depend on a healthy planet to survive as species.

PARTNERSHIPS AND THE TRIPLE BOTTOM LINE

The most successful zoo- and aquarium-based conservation programs have formed mutually beneficial partnerships with other conservation-minded organizations to leverage their contributions into much larger packages of support. The European Association of Zoos and Aquaria (EAZA) has developed some very successful conservation campaigns that all its members can participate in. These EAZA campaigns raise several hundred thousand dollars a year for global species conservation. The WAZA "Biodiversity Is Us" outreach project is yet another example of the collective power of zoological institutions to raise awareness, promote action, and save animals in the wild (WAZA 2016).

The Association of Zoos and Aquariums (AZA) SAFE: Saving Animals from Extinction is also notable (AZA 2016a; Grow, Luke, and Ogden, this volume). The mission of SAFE is to combine the power of visitors to zoological institutions with the resources and collective expertise of AZA members and partners to save animals from extinction. This mission is achievable because of the hands-on expertise of wildlife managers and scientists at zoos and aquariums who are actively applying their talents and resources to work in partnership with field conservationists to save populations of wild animals.

Creating additional revenue streams for conservation funding should be a major part of a business and fund-raising plan. Increasing allocations for conservation programs can be partially achieved through external restricted donations rather than from operating revenues. Roughly 50 percent of the $3 million the Houston Zoo spends annually on field conservation is generated from conservation donations raised each year through various fund-raising events, bequests, and conservation merchandise (Houston Zoo, pers. comm.). Some zoological institutions in the United States charge an extra "conservation fee" on all admission tickets and annual memberships (Zoo Boise, Idaho, pers. comm.). Zoo Boise generates over $130,000 a year with this added admission fee, all of it restricted to conservation programs. The American Association of Zoo Keepers (AAZK) has regional chapters in many zoological institutions that conduct conservation-related activities. The AAZK chapter of the Los Angeles Zoo raised over $55,000 from their "Bowling for Rhinos" event in 2015 (Los Angeles Zoo, pers. comm.). The Houston Zoo holds an annual Conservation Gala featuring inspiring conservation partners and unique auction items, raising over $1 million (net) for conservation programs. Although this money does not go to the operating budget of the zoo, it supports its core mission and has a long-term strategic value to the overall zoo brand.

In the business world this strategy of raising funds for programs that do not directly help the traditional bottom line of profit and loss is sometimes referred to as the triple bottom line (Savitz 2013). This type of accounting factors social responsibility and environmental sustainability into the equation for long-term financial performance and overall business operations. Corporate social responsibility is nothing but maximizing the value of your company over a long period, because in the long term social and environmental issues become financial issues (Sorensen 2015). The diagram in figure 8.4 is a visual interpretation of how triple bottom line accounting will enhance the reputation and brand of your institution.

There is also the sustainability argument of managing for the long-term. Pharmaceutical companies, for example, are always setting aside money for

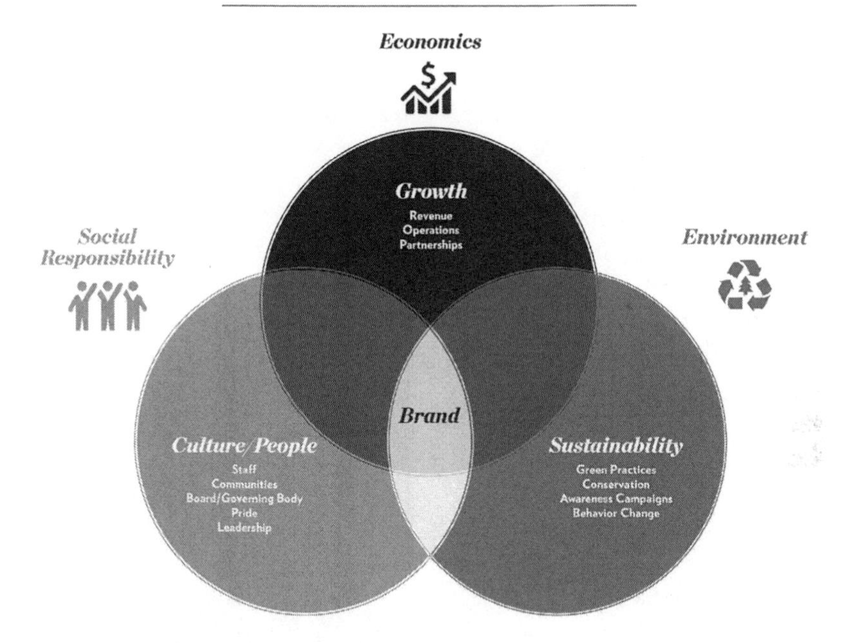

FIGURE 8.4. The Triple Bottom Line.

research and development (R&D) to develop a pipeline of new drugs as their patents for older drugs expire. The animals in zoos and aquariums are not patents, but investing in conservation is akin to investing in R&D. If zoos and aquariums ignore the animal extinction crisis, they will become living museums and will have failed in their primary mission. The end result is the same: if you keep extracting without replacing, you will not survive.

WHERE WILL THE NEXT CONSERVATION CHAMPIONS COME FROM?

No matter how good a conservation strategy or program is, it will fail without the right people on the front lines, who implement the goals and then become public spokespersons. One area that has not received much attention is individual efforts by exceptional people. Most successful endangered species programs are closely aligned with a lifelong champion. To name just a few of the many examples: Rodney Jackson (snow leopards), Raoul du Toit (rhinos), Laurie Marker (cheetahs), William Oliver (pygmy hogs), Merlin Tuttle (bats), Carl Jones (pink pigeons), George Archibald (cranes),

Cynthia Drabek (tree kangaroos), Patricia Medici (tapirs), Colleen and Keith Begg (lions), Russ Mittermeier (primates), Iain Douglas-Hamilton (African elephants), Ian Player (white rhinos), and of course Jane Goodall (chimpanzees).

In my forty-five years working in the zoo and conservation community I have met some incredible people who have devoted their entire careers to caring for and saving one species or one habitat. There are even a few former zoo directors in this group, such as William Conway, George Rabb, Bernhard Grizmek, and Heini Hediger. As these individuals die or get too old to continue the fight, we need a pipeline of young champions to take their places. Money, research, and education are prime ingredients for a conservation program, but without champions on the ground, working and living with the local communities and their wildlife, prospects for success are greatly reduced.

Some zoo zoologists, including me, are concerned that we are losing more conservation champions than we are gaining. Zoos and aquariums are often the first exposure to wildlife for future conservation champions. So while supporting conservation activities is critical, they must also continue to be a gateway to the wild for aspiring young biologists. This also holds true for individuals working in captive wildlife facilities, as they can be indispensable for researchers working with zoo and aquarium animals.

Working with captive wildlife has its challenges and requires an experienced animal staff committed to the welfare of their animals as well as the conservation of the species. Most researchers would be the first to confirm that they could not have obtained their data without the full cooperation of the animal care and veterinary teams. Good animal care specialists develop a unique rapport and trust with the animals under their care. This bond between keeper and animal allows for much greater flexibility with procedures such as obtaining biological samples, testing satellite collars, or taking other measurements.

MAXIMIZING POTENTIAL

Since the publication of the 2005 Conservation Strategy, WAZA zoos and aquariums have dramatically increased their conservation commitments in terms of both financial donations and pro-environmental behavior change programs. That is the good news. However, they are still performing at only a fraction of their potential in raising funds and awareness for the extinction crisis. While most zoos and aquariums state "conservation" as a core mission, they have not yet developed business and fund-raising plans that allow them to contribute more than 1 or 2 percent of their operating budgets to field conservation efforts. The 2015 Conservation Strategy is more than the usual call to arms: it provides a business plan for integrating conservation within all levels of an organization. It provides practical guidelines and credible examples of how to increase the intellectual and financial investments in conservation-related activities of your own institutions.

Every zoo and aquarium is at a different stage of evolution in supporting conservation of animals in the wild. Some are already recognized leaders in zoo- and aquarium-based conservation, such as the Bronx Zoo / Wildlife Conservation Society (New York, US), Monterey Bay Aquarium (California, US), Disney's Animal Kingdom (Orlando, Florida, US), Houston Zoo (Texas, US), Lincoln Park Zoo (Chicago, Illinois, US), St. Louis Zoo (Missouri, US), Zoo Boise (Idaho, US), San Diego Zoo (California, US), Zoos Victoria (Melbourne, Australia), Chester Zoo (England), ZSL London Zoo (England), Durrell Wildlife Park, Jersey (Channel Islands, UK), Perth Zoo (Australia), Zoo Zurich (Switzerland), Zoo Leipzig (Germany), Nordens Ark (Sweden), Singapore Wildlife Reserves (Singapore), and Temaikén Zoo (Buenos Aires, Argentina), are a few great role models. It is important to note that this sample includes both nonprofit and for-profit facilities as well as government-funded entities. The excuse that local tax dollars cannot be used for international conservation programs has been effectively challenged with public awareness campaigns and the education of governing bodies. When governing officials understand (and can celebrate) the important role their institutions play in helping wildlife and people on a global scale, they are more supportive of allocating public funds for these endeavors.

REFLECTIONS ON THE FUTURE

I wanted to end this chapter on an optimistic note, but despite all the great progress and examples cited, we are still losing more species and wild habitats every year. The most prominent ones are the ninety-six elephants and three rhinos lost each day ("Ninety-Six Elephants" 2016; Save the Rhino 2016), but how many species of sharks and frogs are going extinct on our watch? The sixth extinction is upon us, and we are still debating what to do and what not to do.

The primary purpose of this chapter has been to demonstrate and document the critical role *some* zoos and aquariums are playing in fighting the extinction crisis. Unfortunately, *most* public zoological institutions are still making minimal financial commitments and prefer to claim that they inspire people to action just by a visit. While a visit to your local zoo or aquarium can be enlightening and may lead to some type of conservation action, it is difficult to quantify and frankly, in my opinion, is not enough.

I make this statement even though the experience was enough for me. I still remember my first visit to the Bronx Zoo in 1959 with my second-grade class. That experience helped shape my future career choices. It was easy to impress a seven-year-old city kid, especially in an era of three-channel black-and-white TV and no computers. Today is a much different situation, and a visit to a zoo or aquarium, while more entertaining, does not usually connect the captive animals to their plight in the wild.

The zoological community likes to say their captive-born animals are ambassadors for their counterparts in nature, but that can backfire in terms of public perception. No matter how naturalistic and spacious a zoo exhibit or aquarium tank, some folks still believe the animals would be better off in the wild. The recent successful importation of seventeen wild-born African elephants from Swaziland, South Africa, to three AZA-accredited zoos in the United States set off a massive campaign by animal rights groups and some elephant field workers to prevent this well-researched and well-vetted operation (Ramirez 2016). Even though these elephants were in a severely degraded fenced-in habitat in Africa and due to be culled, there still was a strong sentiment to prevent these "wild" elephants from leaving for smaller enclosures in zoos. What many would agree was a rescue operation was labeled cruelty by others with a different perception of what is truly wild. Some extreme critics will even support the phrase "better off dead than captive-bred." Of course, if animal rights advocates really understood what was going on in "the wild," they would want to shut down nature, not zoos.

The best way to change people's perception is to be proactive rather than defensive. Zoos and aquariums must continue to improve their conservation messages and to increase their direct support of conservation activities. If all our zoological institutions can commit to the principles and guidelines set forth in the WAZA 2015 Conservation Strategy, and stay the course, they will ensure not only the success of their institutions but also the survival of our planet. They may even be able to change the definition of these facilities. Imagine a new universally accepted definition for a zoo or aquarium that affirms "places where people come to see and learn about animals, which play a major role in saving populations of species in the wild."

But again, as some of the examples cited in this chapter and by other contributors to this volume confirm, zoos must be part of the story rather than just writing a check once a year. I believe the WAZA Conservation Strategy, along with so many other conservation publications, provides the tools and incentives to motivate all zoos and aquariums to increase their commitment. Wildlife conservation is core to the mission of every zoological facility, and each one needs to "walk the talk"—or close its gates.

CHAPTER NINE

Saving Animals from Extinction (SAFE): Unifying the Conservation Approach of AZA-Accredited Zoos and Aquariums

Shelly Grow, Debborah Luke, and Jackie Ogden

COMMITTED TO CONSERVATION

Conservation challenges loom large, and they are larger than what any single institution can overcome on its own, indicating a need for unification around shared goals (Cerezo and Kapsar, this volume). The Association of Zoos and Aquariums (AZA) and its members envision a world where, as a result of the work of AZA-accredited zoos and aquariums, all people respect, value, and conserve wildlife and wild places. Historically, AZA and its individual members have each dedicated themselves to this vision independently by organization or in small groups. In 2014 AZA SAFE: Saving Animals from Extinction was developed to provide a scope broad enough to unify the diverse AZA community yet focused enough to achieve measurable conservation outcomes. By focusing on the AZA community's strengths, AZA SAFE seeks to unite conservation partners, AZA professionals, and AZA's 180 million annual visitors to achieve biological conservation.

The history of zoological commitment to conservation traces its roots to 1907, when fifteen American bison from the New York Zoological Society, now the Bronx Zoo, were reintroduced to the Wichita National Forest Preserve to prevent what was expected to be the species' extinction (US Fish and Wildlife Service, n.d.; Barrow, this volume). This and subsequent breeding and release efforts are some of the more widely recognized conservation successes of what are now AZA-accredited zoos and aquariums (see, e.g., Allard and Wells, this volume). More recent breeding and reintroduction initiatives have been linked to thirteen (Conde et al. 2011b) of the sixty-eight vertebrate species whose threat level has been reduced since 1980 (Hoffman et al. 2010), and many of these included the active engagement of AZA-accredited facilities. Ex situ propagation and release efforts account for just a fraction of AZA-accredited zoo- and aquarium-led conservation accomplishments. In 2014 the keywords "reintroduction" or "assurance population" were used to describe only 11 percent of field conservation projects reported to AZA, and they accounted for less than 10 percent of the funds spent on field conservation (AZA 2015a).

The range of vital field conservation and species recovery initiatives members support and lead includes population biology/monitoring, in-range conservation/environmental education, community participation, capacity building/training, conservation biology, and habitat protection (AZA 2015a). Furthermore, the conservation commitments of AZA-accredited zoos and aquariums include the delivery of local conservation education programs that link personal behavior to reducing threats to wildlife, leading or facilitating important scientific research, and greening their business operations to conserve natural resources.

AZA surveys its members annually to understand the full suite of each organization's conservation activities. While general conservation-related member surveys have been conducted since 1990, their foci have been sharpened since 2010 to create surveys that provide specific metrics related to the areas of field conservation, mission-focused research, education programs, and green business operations. These newer surveys are accompanied by definitions that maximize reporting consistency across all AZA member facilities, and submissions are reviewed by experts in each field.

As defined by the AZA community, field conservation focuses on projects that have a direct impact on animals or habitat in the wild (AZA 2013a). In line with surveys undertaken by the World Association of Zoos and Aquariums (Gusset and Dick 2010), AZA surveys have found the AZA community to be among the significant funders of field conservation activities relative to major international conservation organizations. From 2010 to 2014, AZA members reported field conservation expenditures from $130 million to $160 million annually (AZA 2015a). During that same time, AZA conservation projects spanned the globe and took place in 112 to 131 countries each year, with almost half occurring in North America, mainly the United States. Every year these projects were implemented in partnership with federal, state, and local government agencies, along with universities and nongovernmental organizations, and their benefits have been felt across more than 1,100 species and subspecies. In 2014 alone, 222 species listed as either threatened or endangered under the US Endangered Species Act, 223 species listed as either critically endangered or endangered by the International Union for the Conservation of Nature's Red List of Threatened Species, and five species categorized by the IUCN as extinct in the wild—the Guam rail, Kihansi spray toad, scimitar-horned oryx, socorro dove, and Wyoming toad—benefited from these efforts.

The AZA community's conservation contributions also include activities that have an indirect, but important, effect on animals and habitats in the wild. The IUCN-Conservation Measures Partnership's classification of conservation actions (Salafsky et al. 2008) includes education and awareness, and AZA accreditation standards require that education be a key component of the member's mission (AZA 2015b). When experienced at an AZA-accredited facility, education is reinforced through social learning with family, friends, and trained interpreters and helps build connections between people and animals, raises awareness about people's effect on the animals, and capitalizes on empathy and knowledge to influence pro-environmental behavior (Clayton and Le Nguyen, this volume; Grajal, Luebke, and Kelly, this volume). Members' education programs repertoire includes conservation education programs, which specifically incorporate both awareness building and promoting conservation action(s) (AZA 2013b). In 2014 there were more than eighty-two million participants in the AZA community's formal conservation education programs (AZA 2015a).

Further contributing to conservation, AZA-accredited zoos and aquariums understand that conservation of wildlife requires the conservation of natural and man-made resources. Many members are embracing the "triple bottom line" concept and its environmental, social, and economic spheres of sustainability (AZA 2013c). As institutions adopt conservation behavior in their daily operations, they support their core business operations, decrease their environmental impact, and model ways guests can become part of the solution to the environmental problems at the root of the current extinction crisis. Each year, AZA-accredited zoos and aquariums are purchasing and generating renewable energy on-site, working toward zero-waste or carbon-neutral goals, receiving third-party certifications for their environmental management systems and building construction, and working with vendors to source local and sustainable foods (AZA 2015a). In 2014, eighteen AZA-accredited zoos and aquariums reported generating or purchasing sixteen megawatts of renewable energy, and nineteen reported having buildings that were LEED-certified by the US Green Building Council (AZA 2015a). In 2015 Utah's Hogle Zoo hosted a zero-waste icebreaker for the 2,500 attendees of AZA's Annual Conference, the first time a zero-waste goal was set for an AZA conference event. In 2016 Omaha's Henry Doorly Zoo and Aquarium followed suit by hosting a zero-waste Zoo Day event for over a thousand attendees at AZA's midyear meeting.

Finally, AZA-accredited zoos and aquariums are science-based organizations that both apply science to their operations and actively contribute to the original research literature. Surveys circulated since 2013 indicate that species and habitat conservation are members' second most common focus of research, behind animal care, health, and welfare (AZA 2015a). Research, including that dedicated to detecting, diagnosing, and halting population declines, is being conducted at an increasing number of AZA-accredited zoos and aquariums and may be particularly compelling because these organizations can link research to a conservation issue and bring it to millions of visitors (Knapp, this volume).

These examples demonstrate the conservation reach of the AZA community and highlight the depth of members' commitment. However, these examples focus on *outputs*—the number of species, the actions undertaken, the project locations, the partners and target audiences, and the amount of money spent—rather than on conservation *outcomes*. Furthermore, they indicate a diffuse approach to the AZA community's engagement in conservation, suggesting that opportunities may exist for a more focused approach.

AZA-accredited zoos and aquariums became increasingly collaborative during the latter part of the twentieth century, but much of that cooperation was focused on animal population management and was exemplified by the establishment of the Species Survival Plan (SSP) program in the 1980s (Conway 1982; see also Henson, this volume; Kisling, this volume). Over the years, various AZA- and member-driven efforts have arisen to unify and coordinate field conservation efforts as well. Some Animal Programs, specifically the SSP programs and Taxon Advisory Groups, have created and led initiatives that have encouraged the AZA community to work together on conservation goals. Some of these programs have been in place for decades, and several of the first SSP programs were established to complement active conservation efforts (see, e.g., Johnson 1990; Ballou et al. 2002). Other Animal Program–led initiatives are newer (see, e.g., Lukas and Stoinski, this volume). In 2014 almost thirty Animal Program–led field conservation initiatives were actively supported by members (AZA 2015a). Additional AZA-driven efforts to unify the AZA community around specific species or issues include establishment of Fauna Interest Groups in 1991 (Hutchins and Wiese 1991), the Bushmeat Crisis Task Force in 1999 (Eves, Hutchins, and Bailey 2008), the Butterfly Conservation Initiative in 2001 (Grow, Allard, and Luke 2015), and the Year of the Frog in 2008 (Grow and Allard 2008).

However, these efforts have not been broad enough in scope to unify the diverse zoological community around field conservation to the same degree that the SSP programs have coordinated population management.

AZA SAFE: SAVING ANIMALS FROM EXTINCTION

Despite massive conservation efforts around the world, the threat of extinction continues to increase for many species, and some scientists suggest we may be on the verge of a sixth mass extinction (Wake and Vredenberg 2008; Barnosky et al. 2011). The scope of the conservation crisis is evident (Pimm et al. 2014; De Vos et al. 2015), and a new approach for strategic, collaborative, and coordinated engagement is critically needed. To help meet the challenge, AZA SAFE: Saving Animals from Extinction was developed in 2014 to unite conservationists, AZA professionals, and AZA's 180 million annual visitors in their efforts to achieve biological conservation. External funders are also invited to partner with SAFE, while member facilities coordinate and participate in projects prioritized and vetted by the broader conservation community.

Key elements of the AZA SAFE model are these:

- Use of the One Plan Approach to species conservation, where management strategies and conservation actions are developed by those responsible for all populations of a species, both inside and outside their natural range (Byers et al. 2013; Traylor-Holzer, Leus, and Byers, this volume).
- Use of an adaptive management framework, such as that described in the Open Standards for the Practice of Conservation, which recommends common language and processes and has been adopted by many conservation organizations (Conservation Measures Partnership 2015).
- Application of social science tools to build on positive experiences at AZA-accredited zoos and aquariums that promote conservation actions aimed at reducing the impact people have on wild populations (Barongi et al. 2015).
- Engagement of all responsible parties in integrated species conservation planning, including nongovernmental organizations, government agencies, and universities working with that species as well as AZA Animal Program leaders with relevant animal management, care, research, and conservation expertise. These groups identify the threats to that species and prioritize the conservation actions needed to address each threat.

- Development of three-year AZA SAFE Conservation Action Plans (CAPs), which build on these planning efforts and identify projects that align AZA strengths and expertise with a subset of the prioritized actions needed.

The CAPs ensure that the conservation projects included have been identified as a priority by the entire group of species-specific conservationists; include measurable objectives, budgets, and timelines; provide AZA-accredited facilities of all sizes the opportunity to coordinate, collaborate on, or support crucial species conservation efforts; and afford an opportunity for philanthropists to support projects in a way that elevates their local zoo or aquarium's conservation leadership role.

The AZA SAFE model was developed and tested in 2014 and 2015 on four signature species (African penguin, cheetah, vaquita, and western pond turtle). To date, each has benefited from a meeting of stakeholders in-range where needed conservation actions were prioritized. AZA staff members and Animal Program leaders reviewed those needs and aligned them with the community's strengths to highlight areas that could be addressed by AZA-accredited zoos and aquariums. These analyses were shared with the original group of stakeholders for additional feedback and to ensure that everybody remained clear about and supportive of the final conservation areas to include in the CAP. The CAPs are then shared with the AZA community to identify members that are able to coordinate or collaborate on the projects. AZA staff members will continue to coordinate communication among all stakeholders to ensure progress and transparency, as well as evaluate both on the ground conservation progress and the AZA SAFE process. Six additional species or taxonomic groups have been identified as SAFE signature species, including the Asian elephant, black rhinoceros, gorilla, sea turtle, sharks, and whooping crane. Moreover, the criteria and processes needed to implement a complementary conservation model for additional species are being developed for AZA member zoos and aquariums.

MOVING THE COMMUNITY—AND THE NEEDLE—ON CONSERVATION

Since the 1980s, AZA has supported an evolving organizational infrastructure intended to aid and promote collective field conservation (Hutchins and Conway 1995). In 2013 AZA's board of directors passed a resolution that, in part, "urges all AZA-accredited institutions to increase their commitment to conservation, achieving a level of support that will allow them to be defined individually and collectively as organizations committed to conservation of species in the wild" (AZA 2013d). AZA SAFE will elevate the conservation capacity and reach of the AZA community and achieve clear conservation outcomes. SAFE provides a unique, multidisciplinary approach for collaborative conservation that will address the conservation crisis to help save vulnerable wildlife species from extinction and protect them for future generations.

CHAPTER TEN

Integrating Ex Situ Management Options as Part of a One Plan Approach to Species Conservation

Kathy Traylor-Holzer, Kristin Leus, and Onnie Byers

ONE PLAN FOR A SPECIES

All too often, species conservation plans are developed by a relatively small group of individuals, typically wildlife managers and field biologists, with limited consideration of the broad range of conservation options offered by ex situ tools. Similarly, zoos, aquariums, and other ex situ facilities often develop species management plans and collection plans without collaborating with the field community to address the real conservation needs of the species. This schism in species conservation planning leads to developing plans for wild and captive populations in isolation rather than within an integrated metapopulation approach. Subsequently, ex situ populations often are not structured or used to the best conservation advantage.

The realization of this situation began to surface about 2010 as a series of analyses outlined the lack of "sustainability" of most of the world zoos' breeding programs (e.g., Lees and Wilcken 2009; Leus et al. 2011; Long, Dorsey, and Boyle 2011). Sustainability for zoo populations is often defined as positive population growth through reproduction and the ability to retain at least 90 percent of genetic diversity for one hundred years (90/100 rule), a genetic goal modified from that proposed by Soulé et al. (1986) for long-term assurance populations ("arks"). For decades the 90/100 rule served as the gold standard against which zoo breeding programs were measured, rather than programs being structured and evaluated against specific goals targeting conservation or other needs.

In response to increasing concerns over the sustainability of zoo populations and their contributions to species conservation, an international workshop was convened in 2010 that focused on intensively managed populations (IMPs) for conservation (CBSG 2011). Professionals from eight regional zoo associations, academics, and field biologists from twelve countries addressed these issues and produced the following vision:

> To preserve biodiversity, the global conservation community commits to providing the level of intervention necessary to prevent the extinction of species. Intensive population management (including, but not limited to management within zoos and aquariums, botanic gardens, other propagation centers, closely managed reserves, and genome banks) is effective (only) when integrated with other conservation measures within an overall species conservation strategy that fully addresses the threats to the species, using the best available science, technology, and practices. We work toward a world in which all species can live within healthy ecosystems as part of evolving communities, without the need for continued human intervention.

This vision solidified the zoo community's recognition of the need to develop species-specific, goal-targeted conservation programs in the context of a single, comprehensive conservation plan for a species. This requires close collaboration between the in situ and ex situ communities managing a species.

A One Plan Approach to species conservation promotes joint development of management strategies and conservation actions for all populations of a species by all responsible parties to produce a single, comprehensive conservation plan for that species. The ultimate goal is to support the species' conservation in the wild (Byers et al. 2013). This strategy evaluates all available methods to support a viable metapopulation, with input and buy-in from diverse responsible parties through collaborative efforts. The resulting plan is likely to incorporate those management strategies that will be most efficient and effective in contributing to the species' conservation. While the concept and practice of integrated conservation planning is not new, it is far too rare in practice. In 2011 the IUCN SSC Conservation Breeding Specialist Group coined the term One Plan Approach (or OPA) and has adopted this approach more consistently into its species planning efforts. Since then this concept has spread rapidly throughout the zoo and aquarium community and become the focus of regional and global zoo association conferences and initiatives around the world. Most recently, the One Plan Approach provides a guiding framework for the 2015 *World Zoo and Aquarium Conservation Strategy* (Barongi et al. 2015; Barongi, this volume).

IMPORTANCE OF A COMPREHENSIVE APPROACH

The vision created at the 2010 IMP workshop recognized the need for species-specific conservation actions that might involve a diversity of management goals and actions to effectively address threats faced by each species. As the status of wild populations changes over time, new threats may emerge, requiring new population management strategies.

The increasing challenges of climate change, habitat loss, exploitation, invasive species, and infectious disease epidemics result in many wildlife populations' becoming smaller and more fragmented. This leads to a new threat—the added vulnerability of small populations to random stochastic processes (e.g., environmental variation, catastrophes, random variation in survival and reproduction, skewed sex ratio, genetic drift, inbreeding) that can feed back into each other, causing a species to be caught in an *extinction vortex* (Gilpin and Soulé 1986). Once caught in this downward spiral, populations may be destined for extinction even if the original primary threats are removed.

Conservation strategies that deal only with the primary threats of small threatened populations run the risk that these remnant populations of a species will go extinct before these driving threats are eliminated. Conversely, population management strategies that concentrate only on ameliorating stochastic threats through periodic demographic or genetic reinforcement of wild populations will likely perpetuate a continuing need for intensive management to offset population instability as long as the primary threats remain, reducing conservation resources available to other threatened species and never realizing a fully conserved, self-sustaining species (Redford et al. 2011). To maximize conservation outcomes, conservation planning needs to consider all threats to species survival, including threats that are long-term as well as immediate and stochastic as well as deterministic (fig. 10.1). A shining example is the Channel Island fox (*Urocyon littoralis*) recovery program, which combined short-term ex situ management with eradication of primary threats (removal of nonnative golden eagles and prey species) and subsequent releases to reverse severe decline and establish a robust wild population within fifteen years that is now proposed for delisting (USFWS 2016).

Human development and activities are changing the world's landscape—increasingly, "the wild" is no longer wild (Minteer, this volume). As the biodiversity crisis intensifies and wildlife populations dwindle, to avoid extinction more species will likely require some form of intensive population

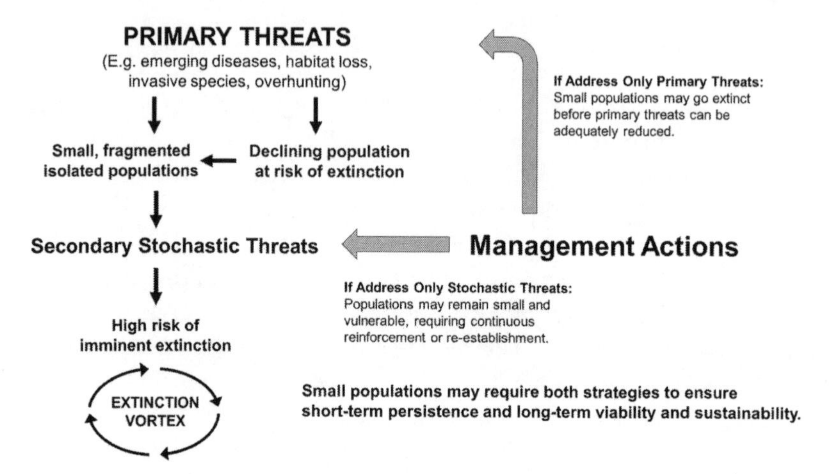

FIGURE 10.1. Management actions needed to address primary and secondary (stochastic) threats to wild population viability.

management across a broad in situ–ex situ continuum. Zoos and aquariums must commit to the conservation of species (and ecosystems) through all means possible, whether through ex situ management, support of in situ conservation efforts, or both.

HOW EX SITU ACTIVITIES CAN SUPPORT CONSERVATION

The One Plan Approach ensures that ex situ activities are assessed for their potential as effective conservation tools to preserve a species. In some cases ex situ management can play a significant role in species conservation and therefore should be assessed as a possible tool in species conservation planning. Ex situ activities are or have been an important component of conservation action for many species across a broad geographic and taxonomic range. Captive breeding was identified as playing a major role in the recovery of sixteen of sixty-four species for which the threat level was reduced (Hoffman et al. 2010). Table 10.1 provides a few examples of various ex situ conservation activities, some of which contributed to downlisting species by one or more categories on the IUCN Red List. Ex situ conservation activities can support species conservation and prevent extinction (Traylor-Holzer, Leus, and McGowan 2013) in the following ways:

1. Offsetting the impact of threats. Ex situ activities can improve the demographic or genetic viability of a wild population by counteracting the impacts of primary or stochastic threats to the population, such as reduced survival, poor reproduction, and genetic isolation—for example, through head-start programs that remove juveniles from the wild for ex situ care and return them once they are less vulnerable, or by cross-fostering captive-born neonates to wild parents.
2. Addressing the causes of primary threats. Ex situ activities can help reduce primary threats such as habitat loss, exploitation, invasive species, or disease through specifically designed research, training, or conservation education activities that directly and effectively addresses the causes of these threats—for example, through ex situ research to detect, combat, or treat infectious diseases.
3. Buying time. Establishing a diverse and sustainable ex situ rescue or assurance population may be critical in preventing species extinction when the wild population is declining and primary threats are not under control—for example, populations facing widespread infectious disease epidemics or decimation by invasive species (see, e.g., Mendelson, this volume).
4. Restoring wild populations. Once the primary threats have been sufficiently addressed, ex situ populations can be used as a source to reestablish wild populations (e.g., Arabian oryx, discussed by Allard and Wells, this volume).

Ex situ conservation activities extend beyond the traditional long-term managed zoo breeding program approach to retain 90 percent of genetic diversity for one hundred years. Rather, they may be short- or long-term, may or may not involve a breeding population, may consist of biosamples (e.g., a genome resource bank; see Ryder, this volume) rather than living individuals, and may expand beyond zoos and aquariums to a broader range of ex situ facilities and organizations (e.g., gorilla conservation efforts; see Lukas and Stoinski, this volume). Ex situ activities may provide more options for integrating with in situ conservation needs when located within the range state of the species. In a conservation-oriented approach, ex situ programs are tailored to the specific needs of the species and may vary widely in their population goals and timelines. Conversely, a true metapopulation approach for a species (linking all captive and wild populations of a species) can improve the viability and sustainability of long-term ex situ populations, if appropriate (Lacy, Traylor-Holzer, and Ballou 2013). The myriad

TABLE 10.1. Examples of species with ex situ programs that serve conservation roles

		Ex situ conservation activities			
Species (RL status)	*Threat(s)*	*Address primary threat*	*Buy time in emergency*	*Offset instability*	*Restore wild population(s)*
Tasmanian devil (EN)	Devil facial tumor disease (DFTD)	Research on DFTD (genetic resistance)	Assurance population against extinction in the wild by DFTD	Potential source for reinforcement in wild once DFTD is controlled	Potential source population once DFTD is controlled
Sumatran tiger (CR)	Habitat loss and fragmentation, human-tiger conflict, poaching	Training of conflict tiger teams (tiger health assessment, immobilization)	Assurance population against extinction		
Przewalski's horse (EN)	Extinction in wild in 1960s; small size, low genetic diversity, disease		Assurance population (prevented species extinction in the past)	Source population for reinforcement of wild populations	Source for reestablishing wild populations
Golden lion tamarin (EN)	Habitat loss, fragmentation, trade		Assurance population against extinction	Source population for reinforcement of wild populations	Source for reestablishing wild subpopulations in original range
Western pond turtle (VU)	Predation by invasive bullfrog, shell disease, habitat loss	Research on ulcerative shell disease		Head start wild juvenile turtles and return to wild when less vulnerable to predation	Potential source for establishment of new wild populations
Kihansi spray toad (EW)	Loss of specialized habitat		Rescue/assurance population against extinction in the wild		Source population for reestablishment of wild population
American burying beetle (CR)	Habitat loss, fragmentation, pesticides, reduced food availability	Research on life history traits, health assessment, DNA preservation protocols	Reservoir against genetic loss in the wild	Source population for reinforcement of wild populations	Source for establishment of new wild populations
Albany cycad (CR)	Small population, illegal harvesting, trade	Ex situ production to reduce harvest pressure	Assurance population against extinction (not viable in the wild)	Source population for reinforcement of wild populations	

Note: Status indicates current IUCN Red List category.

of population management options, from intensive management of in situ populations, to head-start and release programs, to ex situ breeding programs, blurs the artificial dichotomy of captive and wild (Redford, Jensen, and Breheny 2012).

EVALUATING EX SITU OPTIONS

To effectively consider ex situ options within species conservation planning, there needs to be a systematic evaluation of the potential benefits, costs, risks, and feasibility of such options as part of the process. The Species Survival Commission (SSC) of the International Union for Conservation of Nature (IUCN) recently revised its guidelines on the use of ex situ management for species conservation (IUCN/SSC 2014), an effort coordinated by the Conservation Breeding Specialist Group (CBSG; now the Conservation Planning Specialist Group, or CPSG). These guidelines outline five steps for more formal, informed, and transparent decision making process to provide guidance on whether ex situ activities are a beneficial and appropriate component of an overall species conservation strategy as well as on the nature of any recommended ex situ programs (Traylor-Holzer, Leus, and McGowan 2013):

1. Conduct a thorough status assessment and threat analysis. The status of both in situ and any existing ex situ populations should be assessed. In addition, a thorough and detailed examination of the causes leading to the primary threats, as well as the precise effects that both the primary and stochastic threats have on reproduction, survival, and distribution of the species, allows the identification of potential intervention points for conservation action.
2. Identify potential ex situ conservation roles. Potential intervention points can guide the identification of ex situ activities (as well as other conservation actions) that can prevent or reduce negative effects on the wild population. This will define any potential conservation roles that ex situ management can play in the overall conservation of the species.
3. Define the required program structure. The purpose and function of an ex situ population or activity will determine the characteristics and dimensions of the program needed to fulfill the identified conservation role(s). This includes appropriate demographic or genetic management, or both, to meet program goals (e.g., minimizing genetic adaptation to captivity).
4. Assess program feasibility and risks. It is important to assess not only the potential value of an ex situ program, but also resources and expertise

needed for such ex situ activities, costs and risks associated with them, and feasibility and likelihood of success.

5. Make a decision on ex situ options. The information above is used to reach a decision regarding whether ex situ activities, and which ones, are included in the overall strategy for conservation, and to ensure that this decision is informed and transparent (that it demonstrates how and why the decision was taken)

In this process, the conservation need for ex situ activities is not automatically dismissed, nor is it automatically included; rather, a thorough and informed evaluation ensures that any recommended ex situ activities are carefully tailored to the conservation needs of the species. These guidelines are applicable whether or not an ex situ program is already under way, since their application may recommend establishing an ex situ program (or not), or may suggest a revision in the structure and goals of an existing program. Amphibian Ark adopted a similar decision-making process through development and implementation of its Conservation Needs Assessment Tool, which has proved valuable in prioritizing amphibian ex situ conservation needs (Zippel et al. 2011; Carrillo, Johnson, and Mendelson 2015). The botanical community is also adopting this approach of assessing threats and considering all available conservation options across the in situ—ex situ spectrum (Oldfield and Newton 2012).

INTEGRATING EX SITU OPTIONS INTO SPECIES CONSERVATION PLANNING

Applying the IUCN *Guidelines for the Use of Ex Situ Management for Species Conservation* ideally should become an integral part of existing species conservation planning tools such as CPSG's Population and Habitat Viability Assessment (or PHVA) process and IUCN SSC's recommended planning process as outlined in its handbook *Strategic Planning for Species Conservation* (IUCN/SSC 2008). Not all species conservation strategies will require an ex situ component, in the same way that other management interventions such as antipoaching efforts or habitat corridors may or may not be required to conserve a species. In some cases ex situ management will be central to a species conservation strategy, in others it will be secondary, and for some species it will not feature at all. The five decision steps outlined in the IUCN guidelines can integrate well into the broader conservation planning process by linking with the threat analysis to suggest potential man-

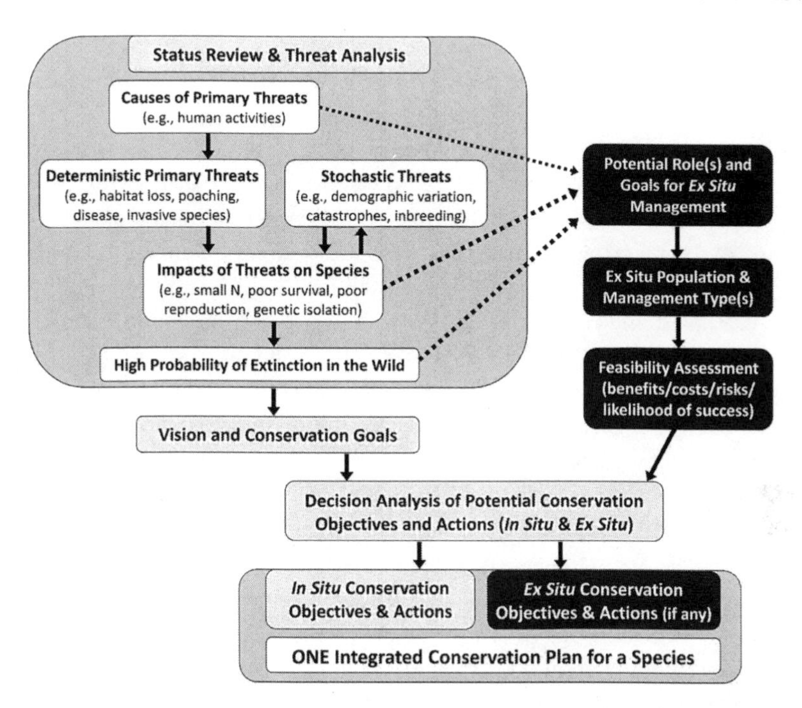

FIGURE 10.2. Incorporation of the decision process proposed in the revised IUCN guidelines (dark boxes) into the species conservation planning process to develop one integrated plan for the species. IUCN 2014.

agement strategies that can be evaluated and subsequently incorporated or not, as appropriate, into the final species conservation strategy (fig. 10.2). This can be done within a more comprehensive species conservation planning exercise (e.g., greater sage grouse PHVA; Lloyd et al. 2014) or as a separate but complementary exercise that focuses specifically on assessing ex situ options (e.g., prairie butterfly ex situ planning workshop; CBSG 2016) as part of a broader conservation planning process.

The Okinawa rail (*Hypotaenidia okinawae*) offers a case study for integrating ex situ options into species conservation planning. This bird species is endemic to the island of Okinawa and classified as Endangered on the IUCN Red List and as an Alliance for Zero Extinction (AZE) species owing to its small population in a single restricted site. In 2006 CBSG facilitated a stakeholder-diverse conservation planning process for this species, which included experts involved in conservation of the Guam rail (a closely related species facing similar threats). A primary threat to the Okinawa rail is the spread of invasive Indian mongoose across the island, along with lesser threats of roadkill mortality and feral cat predation. Population viability analysis projected a rapid decline to extinction in about fifteen years if mongoose expansion was unabated. No ex situ population existed for the species at that time, while approximately a thousand birds remained in the wild.

The conservation planning process identified management actions aimed at the primary threats to the wild population. As an additional measure against imminent extinction, the establishment of a captive population was evaluated and recommended, with the conservation goal of serving as an assurance population against species extinction and as a potential source population for future reinforcement or reintroduction to the wild once mongoose were controlled or eradicated (CBSG 2006). Subsequently an ex situ breeding population of about seventy-seven rails (2013 population count) has been established and is augmented by rescued eggs. Mongoose control efforts stopped the spread of this invasive predator and expanded the mongoose-free zone, leading to the expansion of the Okinawa rail in range and numbers to an estimated fifteen hundred rails by 2013. Techniques are being developed to successfully release captive-hatched rails into the wild to support population expansion while maintaining a viable ex situ assurance population. Local support for the rail is promoted through an ecological exhibition and education center housing rails for display, which encourages public awareness to reduce roadkills and control cat predation.

INTEGRATING CONSERVATION INTO EX SITU COLLECTION PLANNING

Ideally, every species at risk would have an integrated species conservation plan that delineates any appropriate ex situ conservation activities. In reality, few OPA plans exist. Meanwhile, zoos and aquariums manage living animals and must decide on the best use of limited space and resources. It is not appropriate to assume that all threatened species will benefit from an ex situ program, or that programs for nonthreatened species have no value (Bowkett 2014); a more detailed assessment beyond the Red List status of threat category is required to set the best priorities for ex situ management for species. Birdlife International conducted a prioritized assessment of avian species that are recommended to be considered for ex situ breeding, which may be one source of helpful information in zoo collection planning for avian species (Collar and Butchart 2014). Such assessments do not exist for most taxa, however, and will require more effort to derive.

Elements of the IUCN guidelines for ex situ management and a more integrated approach can be applied to the process of regional collection planning undertaken by Taxon Advisory Groups (TAGs) to better align zoo and aquarium collections and management with species conservation needs. This could include a modified and expanded threat analysis and status review based on existing data sources; collaboration with taxon-based IUCN Specialist Groups and ongoing field conservation and research projects to identify conservation needs; careful tailoring of program form and goals to these needs; and careful consideration of the distribution of space and resources among species within a region and among regions for greater conservation benefit. This process can be applied to both threatened and nonthreatened species, since even nonthreatened ex situ populations may serve a conservation role as a surrogate species. It is recognized that species may also serve nonconservation purposes, which may need to be considered within this framework. A similar modified process (threat analysis and evaluation of ex situ conservation roles) has been adopted by several Global Species Management Plans of the World Association of Zoos and Aquariums (WAZA).

The Callitrichid (marmosets and tamarins) TAG of the European Association of Zoos and Aquaria (EAZA) applied the OPA philosophy and IUCN guidelines to its 2014 Regional Collection Plan (see Wormell et al. 2014 for details), providing a model for this process. Status and threat information for forty-two callitrichid species was gleaned from the IUCN Red List assessments, species action plans, and other publications for wild populations. Information was also compiled on the status of any existing ex situ populations by other regional zoo associations or in sanctuaries or rescue centers. The TAG worked closely with specialist group and field conservation representatives to identify potential conservation roles for callitrichid ex situ programs, especially on a regional level. Because in situ specialists are not always aware of the breadth of ex situ management options, they were not simply asked, "Do you think there is a need for a captive breeding program for this species in EAZA zoos?" Rather, they were introduced to the four ways ex situ activities can further species conservation and were given descriptions of potential roles an ex situ program might serve, as well as information on the status of any existing EAZA populations so they could contribute to an informed decision about potential ex situ management roles.

The TAG found that the five-step process described in the IUCN guidelines can be applied equally well to species without direct conservation roles but with other potentially valuable roles for zoos. The required ex situ program structure to reach conservation or other roles was considered as well as the feasibility of reaching this status regionally or globally, and weighed against competing needs for space and other resources by other callitrichid species.

The TAG divided the forty-two species into four categories for assessment based on the species' threat status (threatened versus nonthreatened) and current ex situ status (held in the region versus not held in the region). Nonthreatened species not present in the region were designated "do not obtain." Priority was given to threatened species that would benefit from ex situ management, whether or not such a program currently exists. This led to developing a list of currently held nonthreatened species that could potentially be replaced by threatened species with conservation need. Nonthreatened species were prioritized based on their research, education, husbandry model, or display importance as well as their evolutionary uniqueness (Isaac et al. 2007). This Regional Collection Plan evaluation process led to the realization that the program structure and conservation goals differed among species and do not always align directly with current zoo association program categories. This suggested a need for greater flexibility in such types of management programs to adequately meet conservation needs. EAZA is currently restructuring its species management program to reflect these needs.

This experimental foray into a new collection planning method led to development of the new CPSG-facilitated process of Integrated Collection Assessment and Planning (or ICAP), initially applied to the global assessment of forty-three canid and hyaenid taxa in 2016. This process relies on significant input by both the in situ and ex situ communities to assess the realistic potential conservation contributions by the zoo and aquarium community and to inform the regional TAGs in their regional collection planning.

CONCLUSION

There is increasing need to strengthen the integration of in situ and ex situ conservation planning to ensure that, whenever appropriate, ex situ populations and activities support in situ conservation to the greatest extent possible. When used strategically, ex situ management can be a potent tool for species conservation. Integrated species conservation planning must come from both sides and may require a concerted effort. Bridging the schism in species planning between the in situ and ex situ communities will not be without challenges and likely will require a learning period for each side to think outside the box. The field conservation community should better understand and embrace ex situ management options as part of the larger conservation toolbox to address threats to wild populations and avoid extinction. Likewise, the ex situ community should reevaluate its existing programs in terms of the conservation role(s) they can play and structure those programs to achieve those roles effectively (Lacy et al. 2013). In the ideal future, integrated conservation plans developed in concert with all stakeholders would exist for all threatened species, in which potential conservation role(s) for ex situ management have been carefully evaluated for appropriateness, feasibility, and effectiveness.

Participants at the 2010 IMP workshop summarized what they believed needs to be achieved in the following goal: "The world zoo and aquarium communities are, and are acknowledged as, effective conservation partners in the context of integrated conservation strategies that include intensive population management." We hope the rising profile of the One Plan Approach and the newly revised *IUCN SSC Guidelines on the Use of Ex Situ Management for Species Conservation* will serve as catalysts to promote increased integration of planning activities and collaboration between the field and ex situ communities. The result can be a win-win situation for all: zoos and aquariums can benefit through improved collection planning and population sustainability and can develop into and be perceived as true conservation partners; field conservationists can gain new tools and resources for addressing both short-term and long-term threats to species in the wild; and threatened species themselves may benefit from the buffer against extinction provided by appropriately structured ex situ activities. By including all stakeholders and evaluating all possible management options, the One Plan Approach ensures that all potential conservation efforts are used to save a species from extinction.

CHAPTER ELEVEN

Zoos and Gorilla Conservation: Have We Moved beyond a Piecemeal Approach?

Kristen E. Lukas and Tara S. Stoinski

Zoos and aquariums are places where the general public convenes to view wildlife, spend time with friends and family, and learn about the natural world. Informed and realistic about the myriad challenges facing the natural world, good zoos and aquariums are making increasingly stronger commitments to wildlife conservation and becoming places where the general public can connect to what's happening in nature and act to be part of the solution. However, most zoos and aquariums are nonprofit organizations, and many struggle to cover costs of basic operations (animal care and welfare, facilities and exhibits, guest services) while also prioritizing and raising funds for important mission-based activities such as conservation, education, and research. With hundreds of species and thousands of animals under human care, determining the most efficient and effective mechanisms for supporting wildlife conservation requires special consideration. Fortunately, many mechanisms have arisen in the past few years to ensure that zoos and zoo visitors are connected to wildlife conservation, particularly for charismatic species such as gorillas. Wildlife conservation is a central tenet of the continued justification for the existence of modern zoos. But are zoos doing enough—and acting quickly enough—to have a meaningful effect on the long-term preservation of wildlife and wild places? If the opportunity to look directly into the brown eyes of a zoo-housed gorilla doesn't generate the inspired vocation or sizable donation that secures a future for its wild counterparts, is the vision that zoos are conservation agencies nothing more than a mirage? Although it may not be easy for zoos to garner support for all species, gorillas may present a model for how zoos are leveraging zoo-based populations to ensure the long-term survival of wild populations.

There are two species of gorillas: western gorillas (*Gorilla gorilla*) and eastern gorillas (*Gorilla beringei*). The International Union for Conservation of Nature's Red List of Threatened Species (http://www.iucnredlist.org/) currently lists all four subspecies as either endangered or critically endangered in the wild. Eastern gorillas are threatened with human encroachment, emerging diseases, poaching, and political instability. Although critically endangered, mountain gorillas (*Gorilla beringei beringei*) are the only ape population in the world that is increasing in size. However, the population numbers fewer than nine hundred individuals and continues to face significant human pressures, requiring an "extreme conservation" approach that includes daily protection and monitoring of every individual gorilla by interdisciplinary teams of skilled people (Robbins et al. 2011). Grauer's gorillas (*Gorilla beringei graueri*) are found only in Democratic Republic of Congo and have faced drastic reductions in numbers; they are currently listed as one of the top twenty-five most endangered primates (Schwitzer et al. 2015). Western gorillas face similar threats of illegal hunting, habitat loss, and emerging wildlife diseases such as ebola. Western lowland gorillas (*Gorilla gorilla gorilla*) number more than 125,000 and, in stark contrast, Cross River gorillas (*Gorilla gorilla diehli*) number fewer than three hundred; both subspecies are critically endangered.

In zoos, the global population of western lowland gorillas (*Gorilla gorilla gorilla*) is managed by five regions and currently consists of 879 gorillas in 145 zoos (Köhler and Bender 2016). Across North America, approximately 350 western lowland gorillas live in 50 zoos accredited by the Association of Zoos and Aquariums (AZA) and are cooperatively managed as part of the Gorilla Species Survival Plan (SSP) (Lukas, Elsner, and Long 2015). Because gorillas are very popular and appeal widely to zoo visitors, they are an excellent focal point for examining how zoos can create conservation programming around flagship species. Zoo guests are attracted to gorillas' appearance and behavior, and that attention creates an opportunity for zoos to tell the story of why gorilla populations are critically endangered in the wild and how they are benefiting from the support of zoos—but only if zoos are actively part of the solution, which isn't always the case.

Nearly ten years ago Stoinski, Lukas, and Hutchins (2008) conducted and published a review of ways institutions accredited by AZA contributed to in situ conservation of gorillas from 1997 to 2005. The results revealed that 58 percent of Gorilla SSP zoos were involved with or supported gorilla conservation and that efforts were largely short-term, sporadic, and unevenly distributed among institutions. We therefore called for a new approach to zoo support for gorilla conservation. Specifically, we recommended a more cooperative and comprehensive program to plan and raise funds for longer time frames. To that end, the Ape TAG Conservation Initiative was proposed at the level of the AZA Ape Taxon Advisory Group (TAG), representing each of the six great ape species (eastern gorillas, western gorillas, Sumatran orangutans, Bornean orangutans, chimpanzees, bonobos) and two gibbon species.

This chapter examines how zoos responded to the call to increase the depth and extent of commitment to in situ conservation of gorillas, either through cooperative mechanisms such as the Ape TAG Conservation Initiative or through targeted efforts to connect individual zoos with specific field sites.

THE APE TAG CONSERVATION INITIATIVE

The Ape TAG Conservation Initiative (http://www.apetag.org/Conservation Initiative.html) (fig. 11.1) was launched in 2010 with the goals to provide multiyear support for high-priority ape populations and sites; to increase the number of zoos contributing to in situ conservation of apes; to increase the presence of the zoo community in ape conservation; to encourage law enforcement and in situ education through support of sanctuaries; and to provide zoos with resources to convey ape conservation messages to the public and promote their support for in situ conservation. In each funding cycle, eight projects (western gorillas, eastern gorillas, Bornean orangutans, Sumatran orangutans, chimpanzees, bonobos, gibbons, and siamangs) are selected through a rigorous two-tier process that involves representatives from the international ape conservation community, AZA, United States Fish and Wildlife Service, IUCN Primate Specialist Group, and United Nations Environment Programme's Great Ape Survival Program. Zoos join at one of three levels by committing to a contribution of $1,000 ("silver"), $5,000 ("gold"), or $10,000 ("platinum") per year for three years minimum. Of the funds generated, 95 percent is split equally among eight projects and 5 percent is given to in situ sanctuaries.

Since the beginning of phase 1 (2011–13) of the Ape TAG Conservation Initiative, more than $500,000 has been distributed to sixteen ape conservation projects and sanctuaries in Africa and Asia. Phase 1 contributions were strengthened after the Arcus Foundation awarded a matching grant

FIGURE 11.1. The Ape TAG Conservation Initiative coordinates long-term, substantial support for field projects conserving western gorillas and eastern gorillas, among other ape species. K. E. Lukas.

of $100,000, illustrating the power of this multi-institutional collaboration to leverage additional funding. More than forty AZA facilities participated in the Ape TAG Conservation Initiative in phase 1, representing one-third of the AZA facilities housing apes. Accomplishments from phase 1 relevant to gorilla conservation include recognition as a National Park of the Goualougo Triangle, which is home to significant populations of chimpanzees and western lowland gorillas in the Republic of Congo, and yearly antipoaching patrols covering more than 1,250 square kilometers of bonobo, chimpanzee, and Grauer's gorilla habitat in the neighboring Democratic Republic of Congo (DRC). The awards also supported capacity building for national park staff members working to protect gorilla habitat and community engagement activities that reached thousands of individuals living around national parks in multiple countries in Africa. Phase 2 (2014–16) of the Ape TAG Conservation Initiative was launched in early 2014, and 30 percent of ape-holding institutions have made multiyear commitments to the project. The AZA community considers this program a bright spot for Animal Programs with respect to conservation, and it is a model that is being replicated for other species, including tigers and lions.

INDIVIDUAL ZOOS CONNECT AND COMMIT TO GORILLA CONSERVATION

Another measure of zoo support for gorilla conservation is provided by AZA's Annual Report on Conservation and Science, in which zoos reported spending $705,545, $893,844, and $1,700,569 on gorilla conservation in 2012, 2013, and 2014, respectively, for a total of $3.5 million over a three-year period. Going back even further to 2010, more than $4.5 million was directed toward gorilla conservation over a five-year period. It is essential to recognize the substantial investment made by the Wildlife Conservation Society (WCS), which operates an aquarium and four zoos, including the Bronx Zoo. WCS is a world leader in the conservation of all four gorilla subspecies, and its financial commitment to conservation makes up the vast majority of funds reported to AZA each year. WCS currently has seven projects in central Cameroon, the Congo Basin Coast, Republic of Congo, and Gabon that protect critically endangered western lowland gorillas. A project in Cameroon and Nigeria protects critically endangered Cross River gorillas. WCS also works to conserve mountain gorillas in the Greater Virunga region spanning DRC, Rwanda, and Uganda, as well as Grauer's gorillas in the Maiko-Kahuzi landscape in DRC. One way the Bronx Zoo, one of the WCS institutions, raises awareness of, and funds for, WCS gorilla conservation projects is through the Bronx Zoo Congo Gorilla Forest exhibit, which has raised more than $20 million for conservation projects in the Congo basin since opening in 1999. The collaboration of WCS and the Bronx Zoo is an extraordinary illustration of how a zoo can capitalize on the visitor experience to raise awareness and significant support for conservation.

Although it is somewhat valuable to think of zoos' commitments in terms of dollars spent, it is also important to evaluate the extent to which zoos make meaningful and lasting connections to field initiatives aimed at studying, protecting, or supporting gorilla conservation. Several zoos have responded to the call to move beyond a piecemeal approach by making long-term commitments to field conservation projects.

Notable examples include the following eight zoos, which have all moved beyond one-off grants to make long-term investments in projects protecting wild gorillas.

Cincinnati Zoo and Botanical Gardens: Mbeli Bai Project in Nouabalé-Ndoki National Park (Western Lowland Gorillas, Republic of Congo)

Established in 1995, Mbeli Bai in Nouabalé-Ndoki National Park (NNNP), Republic of Congo, is the longest-running field study of western lowland gorillas. Cincinnati Zoo and Botanical Gardens has supported the Mbeli Bai project, the Mondika gorilla habituation, tourism, and research project, and Club Ebobo conservation education program since 2003. Cincinnati Zoo's curator of primates, Ron Evans, made an on-site visit in 2007, and Cincinnati Zoo subsequently conducted several projects linking the zoo's gorilla exhibit with the in situ conservation program at NNNP. Specifically, Cincinnati Zoo developed and tested the influence of videos that integrated Western concepts and zoo management of gorillas with practical in situ conservation messages in an effort to convey broader perceptions of gorillas to the local people. The videos were evaluated in conjunction with evaluations of the effect of Club Ebobo programming in several villages. Support for these programs is ongoing, and a memorandum of understanding was developed between the Cincinnati Zoo and WCS in 2014 in which Cincinnati Zoo pledged three years of support for Nouabalé-Ndoki National Park's Mbeli Bai study/Mondika Research Center and Club Ebobo conservation projects.

Cleveland Metroparks Zoo and DFGFI (Mountain Gorillas, Rwanda)

In 2007, Cleveland Metroparks Zoo awarded a seed grant to one of Dian Fossey Gorilla Fund International's (DFGFI, https://gorillafund.org/) community health projects and in doing so established a foundation for providing ongoing support to the Fossey Fund's core program areas. The zoo subsequently awarded one-time grants for biodiversity research and professional development of Fossey Fund staff members. In 2012 one of us (KL) coorganized and attended the second Gorillas across Africa workshop in Rwanda and observed firsthand how the missions of the two organizations—Cleveland Metroparks Zoo and the Fossey Fund—aligned perfectly, with core programs dedicated first to the animals themselves, but with an equally strong focus on capacity building, research, and community outreach. In 2014 the Cleveland Metroparks Zoo established a long-term partnership with the Fossey Fund to expand the reach of its university programs. Cleveland Metroparks Zoo now funds and participates in the training of biology majors from the University of Rwanda, who conduct senior thesis projects at Karisoke Research Center, and it also provides professional development opportunities for Fossey Fund staff members and core support for the organization. Cleveland Metroparks Zoo spent more than $75,000 on its Gorilla Conservation Program in 2015 and has developed a five-year plan in partnership with the Fossey Fund that will triple the zoo's contributions to gorilla conservation by 2020. Partnerships like these (see Spring, this volume; Mendelson, this volume) are essential for leveraging resources to maximize conservation impact.

Columbus Zoo's Partners in Conservation (Mountain Gorillas, Rwanda)

Columbus zookeeper Charlene Jendry traveled to Rwanda in 1991 and, on returning, established Partners in Conservation (PIC) with the help and support of zoo director Jack Hanna and several zoo docents (https://globalimpact.columbuszoo.org/about/partners-in-conservation). PIC's mission is to provide funding to people in central Africa to support sustainable livelihoods and wildlife conservation. A number of organizations dedicated to the conservation of mountain gorillas across their range have benefited from PIC support, including the Gorilla Doctors (for more than twenty-five years), DFGFI (for twenty-eight years), and the International Gorilla Conservation Programme (IGCP). In addition, PIC has provided substantial support to a range of community-based conservation projects in Rwanda and DRC including the "Strong Roots" tree-planting initiative in DRC and the "Cleaning of the Virungas" project where more than a thousand former poachers have been hired to remove plastic, invasive plants, and metal from gorilla habitat. PIC's twenty-year-old Artisan Project has provided four hundred men and women from sixteen cooperatives in Rwanda with an economic alternative to poaching. PIC is currently funding the 2015/2016 gorilla census in partnership with IGCP.

Disney's Animal Kingdom

Since 1995, Disney's Animal Kingdom has contributed more than $1.3 million to gorilla conservation in seven countries. In addition, Disney's Conservation Fund has awarded more than a dozen Rapid Response Fund emergency grants to deliver veterinary services, food, and rehabilitation services to gorillas in need. Disney provides more than just funds to gorilla conservation efforts around the world. For example, the GRACE Sanctuary (http://gracegorillas.org/) for Grauer's gorillas in the eastern Democratic

Republic of Congo has benefited greatly from the leadership and knowledge provided by animal care, exhibit, and veterinary professionals from Disney's Animal Kingdom (and other zoos) who shared expertise and guidance as the sanctuary was built and established. Most recently, Disney's gorilla conservation efforts are centered on "reversing the decline," or implementing a strategic plan to reverse the decline of Grauer's gorillas in the eastern Democratic Republic of Congo. With strong partners such as the Jane Goodall Institute, WCS, AZA, DFGFI, Flora and Fauna International, and the African Wildlife Foundation, Disney is decreasing the rate of habitat loss, mitigating threats, and implementing community-based conservation across the landscape to protect Grauer's gorillas.

Houston Zoo and Gorilla Doctors (Mountain Gorillas, Rwanda, Uganda, and Democratic Republic of Congo)

The Houston Zoo began supporting the Gorilla Doctors (formerly the Mountain Gorilla Veterinary Project) in 2009. The Gorilla Doctors (http://www.gorilladoctors.org/) is an international team of veterinary professionals who work with government and nongovernmental agencies to conduct medical interventions that provide care and support as needed to eastern gorillas in Rwanda, Uganda, and Democratic Republic of Congo. The zoo has provided both financial backing and in-kind support through the involvement of its director of conservation, Peter Riger, as a member of the board of directors. The Houston Zoo is currently growing its support for training African veterinarians to help build regional capacity, reflecting the zoo's investment in creating a stable future for the long-term care and protection of eastern gorillas. The zoo has made a formal commitment through 2019. In addition to its commitment to Gorilla Doctors, the Houston Zoo has a comprehensive portfolio of projects that includes funding and in-kind support for the Conservation Heritage—Turambe mountain gorilla conservation education program in Rwanda (since 2012) and GRACE, the Grauer's gorilla sanctuary in Democratic Republic of Congo (since 2013).

Lincoln Park Zoo and the Goualougo Triangle Ape Project (Western Lowland Gorillas, Republic of Congo)

The Goualougo Triangle Ape Project (GTAP, http://www.congo-apes.org/) is a conservation initiative in the Republic of Congo that aims to study and protect sympatric chimpanzees and western lowland gorillas. GTAP first received a small grant from the zoo's Conservation Fund in 2003, and the zoo subsequently funded some field expenses for several years. In 2008 Lincoln Park Zoo made a long-term investment in the project and hired codirector Dr. David Morgan as a full-time employee in the Lester E. Fisher Center for the Study and Conservation of Apes. Lincoln Park Zoo continues to provide salary and travel support, cover field expenses, and contribute in-kind administrative support for the camera trap video library (Goualougo Video Lab) and in public relations, fund-raising, and accounting. In addition, the zoo has sent staff members including scientists, curators, and marketing staff to the field site to lend expertise and develop new communication tools to increase awareness of and support for the program.

North Carolina Zoo and Cross River Gorilla Conservation (Cross River Gorillas, Nigeria and Cameroon)

The Cross River gorilla is one of the most endangered apes. Found only in West Africa along the border of Nigeria and Cameroon, the population is estimated to number fewer than three hundred individuals. In 2007 North Carolina Zoo established a partnership with the Wildlife Conservation Society to help conserve Cross River gorillas (http://www.nczoo.com/conservationanimals/savingwildspecies/crossrivergorilla.aspx). Specifically, Dr. Rich Bergl, curator of conservation and research, and others have developed and applied a range of geographic information system (GIS) and global positioning system (GPS) technologies to help in-country teams locate the elusive gorillas and document illegal activities that threaten them. As a result of this monitoring and other associated conservation efforts, several of the Cross River gorilla sites have been able to document reductions in threat levels, increased conservation effort, and stable wildlife populations (Bergl et al. 2014), something seen at few other locations in equatorial Africa. North Carolina Zoo continues to work with WCS and other conservationists in Nigeria and Cameroon to provide training, equipment, support, and long-term planning to secure this population of gorillas (fig. 11.2). In 2015 AZA recognized the work of North Carolina Zoo and WCS with a Significant Achievement in Conservation award.

Zoo Atlanta and DFGFI (Mountain Gorillas and Grauer's Gorillas in Rwanda and Democratic Republic of Congo, Respectively)

DFGFI's work focuses on conserving both subspecies of eastern gorillas—the Virunga mountain gorillas (*Gorilla beringei beringei*) and Grauer's go-

FIGURE 11.2. Cross River gorillas are protected by conservationists trained and equipped by the North Carolina Zoo and other partners. Photo courtesy of Rich Bergl, North Carolina Zoo.

rillas (*Gorilla beringei graueri*). Since 1995, Zoo Atlanta has hosted the international headquarters of the Fossey Fund on-site at the zoo in Atlanta, Georgia. This has included providing space, technology support, human resources support, benefits, security, maintenance, and so forth. In addition, the zoo and the Fossey Fund cofunded a staff person (TS, one of the authors of this chapter) for twelve years, and recently the zoo has provided partial salary support for a second scientific position. By providing these essential functions and resources, the zoo has enabled the Fossey Fund to put more of its critical resources toward its gorilla conservation work in the field. The total amount of in-kind and direct support the zoo has provided since 1995 is $1,320,000, and the support provided by more than fifty additional zoo partners since the late 1980s exceeds $1,240,000.

It is important to note that these nine zoos (including WCS) may provide additional support for other gorilla conservation initiatives not described above. In addition, there are other zoos not featured here that have made significant and long-term commitments to gorilla conservation, including Dallas Zoo (DFGFI and GRACE), Indianapolis Zoo (DFGFI and GTAP), and Woodland Park Zoo in Seattle (Mbeli Bai). Other zoos making substantial contributions to gorilla conservation in the past five years include Denver Zoo, Detroit Zoo, Nashville Zoo, Oklahoma City Zoo, and Sacramento Zoo.

AZA SAFE: GORILLAS

In 2015 AZA announced the launch of the AZA SAFE: Saving Animals from Extinction (SAFE) initiative (https://www.aza.org/SAFE/; see Grow, Luke, and Ogden, this volume). With 231 accredited zoos and aquariums, AZA is in a position to leverage the collective expertise and fund-raising capacity of its institutions to significantly increase awareness of and support for wildlife conservation while inspiring and engaging the 180 million visitors who seek us out annually. Only ten species have been selected as areas of focus for AZA's awareness-building and financial support. Gorillas, as a group, are one of them. As we noted above, all four subspecies of gorilla face significant threats and need immediate and significant support to ensure their continued survival. All gorillas can benefit from the expertise and skill sets in place at AZA zoos, including those that can influence ecological research, capacity building, conservation technologies, captive care, education and outreach, university partnerships, social media campaigns, legislation, and policy change.

AZA organizations breed gorillas exceptionally well, and these individuals can serve as ambassadors for their wild counterparts. Currently there are more gorillas in AZA zoos than any other primate species. In addition, AZA possesses collective expertise in husbandry and veterinary medicine that can play a major role in reducing health threats in natural gorilla populations. Because so many are habituated, medical interventions are much more possible and prevalent in mountain gorillas than in many other populations of wild animals. For example, after a measles outbreak in the 1980s, the population was effectively vaccinated against measles. Gorillas occur in regions that harbor both established and emerging diseases and have a high risk of human-induced disease outbreaks. These outbreaks could have a catastrophic impact on this extremely small population. We need to better understand the potential role of disease in gorilla conservation and continually develop health protocols to reduce pathogen transmission—both are areas where AZA zoos can play a role. In addition, AZA members are specialists in small population management, which is likely to become increasingly important as the population continues to grow and potentially reach carrying capacity.

Great apes are our closest relatives and, as such, people relate to and connect with them (fig. 11.3). They should be one of the easiest species for

FIGURE 11.3. In addition to benefiting from the collaborative Ape TAG Conservation Initiative, mountain gorilla (Gorilla berengei berengei) conservation efforts are strengthened by long-term commitments from eight AZA zoos. Photo courtesy of Peter Riger, Houston Zoo.

which to garner widespread public support. The conservation community is extremely lucky to have designated United States tax dollars available for ape conservation through the Great Apes Conservation Act of 2000. However, the release of the Multinational Species Conservation Funds must be approved by Congress each year. A consortium of NGOs, including AZA, works with both the Senate and the House to provide letters of support for these funds—unfortunately, this year we were unable to secure support from the Senate. This is truly regrettable, particularly because the funding request also covers conservation measures for rhinos, tigers, and elephants. And these funds not only support conservation but also directly help US interests. For example, elephant poaching fuels terrorist organizations, and many conservation NGOs work toward United Nations Millennium Development goals. If an informed public were in contact with its elected officials, we could ensure and probably increase the funds available for conservation. In addition, one of the most significant rising threats to African apes is palm oil. AZA has already championed public awareness campaigns to mitigate threats to orangutans posed by the palm oil industry. Now we need to expand this to include African apes. Through the 180 million people who visit us each year, we have incredible power to create an informed, mobilized electorate and influence consumers' decisions.

The potential of this growing attention to gorillas and gorilla conservation by North American zoos through AZA SAFE is significant, and when we succeed, we will create the world's largest gorilla conservation movement and establish a secure future for ape populations across Africa. As we noted above, more than 15 percent of the zoos housing gorillas in North America have made significant, long-term commitments to field conservation programs, and both the Ape TAG Conservation Initiative and AZA SAFE provide immediate opportunities for engaging the remaining zoos in conservation. By leveraging staff expertise, along with current conservation connections, AZA SAFE can help zoos and aquariums engage millions of visitors annually in conservation and not only make a meaningful contribution to gorilla conservation but exceed the public's growing demand that zoos do more to conserve wildlife in nature.

CONCLUSION

We are fortunate to have the WAZA World Conservation Strategy (see Barongi, this volume) that provides us with leadership and vision for the role of zoos and aquariums in international wildlife conservation. There are many models zoos can adopt to increase support for field conservation. All require leadership and unwavering commitment from zoo leaders (including governing boards) and involve making reliable, long-term investments. It may not be easy for zoos and aquariums to maintain steadfast dedication to the conservation mission while also ensuring optimal animal welfare and providing an engaging day out for visitors. But it is an incredibly special opportunity when a visitor can look directly into a gorilla's eyes and have that gorilla return its gaze for even a few moments. There's no mirage there—only the vision, pure and simple, that such connections inspire and promote conservation action (see Grajal, Luebke, and Kelly, this volume).

Have we moved beyond a piecemeal approach? Yes. The Ape TAG Conservation Initiative, the long-term conservation programs led or supported by individual zoos, and now AZA SAFE are providing the impetus, cohesion, and leadership needed for our profession to become a major force for wildlife conservation. However, two new questions deserve consideration over the next five years:

1. Is every AZA-accredited zoo and aquarium making long-term, meaningful contributions to wildlife conservation? This year, the AZA Field Con-

servation Committee is working to ensure 100 percent participation in wildlife conservation by every one of the 231 accredited zoos and aquariums. Wildlife conservation is a core business for zoos and aquariums, and therefore not even one institution should be able to evade the responsibility we assume as members of the world zoo and aquarium community to ensure a secure future for wild animals in nature.

2. Are AZA zoos and aquariums well on the way to collectively becoming the world's largest wildlife conservation organization? The AZA Field Conservation Committee has posited that if every AZA institution contributed 3 percent of its operating budget to field conservation every five years, AZA would collectively contribute $1 billion to wildlife conservation. Working within the context of IUCN Conservation Action Plans and with the help of many, many partners, *this* is how zoos and aquariums can and must realistically achieve our collective conservation mission.

If we cannot answer yes to both questions, not only will we be unable to defend our profession to detractors, but we will have missed a great opportunity to play a key role in redefining the future for wildlife.

ACKNOWLEDGMENTS

Many thanks to the following individuals who provided information and images for their gorilla conservation work: Elizabeth Bennett, (Bronx Zoo/WCS), Rich Bergl (North Carolina Zoo), Ron Evans (Cincinnati Zoo), Charlene Jendry (Columbus Zoo and Aquarium), Claire Martin (Disney's Animal Kingdom), Peter Riger (Houston Zoo), and Steve Ross (Lincoln Park Zoo).

Ape TAG Conservation Initiative members, including Platinum Supporters Detroit Zoo, Oklahoma City Zoological Park, and San Diego Zoo and Wild Animal Park; Gold Supporters Cheyenne Mountain Zoo, Cleveland Metroparks Zoo, Dallas Zoo, Houston Zoo, Sacramento Zoo, Sedgwick County Zoo, Woodland Park Zoo, and Zoo New England; and Silver Supporters Chimpanzee SSP, Ellen Trout Zoo, Fort Wayne Children's Zoo, Fresno Chaffee Zoo, Gorilla SSP, Lincoln Park Zoo, Little Rock Zoo, Milwaukee County Zoological Gardens, Naples Zoo, North Carolina Zoological Park, Orangutan SSP, Potawatami Zoo, San Antonio Zoo, Tampa's Lowry Park Zoo, Topeka Zoo, Utah's Hogle Zoo, Virginia Zoological Park, and Zoo Atlanta.

CHAPTER TWELVE

Lessons from Thirty-One Years at the Monterey Bay Aquarium and Reflections on Aquariums' Expanding Role in Conservation Action

Margaret Spring

URGENCY FOR ACTION

The ocean, with its vast diversity of habitats, provides humans with countless essential benefits and services. It is our lungs, our pantry, and our playground. The ocean generates half the oxygen we breathe. It also regulates Earth's climate, buffering us from the full impact of greenhouse gas emissions, global temperature rise, and intensifying weather events.

The ocean also provides myriad intangible benefits. Its timeless rhythms and blue horizons help clear our minds, making space for fresh ideas. Its complex living systems inspire creativity and innovation, giving rise to art that enriches us and technology that sustains us. But these critical benefits are eroding as growing human needs strain the ocean's living systems.

Although US fisheries are recovering, mounting demand for seafood is driving global overfishing, destroying marine habitats and disrupting the ocean food web. Half the world's fish stocks are fully exploited, meaning they're at their maximum sustainable catch limits; another quarter are considered overexploited. Illegal fishing and bycatch threaten the sustained survival of tuna, sharks, seabirds, and sea turtles. If this trend continues unchecked, it may become impossible for wild fish stocks to help feed the world's human population–now over seven billion and growing (World Bank 2014).

Agricultural and urban runoff contaminates coastal waters and creates ocean dead zones. Plastic pollution ensnares, chokes, and poisons marine life while adding a side of toxins to our seafood. Oil spills and unsustainable coastal development pose additional threats to marine ecosystems. The human-caused changes to the ocean's chemistry compromise the foundations of marine ecosystems and our food supply (IPCC 2014a).

Humanity's continued emission of heat-trapping gases—the overarching environmental challenge of this century—compounds the effects of each stressor by affecting sea level, ocean temperature, species distributions, oxygen levels, and ocean acidity (United Nations 2016).

AQUARIUMS JOIN A GROWING GLOBAL RESPONSE

The ocean wildlife that enthralls our visitors and migrates along our shores is clearly under increasing stress. But aquariums are well positioned to lend their trusted voices to help turn the tide (Pew Oceans Commission 2003; US Commission on Ocean Policy 2004; National Ocean Council 2016).

Market research indicates that the general public looks to, and expects, aquariums and zoos to provide credible and actionable information on complex topics such as climate change, pollution, and seafood sustainability (Meyer, Isakower, and Mott 2015).

In response, aquariums worldwide are evolving in their missions to take on more active conservation roles. Many have launched initiatives to promote public awareness of environmental issues and undertake field conservation work (Muka, this volume). Some are moving from informing people to mobilizing them, and a few, including the Monterey Bay Aquarium, are taking direct action to promote changes, both in policy and in the marketplace.

CONSERVATION AT THE MONTEREY BAY AQUARIUM

In the late 1970s a group of marine scientists affiliated with Stanford University and Silicon Valley crafted a vision. They imagined a small aquarium devoted to the diversity and beauty of life in the waters just offshore, built inside a now derelict cannery from the industrial heyday of the sardine-packing era.

After six years of planning and funded by a $55 million personal gift from David and Lucile Packard, their vision became reality on October 20, 1984. The Monterey Bay Aquarium opened its doors in a state-of-the-art building on Monterey's historic Cannery Row.

Our aquarium started with a focus on the interconnected web of life in the ecosystems of Monterey Bay. It was a framework developed by marine biologist, ecologist, and philosopher Ed "Doc" Ricketts, a close friend of renowned author John Steinbeck. In less than a decade, Monterey Bay had become the heart of the largest national marine sanctuary in the continental United States, its recovery a model for protection of other marine regions.

Since opening our doors thirty-one years ago, we've explored the boundaries of what we can do to encourage conservation, both within the aquarium and beyond our walls. From the beginning, we've witnessed the power of up close and personal experiences with wildlife to inspire our visitors—and in many cases to motivate them. Over the decades, we've become increasingly invested in ocean health in and beyond Monterey Bay, and in 1996 we clarified and simplified the Monterey Bay Aquarium's mission to reflect our conservation-oriented approach. The new mission, "to inspire conservation of the ocean," is a major driver of our work, from exhibits to programs.

VISITOR PROGRAMS

We start with our interpretive programs. Our internal research shows that personal interactions between interpreters and visitors are among the most effective ways to inspire people. Our staff members and volunteer docents are particularly adept at weaving in ocean conservation information to enhance the visitor experience.

In 2009 we joined with other institutions to create an online tool, Climateinterpreter.org, to help communicate more effectively about climate change and to share professional resources. Today a diversity of aquarium users and nonusers—including government agencies, nongovernmental organizations (NGOs), and academic institutions—continually contribute content to the website. We are exploring how this model can be used for other critical ocean issues.

Our conservation mission is also central to our programs for teachers, youth, and students. Each year thousands of teachers and school administrators come to the Monterey Bay Aquarium, many for professional development, learning how to incorporate project-based marine science into their classrooms. We offer award-winning teen programs—extraordinary experiences that develop young people who are ocean literate, inspired, confident, and ready to take action for the ocean. And over the decades we've provided free admission to more than 2.2 million students, from prekindergarten to high school, helping build their marine science knowledge and conservation skills (Meister et al. 2014).

In addition to our education work, our aquarium conducts and supports scientific research, with the following goals: to ensure the safe collection and handling of exhibit animals; to provide clean, high-quality seawater to our exhibits; to accurately assess the sustainability of fisheries rated by our Seafood Watch program; and to inform our conservation policy work. Our Conservation Research staff brings expertise in population biology and ecosystem studies to bear on the conservation of southern sea otters, white sharks, and bluefin tuna—three key species in the North Pacific Ocean. Under the oversight of our science director, our dedicated staff scientists and partners employ state-of-the-art methods to advance our understanding of the animals and ecosystems we study. It's an approach increasingly embraced by our peers, such as the Shedd Aquarium (Knapp, this volume).

At the same time, our Conservation and Science programs are expanding in scope and influence, integrating our research with our growing influence in ocean policy. We're coupling that with our work with businesses and consumers to reshape the global seafood supply chain.

DEVELOPING ACTION-FOCUSED PROGRAMS

Our best-known conservation program began with a single exhibit and has expanded to reach a global audience. "Fishing for Solutions," a special exhibit that opened in 1997, focused on conservation issues surrounding fishing and aquaculture practices. In the course of developing the exhibit, the aquarium committed to serving only seafood from environmentally responsible sources—in our own restaurant and catering services and to our exhibit animals. As we publicized this change, visitors asked for copies of our approved seafood list. In 1999 we launched the Seafood Watch program with a website and our first consumer pocket guide (Seafood Watch 2013).

Today Seafood Watch provides information on the sustainability of seafood products in the North American seafood market. The goal is to empower US consumers, businesses, and suppliers to make purchasing decisions that support a healthy ocean. We've used the best science to make more than eighteen hundred recommendations on which seafood sold in North America is a "best choice" (green), which is a "good alternative" (yellow), and which to "avoid" (red).

In the sixteen years since launching Seafood Watch, we've distributed more than fifty-six million consumer guides, and our mobile app has been downloaded more than 1.5 million times. We've also greatly expanded our sustainable seafood network, which now includes more than 160 aquariums, zoos, and other nonprofit conservation partners, to extend Seafood Watch's reach (Seafood Watch 2016). Our business program encourages restaurants, distributors, retailers, food service companies, and chefs to commit to sustainable seafood sourcing.

This market-based shift to sustainable seafood is growing quickly. According to the Conservation Alliance for Seafood Solutions, of which the Monterey Bay Aquarium's Seafood Watch is a member, "more than 80 percent of the North American retail and institutional food service markets have adopted sustainable seafood policies" (Conservation Alliance for Sustainable Seafood 2014).

The impact is global. The United States imports 90 percent of the seafood Americans eat (NOAA 2014). By shifting seafood sourcing to more environmentally responsible options on a national scale, our corporate commitments are, in turn, promoting sustainable fishing and aquaculture practices around the world.

Another major program was the development of an aquarium-supported advocacy arm, focused specifically on conservation policy. The program was launched in 2004 to take our involvement in ocean conservation to the next level. This effort was enabled by our leadership's involvement in a national-level ocean study that recommended changes to US ocean policy and law (Packard 2009).

Aquarium executive director Julie Packard and trustees Leon Panetta and Jane Lubchenco, who served as members of the Pew Oceans Commission, worked with other thought leaders, including those from the US Commission on Ocean Policy, to set a course allowing the aquarium to make unique and valuable contributions to ocean conservation policies.

Over the next decade, the program harnessed the aquarium's influence and credibility to achieve policy-related outcomes. Among its victories was the successful implementation of the California law creating the nation's first statewide network of marine protection areas (MPAs). We mobilized public support for the science-based process that led to the designation of over one hundred new MPAs along the California coast and participated in the MPA network's strategic, science-based design.

Another outcome of this work was President Barack Obama's issuance in 2010 of a presidential executive order establishing national ocean policy that incorporated recommendations from both ocean commissions (Obama 2010). We joined with a coalition of partners at the national level to promote its development and implementation. And we continue to engage in high-level, international ocean policy discussions, including the World Ocean Summit, hosted by the *Economist* (*Economist* 2016); the Our Ocean

Conference, cohosted by the US Department of State (USDS 2015); and the inaugural Bluefin Futures Symposium, which we cohosted with Stanford University in January 2016 (Bluefin Futures 2015).

EXPANDING OUR SPHERE OF INFLUENCE

These milestones mark progress toward new and strengthened policies that are recovering US fisheries, expanding MPAs, and combating illegal fishing on the high seas. In addition, our work with research partners has tightened protections for at-risk ocean wildlife, including southern sea otters, Pacific bluefin tuna, and sharks. For example, we sponsored a state law banning the trade of shark-fin products in California—a major step in the global movement to reverse the devastating decline of shark populations.

Our policy efforts and activities as a conservation organization tell us that aquariums can play a leading role in conservation today—through the lenses of both marine biodiversity conservation, served through vehicles such as MPAs, and sustainable resource management, pursued via market-based approaches and education.

In 2013, with this solid record in hand and a growing urgency to address even more pressing challenges to ocean health, we evaluated our strengths and weaknesses and identified opportunities to focus our policy, markets, and research expertise for a more powerful and relevant conservation agenda. This analysis reinforced the value of our work in advancing sustainable seafood, restoring and protecting the marine life and ecosystems of the California Current, and promoting innovative and effective conservation models at both state and federal levels. It identified ways to leverage the aquarium's assets for greatest effect and also highlighted opportunities to work with partners outside our walls.

We now have incorporated these findings into a three-pronged approach under a strategic plan developed by aquarium staff and approved by our board in 2014. First, we aim to integrate our strengths, including our credibility with key decision makers beyond the general public. This means our Conservation Research, Policy, and Seafood Watch teams collaborate, delivering the best available science and conservation strategies to the government and business leaders who can drive solutions.

For example, the Seafood Watch red-listing of Louisiana shrimp as a product to "avoid," owing to the state's refusal to enforce the federal requirement of turtle-excluder devices, prompted a reported thirteen thousand retailers across the United States to boycott wild shrimp caught by Louisiana fishers (Alexander-Bloch 2015). In July 2015, under pressure from the state's shrimper association, the Louisiana government repealed its law, effectively requiring its fishers to use the federally mandated gear preventing turtles from drowning in shrimp trawls. Seafood Watch promptly upgraded its listing of Louisiana shrimp to "good alternative," opening up an enormous new retail market for the product.

Second, the aquarium's strategic plan reflects our expanding sphere of influence, starting with our own backyard. The Monterey Bay region and greater California constitute our identity and base. Regionally and across the state, we are raising awareness of important ocean issues like MPAs, ocean plastic pollution, and climate change.

Third, we're cultivating ocean champions among decision makers from Sacramento to Washington, DC. Following our key ocean wildlife—sea otters, bluefin tuna, and sharks—along the California Current and across the Pacific, we also advocate for their protection in international fishery management forums, while our Seafood Watch program is quickly becoming a recognized global standard of sustainability.

Given that we do not have unlimited capacity or expertise, we've carefully selected focal areas for action by using the following questions as strategic filters: Is there a clear connection to the aquarium's unique identity and strengths? Does the aquarium add unique value? Is the issue ripe for action, and is there a clear line of sight to conservation impact? Does the aquarium have the capacity—inclusive of partnerships—to be successful?

AQUARIUMS' ROLE IN SHAPING PUBLIC PERCEPTIONS OF CONSERVATION

On the surface, aquariums seem well equipped to deliver ocean conservation messaging and mobilize their audiences to act. But we still face significant hurdles in public perception. We depend largely on survey data to inform our strategies. Our visitor research draws from the roughly two million people who walk through our doors each year. We also continue to track public attitudes about aquariums and ocean conservation so we can meet people where they are, then engage them to take action.

The first part of that formula—meeting people where they are—means understanding how people think about our priority issues. To that end, the Monterey Bay Aquarium was one of the founding partners of the Ocean Project, which has conducted three national surveys since 2009 to track public perceptions of ocean-related issues such as climate change, plastic pollution, and habitat protection.

In collaboration with twelve aquariums and science museums, The Ocean Project surveyed aquarium visitors. One notable finding: Visiting an aquarium tends to ignite a temporary spark of concern for the ocean, inspiring an emotion-based interest in helping reduce its major threats, whether at the individual or the collective level.

According to the Ocean Project's 2014 summary of public opinion research, the US public's knowledge and concern about ocean issues have remained consistently low over the years, with interest spiking only in reaction to major events like the *Deepwater Horizon* oil spill.

But the research reveals some silver linings. Most people want to be a part of the solution to ocean problems, even if they don't view those problems as urgent. They have a very high level of trust in aquariums, zoos, and museums as sources of information—more than they do in nongovernmental organizations, and much more than in government. Respondents had the least trust in government at the federal level.

That's corroborated by a separate 2013 survey by the public opinion research firm FM3, which focused on voters' attitudes toward plastic pollution in California waterways. That survey found that 79 percent of Californian voters trust the Monterey Bay Aquarium as a messenger—more than all other listed government agencies and NGOs (FM3 2013). Likewise, our research has found that people trust aquariums, zoos, and museums more than they do NGOs or government (Meyer, Isakower, and Mott 2015). Perhaps most important, people strongly believe aquariums, zoos, and museums should suggest specific ways they can take action to help solve environmental problems.

MOBILIZING THE NEXT GENERATION FOR OCEAN CONSERVATION

The decline in public confidence in government and NGOs suggests that aquariums have a unique opportunity, and responsibility, to influence the next generation's views about ocean conservation. As a visitor-serving conservation organization, the Monterey Bay Aquarium is attentive to changing circumstances that could affect our mission: to inspire conservation of the ocean.

US demographics are changing rapidly. The success of our ocean conservation efforts depends on our ability to inspire the up-and-coming wave of conservation actors who make up a large part of our audience.

Pew Research reports that more than one in three American workers are Millennials, defined as people born between 1981 and 1997. In 2015 this group eclipsed Generation X as the largest demographic in the US workforce. To effectively reach them, we first need to understand how they communicate and what they value.

One important data source is a national study of youth attitudes and behavior toward ocean conservation, conducted by the David and Lucile Packard Foundation in partnership with Edge Research (Dropkin, Tipton, and Gutekunst 2015).

The study found that almost 60 percent of Millennials are involved in "causes" such as children's health, poverty, and human rights. But of those, only one-quarter name "environment, conservation, or wildlife protection" among those causes they most identify with: "animal welfare" tops the list of causes, at 32 percent. That concern signals an opportunity for aquariums to reach more Millennials, emphasizing our work in rescuing and releasing distressed wildlife. But it's also a signal that we could lose Millennial support if captivity is perceived to harm animal welfare (Palmer, Kasperbauer, and Sandøe, this volume).

In general, Millennials are more attracted to aquariums that are taking action for ocean conservation. That's why we know it's important to "walk the talk" on the issues we target. For example, as we increase our messaging about ocean plastic pollution, we've removed all single-use plastic from our in-house café, and we supported successful state and federal bills restricting the use of microbeads in personal care products (MBA 2015).

It's important to frame calls to action based on what we know motivates people. The Packard study finds that Millennials believe more in the power of individuals, acting together and alone, to make changes. They don't count on governments to do it, and they don't see NGOs as effective. Within this important group, change begins at the peer-to-peer level; Millennials can be encouraged to influence family and friends.

While Millennials have the energy and the passion to make a difference, the Packard study shows, ocean conservation issues are not yet a part of their trending conversations. So we're using close to home issues that resonate, like plastic pollution and marine life protections, as entry points to more complex topics like climate change and overfishing. It's also important to engage Millennials in their "natural habitat" of digital and social media.

Another important demographic, particularly in California, is the growing Latino population. Latinos have become increasingly influential, both in the US marketplace and in political circles, and many see California as an early indicator of trends for the nation. The California Legislature has a new majority of moderate Democrats, many of whom represent majority-Latino constituencies.

This presents both an opportunity and a need to cultivate new ocean champions. We hope to impress on California's new generation of elected leaders the importance of ocean health to the well-being of the communities they represent. These topics include those relevant to quality of life, such as water quality, plastic pollution, and protected open spaces.

THE NEXT WAVE: CONSERVATION ACTION

Every day at the Monterey Bay Aquarium, we see visitors falling in love with the kelp forest, or the sea otters, or the octopuses. According to our visitor research surveys, most leave with a willingness to take action for ocean conservation—such as by using a Seafood Watch pocket guide, sharing materials with their schools, or joining as members.

Given the well-established success of our exhibits and education programs, creating an aquarium-based conservation effort was not without risk. Internally, some have worried this new focus might distract from the core business of ensuring that our aquarium continues to thrive. Our visitors might not welcome the increased focus on conservation messaging in our exhibits and communications. They might not want to respond to calls to action like supporting legislation and funding specific projects. Or they might look to other nonprofit organizations for this guidance. Some of our members and donors might disagree with our points of view, or they might feel we should steer clear of politics, particularly on polarizing issues such as climate change.

A piece in the *New York Times* described how aquariums and zoos are navigating the potential conflict between public trust and climate leadership (Kaufman 2012). The article points to the New England Aquarium as an institution embracing the opportunity to educate visitors about the effects of climate change, and to the Georgia Aquarium as one choosing to avoid the controversial topic so as to not offend conservative visitors. The story underscores the political tension aquariums and zoos must navigate as we attempt to balance our roles as both attractions and conservation authorities.

But without taking that risk we lose relevance and trust with the next generation. At the Monterey Bay Aquarium, we've ventured outside our comfort zone, wading ever deeper into the realm of conservation action.

We've done so strategically, using audience data to build from our core strengths: a strong reputation and scientific credibility.

The results to date have shown we can have a conservation agenda and still keep the public's respect. Survey data, from research conducted for our internal use by IMPACTS Research and Development, indicate a high level of public trust in our aquarium—not only to provide accurate information, but also to take conservation action.

By "action," we mean more than awareness. There's a wide range of behavior that falls under the "action" umbrella. It can be as simple as individual changes like carpooling and recycling more. It can be communitywide efforts like beach cleanups and bike infrastructure improvements. Or it can be the creation and modification of government policies and legislation—at state, federal, and even international levels.

As our credibility as a conservation actor inched up over the past three years, so did the percentage of respondents who consider the Monterey Bay Aquarium the best and most admired aquarium in the United States. The lesson is clear: Not only can we take positions on ocean conservation, but the public expects us to do so.

And we're just one institution, acting on our own. Together, aquariums and zoos can deliver on a larger promise: to raise a collective voice on ocean policy (Falk et al. 2007). Our experiences to date and our plans for the future may serve as a useful road map for other institutions as they consider expanding their roles in conservation action.

A COLLECTIVE VOICE FOR AQUARIUM CONSERVATION

Our survey data show that an increasing number of US aquariums are engaging in ocean conservation in their exhibits, education programs, and research. Together we have great potential to advance a wide range of conservation initiatives, bringing to ocean conservation issues a broader audience appeal that could help bridge political and regional divides.

US aquariums have not historically leveraged our collective expertise to make a broader conservation impact in the policy and business worlds. To that end, the Monterey Bay Aquarium has stepped up as one of several organizers of a voluntary pilot collaboration among aquariums to achieve a greater collective impact than any one institution acting alone.

The Aquarium Conservation Partnership (ACP) aims to focus aquarium action on a discrete set of impact goals while also serving as a "strategic table" around which aquariums can share best practices in growing our individual and collective conservation leadership. The organization was designed to promote aquarium engagement in policy action, in particular, and to leverage our credibility and influence with decision makers and their constituencies to accelerate government action.

ACP has selected plastic pollution as the primary focus of its two-year pilot (2016–17). It serves as a forum for helping aquariums identify, advance, and enact science-based policy solutions that reduce the sources of plastic pollution in ocean and freshwater ecosystems. ACP will also inform and facilitate aquarium policy action toward other conservation goals, including increasing protections for critical ecosystems, conserving threatened shark and ray species, and improving seafood sustainability.

The broader community of US aquariums and zoos is also moving toward conservation action. In 2015 the Association of Aquariums and Zoos (AZA) launched an effort called Saving Animals from Extinction, or SAFE, to "deepen the already substantial science and conservation work on endangered species occurring at AZA-accredited zoos and aquariums by engaging the 180 million annual aquarium and zoo visitors and partners across the world to protect habitat, decrease threats, and restore populations to sustainable levels" (AZA 2015c). AZA has also taken several steps to make ocean conservation more visible, including creating a web page dedicated to the topic (AZA 2014a).

CONSERVATION AS A CORE BUSINESS STRATEGY

In the past we've thought of our conservation outreach and policy work as benefits nested within our education and research programs. Today, guided by our new thinking, conservation work is an essential part of our business strategy. We've remained attentive to our reputation as one of the world's great aquariums, engaging people through our exhibits and continuing our investment in husbandry research and development. But serious conservation work will remain central to what we do—and how we thrive.

If we want to engage more people in our mission, both as contributors and as advocates, we must move the focus far beyond our walls, extending our reach as a force for conservation. This will require a new emphasis on Internet communications, both for marketing and for social networking. We're moving beyond the "radio tower" model, in which an institution broadcasts its own information, to a social-network model of people communicating with each other.

Our ultimate goal is to build a constituency that will work to protect and restore the world's life-sustaining aquatic ecosystems. This vision—to engage and activate people in a meaningful long-term relationship—may take many forms, whether we are asking people to join, to give, or to act. These forms of engagement are mutually reinforcing. Contributions motivate visits. Visits motivate action. Action motivates visits. Doing meaningful conservation work that builds brand loyalty and respect is the best business investment we can make.

Market research clearly shows that the most financially successful nonprofit, visitor-serving organizations are those that pursue their missions aggressively and demonstrate to the public that they are both principled and strongly committed.

Although public attitudes about conservation vary across global cultures, we believe people everywhere seek a common vision of a sustainable future on Earth—one that's practical and attainable, and one in which they can play a part. It's a complex dynamic, of course, that involves factors like preserving biodiversity and improving human livelihoods—goals that can be pitted against one another, or bound together, depending on case-by-base definitions and framing. But we believe the protection of marine ecosystems, and the wise use of their resources, should be embedded in any vision of sustainability. Aquariums can guide audiences around the world to make a difference for people, our planet, and the wildlife we share it with.

With global ecosystems in decline, there is no time to lose. By understanding our audiences, we can craft meaningful ways to speak to their interests and help them be part of the solutions. These are big challenges, and we don't have all the answers. But decades of data analysis and risk taking have taught us that conservation isn't just about fieldwork. Our sights have been set on policy and behavior changes because in the end they will have the lasting impacts we need.

ACKNOWLEDGMENTS

This chapter reflects over thirty years of innovation at the Monterey Bay Aquarium, under the leadership of executive director Julie Packard. I wish particularly to express my appreciation to my writing partner, Kera Panni, who was indispensable in the development, editing, and timely completion of the chapter. I also thank my colleagues Jim Hekkers, Cynthia Vernon, Mimi Hahn, and Aimee David for their contributions both to the chapter and to this body of work.

CHAPTER THIRTEEN

The Phoenix Zoo Story: Building a Legacy of Conservation

Ruth A. Allard and Stuart A. Wells

Modern, professionally managed zoos and aquariums serve many purposes and reach diverse audiences (WAZA 2005; Falk et al. 2007; Falk, Heimlich, and Bronnenkant 2008). Driven by missions that expand far beyond the menageries of centuries past, modern zoological facilities are not content simply to hold exotic animals and display them for public view (Conway 1995a; Hutchins and Smith 2003; Gusset and Dick 2011). Guests come to zoos and aquariums to bond with family and friends, to play, to learn, to get exercise, and of course to see wild animals, among other motivations (Falk et al. 2007; Falk, Heimlich, and Bronnenkant 2008). To keep up with or even outpace guests' expectations and to maintain relevance in today's world, modern zoos and aquariums must consistently deliver high-quality, compelling guest experiences. But today's institutions must also drive home the message that a visit to a professionally managed zoo or aquarium supports a healthy future for animals in the wild (Conway 1995b; Croke 1997; Hutchins, Smith, and Allard 2003; WAZA 2005), which includes a commitment to conservation and education (Hutchins and Conway 1995; WAZA 2005; Patrick et al. 2007; Gusset and Dick 2011; Barongi, this volume; Grajal, Luebke, and Kelly, this volume; Lukas and Stoinski, this volume). Excellent zoos and aquariums take seriously their roles as conservation organizations with a responsibility to inspire guests to care for wildlife and wild places beyond their gates (Maple, McManamon, and Stevens 1995; WAZA 2005; Barongi, this volume; Grajal, Luebke and Kelly, this volume; Lukas and Stoinski, this volume). The Phoenix Zoo is no exception.

LAYING THE FOUNDATION FOR CONSERVATION ACTION

The Phoenix Zoo has been involved in international conservation efforts from its inception, committing to welcoming the "World Herd" of Arabian oryx from the Middle East even before the zoo's gates first opened in November 1962 (Shepherd 1965; Turkowski and Mohney 1971; Stanley Price 1989). Through the efforts of several essential partner organizations, three of the last remaining Arabian oryx in the wild were brought to the Phoenix Zoo as part of "Operation Oryx," a collaborative project developed to save the species from extinction. This vision was unusual for a nascent zoo during that era (Conway and Hutchins 2001), and it has been a source of pride throughout the zoo's history.

Through various acquisitions and transfers, nine animals ultimately came to the Phoenix Zoo to establish the "World Herd" (Shepherd 1965; Grimwood 1967; Turkowski and Mohney 1971; Stanley Price 1989). Descendants of this group were ultimately released back into the wild in Oman beginning in 1982 (Stanley Price 1989). Although the Arabian oryx did become extinct in the wild in 1972, subsequent reintroductions of oryx born ex situ brought this animal back from extinction (1989). In 2011 the Arabian oryx became the first mammal to have its listing improved from critically endangered to vulnerable after being considered extinct in the wild (IUCN/SSC Antelope Specialist Group 2013). The Arabian oryx is now a classic example of a coordinated species reintroduction program (Wiese and Hutchins 1994; Ostrowski et al. 1998; Leader-Williams and Dublin 2000; Al Jahdhami, Al-Mahdhoury, and Al Amri 2011). Species reintroduction is the intentional release into their native range of animals that had been managed ex situ, and per the IUCN Reintroduction and Invasive Species Specialist Groups it "must be intended to yield a measurable conservation benefit at the levels of a population, species or ecosystem, and not only provide benefit to translocated individuals" (IUCN/SSC 2013, viii). These programs also work to identify and address the ultimate threats to the conservation of these population(s) in the wild. The Arabian oryx effort was an early success story in the history of zoo and aquarium reintroduction efforts and has served as a model for many other institutions collaborating on endeavors to save species such as black-footed ferrets (Wiese and Hutchins 1994; Leader-Williams and Dublin 2000).

SUPPORTING GLOBAL CONSERVATION

The Phoenix Zoo's collection animals are ambassadors for their wild counterparts, and zoo leadership is committed to ensuring that a visit to the zoo is an investment in conservation worldwide. The zoo's annual Conservation and Science Grants program is one way to make good on these promises. Since its inception in 2008, the zoo has provided nearly $500,000 in grants and contributions to field conservation efforts across the globe. Priority is given to projects that demonstrate conservation need, build capacity in range countries, include active participation by local stakeholders, and support species sustainability in the wild. Zoo employees are invited to submit proposals to use their expertise to benefit conservation efforts in the field through the zoo's Staff Conservation Grants program. This program has sent zookeepers, managers, and educators around the world to enhance existing animal management programs, to build new programs to improve conditions for animals in human care, and to create and strengthen conservation education programming and partnerships. In 2015 the zoo launched a Sustaining Grants program, building on the success of the Annual Grants. Sustaining Grant awardees submit an initial proposal in year one and streamlined requests in years two and three to facilitate multiyear projects representative of grant program priorities.

The Phoenix Zoo is certainly not alone in our commitment to supporting wildlife conservation at home and in the field. In 2014, 212 of the 241 members of the Association of Zoos and Aquariums (AZA), a US-based nonprofit member services and advancement organization, reported contributing over $154 million to support more than three thousand field conservation projects in 130 countries (AZA 2014b). The AZA Field Conservation Committee is leading the charge to ensure that all 241 member institutions are participating in some form of conservation support, and it is working toward this goal through communication efforts, mentorship programs, and other strategies (AZA / FCC 2016). Accredited zoos and aquariums make substantial financial and in-kind contributions to wildlife conservation internationally each year (WAZA 2005; Gusset and Dick 2011; AZA 2015b; Barongi, this volume; Grow, Luke, and Ogden, this volume).

AZA zoos and aquariums have long been contributing to field efforts internationally, and in recent years they have been increasing their efforts closer to home (Barber 2008; Grow, Allard, and Luke 2015). Of the field conservation projects reported to AZA in 2014, 40 percent focused on projects based in the United States (AZA 2014b). The Phoenix Zoo has been at the front of this trend. Building on our initial commitment to the Arabian oryx, the zoo has been a proud leader in native species recovery partnerships for several decades. This work led the zoo to be named Conservation Organization of the Year by the Arizona Game and Fish Commission in 2008 and awarded an AZA Significant Achievement Award in North American Conservation in 2010 for its work on Chiricahua leopard frog recovery.

CONSERVING LOCAL SPECIES

Working with state and federal agencies, universities, and private landowners, Phoenix Zoo Conservation and Science Department staff members have developed long-standing partnerships for native species recovery throughout the American Southwest (for more on the importance of partnerships in species recovery in this region see Ivanyi and Colodner, this volume). Only six facilities in the world breed the endangered black-footed ferret (*Mustela nigripes*), and the Phoenix Zoo is one of them (Garelle, Marinari, and Lynch 2015). Since the program began in 1991, more than four hundred ferret kits have been born at the zoo and nearly one hundred released to the wild in Arizona alone, with dozens of others introduced to release sites elsewhere in the species' current range. The zoo's work with Chiricahua leopard frogs (*Lithobates [Rana] chiricahuensis*) began in 1995 and celebrated the release of the twenty-thousandth Phoenix Zoo–reared frog released to the wild in 2014. In 2015 the zoo initiated a new effort to monitor the success of these introductions by surveying release sites where zoo-reared animals have been toe-clipped and released. Happily, several frogs observed in recent surveys are "zoo frogs" thriving in the field two years after release. Zoo staff members look forward to continuing these field efforts to gain a clearer idea of how zoo-based recovery work is affecting wild Chiricahua leopard frog populations.

Thanks to successes with these initial projects, the zoo has developed a strong reputation within the native species conservation community in Arizona. In 2007, after field biologists noticed a downward trend in narrow-headed gartersnake (*Thamnophis rufipunctatus*) populations (Nowak and Santana-Bendi 2002), members of a multiagency Gartersnake Conservation Working Group reached out to the Phoenix Zoo. A research population of seven animals was brought to the zoo's Arthur L. and Elaine V. Johnson Conservation Center, which was designed for rearing native species for release to the wild and for research supporting species recovery efforts. These specimens would help zoo scientists begin compiling information regarding life history, reproductive biology, behavior, feeding requirements, and more with a goal of propagating these animals for introduction to the wild. Very little of this information was known from field studies. For Phoenix Zoo staff members, this action was a tremendous vote of confidence and understanding regarding the distinctive role zoos and aquariums can play in species recovery.

Phoenix Zoo conservation biologists have developed husbandry and management protocols for narrow-headed gartersnakes, ultimately leading to the first successful propagation of this species in managed care (Phoenix Zoo Conservation and Science Department 2015). The resulting husbandry manual has been distributed to the Gartersnake Conservation Working Group, allowing this important breeding and management information to be widely shared. Quantifying behaviors and activities seen in animals held ex situ, which may be difficult to observe in the wild, may help inform future recovery actions. Zoo staff in Phoenix and throughout the industry are proud of the roles they can play in supplementing in situ studies with thoughtful ex situ scientific research (Hutchins, Smith, and Allard 2003).

A similar story has unfolded with the endangered Three Forks springsnail (*Pyrgulopsis trivialis*). Two of the three known Three Forks springsnail populations in Arizona went extinct locally within a short time, possibly because of fire retardant chemicals used to control wildfires. In 2008 the Phoenix Zoo was asked by the US Fish and Wildlife Service to develop protocols for maintaining these rare snails ex situ, something that had never been attempted. As a result of the careful attention of zoo biologists working with a surrogate species, the Page springsnail (*P. morrisoni*), a clearer understanding of development rates for these animals emerged. The typical timing for annual field censusing may in fact have resulted in low counts, since it coincided with the end of the adults' life span and occurred when the young of the year were likely still too small to be counted reliably in the field (Wells et al. 2012; Pearson et al. 2014). In addition to the narrow-headed gartersnake success, the zoo has celebrated the first propagation of Page springsnails outside their native habitat, which has provided the Conservation staff with additional opportunities to gain insight about these rare species that may in turn inform field conservation science and support recovery.

The Phoenix Zoo is also working with state and federal agencies to support the recovery of Mount Graham red squirrels (*Tamiasciurus hudsonicus grahamensis*), masked bobwhite quail (*Colinus virginianus*), California floater mussels (*Anodonta californiensis*), several species of native desert fish (*Cyprinodon macularius* and *Poeciliopsis occidentalis* in particular), and an endangered plant, the Huachuca water umbel (*Lilaeopsis schaffneriana* spp. *recurva*). In all cases the zoo became involved with these species to fill gaps in knowledge of species' life histories through ex situ research, augment wild populations with animals reared at the zoo, and raise awareness about the need for native species conservation among the 1.4 million guests who visit the zoo annually. Through these partnerships and similar relationships with a wide variety of organizations ranging from local government agencies to academic institutions to international nongovernmental conservation groups, zoological facilities demonstrate their commitment to contributing in meaningful ways to conservation science (Allard 2005; Grow, Allard, and Luke 2015; Ivanyi and Colodner, this volume; Muka, this volume; Norton, this volume).

INTEGRATING CONSERVATION AND EDUCATION

As a conservation organization, the Phoenix Zoo recognizes that its role in conservation messaging is as important as its efforts in conservation science. The zoo weaves conservation stories into its many family educational programs and school field trip programs, and it is committed to fostering the creation of a new generation of conservation stewards through university internships and research fellowships, the development of a ZooTeen Conservation Team volunteer program, cohosting the Miami University/Project Dragonfly Advanced Inquiry master's program for science educators and others interested in community-based conservation and environmental stewardship, and more.

This commitment is woven into the everyday guest experience via the zoo's Safari Train script and extensive on-grounds signage. In fact, one-third of the narrative signage at the zoo touches on conservation-specific topics, and that proportion is rising as new signs are written and previous signs updated. New exhibits at the zoo incorporate strong conservation messages emphasizing what guests can do in their daily lives to make a difference for wildlife and their habitats. Special events highlight conservation themes and activities, including animal awareness days, Earth Day, Endangered Species Day, World Oceans Day, a conservation lecture series, and more. The zoo's annual fund-raising gala event was reenvisioned in 2010 as "Rendez-Zoo: An Evening of Conservation and Cuisine" to focus attention on the zoo's role in wildlife conservation.

AN EVOLVING LEGACY

In November 2014, after months of internal discussion, the Arizona Zoological Society board of trustees voted to change the name of the organization responsible for the zoo's operations. Thus the Arizona Zoological Society was reborn as the Arizona Center for Nature Conservation (ACNC). This change was not just cosmetic; the zoo and its umbrella organization had been ramping up their commitment to conservation and understood that most of the metro Phoenix community did not recognize the zoo as a conservation organization. Rebranding further emphasized the board of trustees' commitment to changing that perception and supporting even greater attention to supporting conservation action, reflecting our founders' hopes for "Operation Oryx" in 1962 and looking forward to ensuring our relevance in the century to come.

There was a time when zoo and aquarium conservation departments were vulnerable to budget cuts and outright elimination, particularly during difficult economic times. The Phoenix Zoo/ACNC leadership made an important commitment to sustaining and expanding the ACNC's reach in 2014 when it approved a measure to invest one dollar from every daytime paid admission and five dollars from every membership sold to build the zoo's Conservation Fund. This is not a surcharge paid by zoo guests; it is a dedicated allotment from operating revenue that resulted in a transfer of $683,500 to the Conservation Fund in its inaugural year (fiscal year 2014–15). The zoo transferred $702,000 to the Conservation Fund in fiscal year 2015–16. Through April of fiscal year 2016-17 the zoo has transferred another $620,000 and should be on track to transfer $60,000 to $70,000 more to the Conservation Fund by fiscal year end as this program continues.

The Phoenix Zoo is one example of many zoos and aquariums "walking the talk": we continually challenge each other to increase the scope and scale of our conservation commitment, adding new chapters to our conservation stories and strengthening our legacies as our organizations continue to evolve (Barongi, this volume; Grow, Luke and Ogden, this volume; Knapp, this volume). To remain relevant in the decades to come, we cannot be content simply to be a place for family recreation and connection. We must also use our resources to ensure a future for the wildlife and wild places that inspire our work.

PART IV

Caring for Nature: Welfare, Wellness, and Natural Connections

CHAPTER FOURTEEN

Bears or Butterflies? How Should Zoos Make Value-Driven Decisions about Their Collections?

Clare Palmer, T. J. Kasperbauer, and Peter Sandøe

INTRODUCTION

Zoos are ethically contested institutions, not only in terms of their existence, but also with respect to their aims, policies, and practices. Many of these aims, policies, and practices are underpinned by commitments to defensible and widely shared values including animal welfare and species conservation. However, these values may be in tension, forcing choices to fulfill some aims at the expense of others, or requiring trade-offs where each aim can be only partially met.

Such tensions are particularly salient with respect to the variety of species in zoo collections. Obviously zoos have limited space; even in combination, zoos can only keep a tiny fraction of existing species, and keeping one species essentially means excluding others. So what should drive the mix of species kept, given the aims and values that zoological associations claim to endorse? And how should zoos respond to tensions and conflicts between these values in terms of their collections?

We begin this chapter by exploring key aims endorsed by three major zoo associations. Then we discuss the values underlying these aims, including animal welfare and competing understandings of conservation. We consider why these values are important and examine the dilemmas and difficulties they pose for decision making about the composition of zoo collections. We conclude with some tentative suggestions about future directions for zoo collections.

THE AIMS EXPRESSED BY ZOO ASSOCIATIONS

In considering zoos' aims, we draw on mission statements and other policies adopted by three major zoo associations: the World Association of Zoos and Aquariums (WAZA), the Association of Zoos and Aquariums (AZA; primarily a US-centered organization), and the European Association of Zoos and Aquaria (EAZA). We assume that in joining these associations, individual zoos endorse these statements and policies. (Our primary focus is on zoos, not aquariums, and we will not discuss zoos operating outside these associations.)

The most prominent shared aim is to promote conservation to protect animal species, populations, and habitats. WAZA's strategy document (Barongi et al. 2015, 16; see also Barongi, this volume), for instance, defines successful conservation as "securing populations of species in natural habitats for the long term." Animals kept in zoos should "play a conservation role that benefits wild counterparts" (Barongi et al. 2015, 17), in particular by linking zoo exhibits with fund-raising for specific in situ projects (for instance, through the "One Plan" approach; see, e.g., Traylor-Holzer, Leus, and Byers, this volume). By 2008, WAZA members collectively contributed over $350 million each year to in situ conservation (Gusset and Dick 2011; Barongi et al. 2015). Zoos may also keep animals for reintroduction, or to serve as an "assurance population" for reintroduction "when conditions are ripe" (AZA 2014f). Zoos also train staff and wildlife veterinarians, while conservation-relevant zoo research may contribute to protecting species in the wild. Alongside *direct* contributions to field conservation, zoos also pursue *indirect* conservation work, in particular public engagement and environmental education aimed at changing knowledge, attitudes, and behavior with respect to conservation (Barongi et al. 2015; AZA 2014g).

AZA highlights animal welfare, a clearly separate aim, in its mission statement: "The AZA provides its members with the services, high standards, best practices and program co-ordination to be leaders in animal welfare, public engagement and the conservation of species" (AZA 2014g). *Caring for Wildlife: The World Zoo and Aquarium Animal Welfare Strategy* (Mellor, Hunt, and Gusset 2015) is a wide-ranging policy statement on the significance of animal welfare, and WAZA notes on its website (n.d.), "The goal of the World Association of Zoos and Aquariums is to guide, encourage and support the zoos . . . of the world in animal care and welfare, environmental education and global conservation." While EAZA does not explicitly mention animal welfare in its mission statement, its 2015 *Code of Ethics* requires members to "promote the interests . . . of animal welfare" (EAZA 2015, 1), and animal welfare is emphasized in most of EAZA's official guidelines and position statements.

It's not clear exactly how promoting animal welfare fits with conservation, however. Animal welfare in EAZA's, WAZA's, and AZA's statements appears to require independent promotion rather than being just a side constraint on the pursuit of conservation; this is certainly how some leading interpreters see it (e.g., Maple and Perdue 2013). However, zoo animal welfare is not usually understood as a goal in the same sense as conservation. Animals are not kept in zoos just to promote animal welfare. For instance, WAZA (Barongi et al. 2015, 59) describes conservation as zoos' core *purpose* and positive welfare as their core *activity*. What these welfare commitments do establish, though, is that when animals are kept in zoos, it's important that their welfare is good—for the animals themselves, for visitors to appreciate, and sometimes as an inspiration for animal welfare initiatives in other areas (see Maple and Segura, this volume).

So what implications do these differing aims have for the composition of zoo collections? Direct conservation goals alone might suggest focusing on threatened species of ongoing ecological significance, where populations may persist with assistance, and where there is some likelihood of successful reintroduction. While conservation education *may* focus on those same species, they might not appeal to visitors and so may not be ideal for public engagement. And in terms of animal welfare, some species may flourish better in a zoo environment than others but not necessarily have high conservation value *or* be highly attractive to visitors (Dubois and Fraser 2013). So, given the relative independence of these aims, zoos may face conflicts or trade-offs in terms of what's kept in their collections.

Some of these tensions are acknowledged in the mission statements or ethical codes of zoological associations. WAZA (2005, 61) is explicit: "In practice there could be a conflict of interest between the conservation of a species or population and the welfare of an individual animal." But it offers little assistance about how to tackle such conflicts other than to say that doing so may involve "weighing competing values" and that "these considerations are complex and often dependent on context." Similar value weighing is proposed in Mellor, Hunt, and Gusset (2015). The WAZA *Code of Ethics* (2003) comments: "Any actions taken in relation to an individual animal, e.g. euthanasia or contraception, must be undertaken with this higher ideal of species survival in mind, but the welfare of the individual animal should not be compromised." This seemingly gives little guidance on how zoos might go about "balancing values," if indeed that is what they should be doing; but if euthanasia and contraception are seen as incompatible with welfare, this statement may suggest that weighing species conservation at the cost of animal welfare is not permissible after all.

Zoo mission and strategy statements don't go into much detail about how to understand the values underpinning zoos' aims. We therefore now turn to the wider literature on conservation, environmental, and animal ethics to help spell out what might be meant by "conservation" and "animal welfare" values.

VALUES UNDERLYING CONSERVATION AND ANIMAL WELFARE

Conservation value can refer to conserving a number of different things, singly or in combination: individuals, populations, species, ecosystems, or more abstract qualities such as "wildness" or "place." Recently—although this is not undisputed—it has been proposed that larger, more encompassing entities such as ecosystems should have conservation priority (see Norton, this volume). Certainly, inasmuch as zoos' conservation goals are tied to either original or potential future habitats of the species they support, a key concern is maintaining the health of (and potentially restoring or even creating) these ecosystems (WAZA 2005, 11). Ecosystems may provide both consumptive values (such as food sources) and nonconsumptive values including places and landscapes that people value for cultural, historical, and aesthetic reasons. On some (admittedly contested) ethical views, it's also argued that we have direct moral responsibilities to ecosystems, independent of their instrumental value (e.g., Johnson 1992). So there are many value-based reasons for supporting zoos' ecosystemic goals.

Zoo conservation may also be committed to the value of species themselves or at least to the value of *some* species. Species can be highly valued for intrinsic qualities such as charisma (for instance, polar bears) or beauty, or for their apparent similarity to human beings, as in the case of gorillas (DePinho et al. 2014; Russow 1981). Philosophers such as Johnson (1992) and Staples and Cafaro (2012) argue that we have direct duties to species as well as to ecosystems. These values persist even if species have no remaining natural habitat, as in the case of Pere David's deer, or when reintroduction seems unlikely, as with members of amphibian species being kept in zoos to avoid extinction from chytrid fungus in the wild (see Mendelson, this volume).

Another distinct value is good animal welfare. Welfare is typically understood in one of three ways: in terms of animals' positive and negative subjective *experiences*, the satisfaction or frustration of their *preferences*, or their ability to perform *natural behaviors* (Appleby and Sandøe 2002). In the animal welfare literature there's significant disagreement between those who think that only animals' subjective experiences are relevant to welfare and those who maintain that their being able to perform natural behaviors makes an independent contribution to welfare. Concern for both the subjective states of animals and animals' ability to perform natural behavior appears in zoo association documents, and both seem important to animal welfare in zoos. Many animals held in zoo collections are sentient, that is, able to undergo subjective experiences of pain and pleasure and to be in other positive, or aversive, experiential states. WAZA (2005) commits zoos to avoiding both causing and allowing suffering (strong or protracted episodes of pain or other aversive states) in animals under their care and emphasizes the importance of allowing zoo animals to perform natural behaviors. However, it's not clear whether performing natural behaviors here is regarded as independently important or important only in terms of animals' resulting experiences. This may matter when promoting natural behavior conflicts with preventing suffering—for example, when deciding whether male animals should be allowed to fight, or when considering whether to foster "wild behaviors," potentially stressful to captive animals, before reintroductions (Barongi et al. 2015, 60; see Greene, this volume).

Other issues also require clarification. First, some animals kept in zoo collections—particularly invertebrates—may not have a welfare, at least in the sense of having subjective experiences or preferences. It still makes sense to talk about their performing "natural behaviors," but there's a question whether this *matters* in animals lacking sentience.

Second, concern about zoo animal experience has, historically, focused more on avoiding negative welfare (such as suffering) than on providing *positive* welfare—in terms of feelings of pleasure, satisfaction, or excitement (Maple and Perdue 2013) or the opportunity to exercise natural behaviors. The recent emphasis on enriched environments in zoos indicates increasing concern for positive welfare, as does the new emphasis on animal "wellness" (see Maple and Segura, this volume; Mellor, Hunt, and Gusset 2015); but there's still a question about how significant positive welfare actually is in practice for zoos.

Third, while a focus on animal welfare centers on the ongoing *quality* of animals' lives, it can be argued that it doesn't necessarily have implications for the *length* of their lives, and therefore for painless culling of healthy but "surplus" animals. In contrast, however, Kasperbauer and Sandøe (2015) defend the view that painless culling may be a welfare issue.

This brief overview of the values that underlie concern for conservation and animal welfare indicates tensions between different conservation values, between different ideas of animal welfare, and also potential conflicts between conservation values and animal welfare values. This may give rise to complex and contested situations, especially when thinking about a value-driven species composition for zoo collections. We'll now return to this discussion of zoo collections more directly, drawing on debates about the possible future of conservation and on recent empirical studies of conservation and animal welfare.

COMPOSITION OF COLLECTIONS OF ZOOS IN LIGHT OF THEIR DIFFERENT AIMS

Zoo Collections and Direct Conservation Value

Since zoos have limited space and resources, they must operate tactically by using their collections to best serve their own conservation goals. But these will vary depending on what conservation values have priority.

Strategic decisions of this sort are being taken by zoos and zoo associations, for instance as part of IUCN Population Management Strategies. AZA's "Regional Collection Plan Handbook" (2012) outlines the criteria Taxon Advisory Groups should use when making recommendations for which species should be managed by groups of zoos. First among these criteria are "conservation status" and "extinction risk in the wild."

In line with this, Leader-Williams et al. (2007) found that the numbers of threatened species held by zoos increased from 103 in 1993 to 230 in 2003 (mostly nonmammalian species). However, recent studies of current zoo collections have questioned whether these increases are sufficient to meet conservation goals. Conde et al. (2013, 3), for instance, compared the species held in zoos with those listed on the IUCN Red List as vulnerable, endangered, or critically threatened (together called "threatened") and concluded that with a few exceptions "most collections are not distinguishable from what would be expected if the species were selected at random." Martin et al. (2014) compared bird and mammal species held in zoos with the most closely related species not held in zoos and found that (inter alia) zoo species tended to be less, rather than more, threatened with extinction. These and other studies suggest that, given zoos' own commitments, there's room for a greater focus on threatened species in practice.

An earlier study by Conde et al. (2011b) raises another issue about the fit between conservation and the composition of zoo collections: the classes of species held in zoos. While one-fifth to one-quarter of IUCN threatened and near-threatened mammal species are represented in zoos, only 6.2 percent of globally threatened amphibian species are represented (Dawson et al. 2015), even though 41 percent of amphibian species are listed by the IUCN as threatened or extinct in the wild. Taking only conservation value into account, amphibians—and reptiles—seem good choices for zoo collections (as suggested, for instance, by the Amphibian Ark project; see Mendelson, this volume). In addition to being threatened, many of them require little space, while still contributing as much to conservation research and personnel training as larger species.

Thus Keulartz (2015, 346) recently argued, "The most effective strategy to combat the problem of limited space is without any doubt a shift away from the large charismatic mammals towards smaller species, particularly amphibians, invertebrates and some species of fish, which occupy less space, are relatively inexpensive to keep, have a high birth rate and are easy to reintroduce." While these factors may not apply to all species, holding more small-bodied, nonmammalian species may help zoos protect endangered species (without, as we discuss below, negatively affecting visitors' responses).

The issue Keulartz raises about reintroduction is complicated, however. To date, reintroductions from captive populations have not been very successful. Zoos have played a direct role in the recovery and reintroduction of thirteen animal species (though this number is contested in both directions; Balmford et al. 2011; Conde et al. 2011a, 2011b; Hoffmann et al. 2010). Multiple studies have found that reintroductions with captive-bred populations are less successful than reintroductions with relocated wild populations. For instance, Jule, Leaver, and Lea (2008) surveyed reintroductions of forty-six carnivore species and found that 48.5 percent of the reintroduced populations sourced from the wild, and only 19 percent of the populations sourced from captivity, survived six to eighteen months after release.

A further inhibiting factor for zoos' captive breeding programs is that many populations of threatened species in captivity are small and are distributed between zoos (Conde et al. 2013). To ensure sufficient genetic diversity in breeding programs, zoos therefore need to exchange either animals or gametes. Zoos are tackling such difficulties, of course, for instance with collaborations supported by Regional Collection Plans, but moving animals for breeding faces many hurdles (Barongi et al. 2015, 54). However, a more cost-effective strategy may be for different zoos to hold smaller numbers of species in their collections but to increase the number of individual animals in each species, which may also have welfare benefits (Maple and Purdue 2013, 150).

The likely impact of climate change on ecological systems also raises concerns about the possible success of species reintroductions and the purpose of keeping assurance populations. Species that flourish in such changing systems are unlikely to be currently threatened or endangered. For many currently endangered species, changing climate is likely to increase threat and diminish the likelihood of successful reintroduction—at least into their current native habitats—for the foreseeable future (Sandler 2013a).

This doesn't mean efforts aimed at reintroductions should be abandoned. But it does suggest that when species are being kept for reintroduction, or as assurance populations, it's important to consider their resilience in the face of climate change, climate predictions for their native habitats, and the possibility of potential introductions to nonnative locations in the future. If wild reintroduction is a goal but successful reintroduction anywhere in a realistic, climate-changed future looks unlikely, this should count against including a species in a zoo's collection (though some species could still be important for conservation-oriented research; Minteer and Collins 2013). On the other hand, if the purpose is just considered to be conservation of certain species, even when their natural habitats are gone, zoos may still have an important role to play.

As we noted above, some species are valued independently of their wild habitat based on their beauty or charisma—which might contribute to what AZA calls "exhibit value." However, given zoos' own conservation goals, exhibit value does not by itself seem a sufficiently compelling reason for keeping animals in spaces that could be occupied by those of more direct conservation value. But exhibit value may matter when it comes to *indirect* conservation.

Indirect Conservation: Education and Fund-Raising

Alongside direct conservation, zoos generate resources for field conservation and are used for conservation education (see Barongi, this volume). Without an income stream from visitors, most zoos can't contribute to conservation, and of course conservation education requires visitors. Zoo collections thus must attract visitors.

If species best for direct conservation were also best for visitor preference and conservation education, zoos' collection decisions would (in these respects) be relatively simple. And in fact they may be fairly simple, at least in terms of what can be gleaned from current research. After reviewing the literature, we could find little evidence of significant negative effects on visitor numbers or conservation education when zoos *did* primarily select species ideal for direct conservation. On the contrary, there's evidence that people are particularly interested in seeing rare and endangered animals (Whitworth 2012).

It's widely believed, however, that charismatic mammals, and species with large body sizes, are needed to attract visitors. This view may have affected the current composition of zoo collections. Frynta et al. (2010, 2013) found that people's rankings of an animal's beauty, as well as its body size, were good predictors of whether that species would be found in zoos—implying that zoos take these factors into account when forming their collections and that animals of this kind are well represented in zoos.

However, it has been difficult to establish what visitors' preferences actually are regarding the physical features of zoo animals. Balmford (2000), for instance, failed to find any connection between body size and popularity among zoo visitors, looking at both his own data and that of four other studies. More recently, Whitworth (2012) found that small animals were actually more popular than large animals in United Kingdom zoos. He did find a weak positive correlation ($r = .268$) between the popularity of animals held at a zoo and the number of visitors. This suggests that the type of animals a zoo holds does matter somewhat for zoo attendance, but that body size is not a strong factor influencing which animals people want to see. While further research is needed, this at least in principle suggests that keeping animals with smaller body sizes would not reduce (and might increase) the number of zoo visitors (a view perhaps reinforced by Ivanyi and Colodner, this volume).

On the other hand, there's some evidence that mammals are the most popular animals kept in zoos. Moss and Esson (2010), for instance, studied visitors' interest in forty zoo species at the Chester Zoo in the United Kingdom, spanning mammals, birds, reptiles, amphibians, fish, and invertebrates, measuring both the number of visitors at each exhibit and the time visitors spent there ("holding power"). Mammals were significantly more popular than all other groups on both measures. It's not clear what implications this finding might have (assuming it could be confirmed in other studies). It might mean that while mammals are most popular with visitors, other taxonomic groups would be visited more if fewer mammals were available. Or it might suggest that zoos should keep some "flagship" mammals for attracting visitors and as a gateway to conservation education, but that they could in addition increase collections of smaller, more threatened individuals from other taxonomic groups. Another strategy might be to increase conservation value among mammal holdings by seeking threatened mammal species. Smith et al. (2012), for instance, identified 183 candidate species that receive very few conservation resources but nonetheless possess features (e.g., large, forward-facing eyes) that are aesthetically appealing to visitors.

In terms of conservation education, we don't know much about the specific effect of species composition in zoos. Studies have tended to focus on the *way* zoos present information—for instance, whether there's a human interpreter at exhibits and the effects of signage (Routman, Ogden, and Winsten 2010)—and on overall knowledge gained, for instance, about biodiversity (Moss, Jensen, and Gusset 2015). It may be that once people have entered the zoo, how information is presented is more important than species composition.

Although more research is needed here, education and fund-raising considerations don't appear to winnow out or expand the species suggested by direct conservation values alone. So there may not be much tension between achieving the aims of direct and indirect conservation in terms of zoo species composition. However, animal welfare generates more potential for value conflict.

Animal Welfare

One potential consequence of pursuing good animal welfare is narrowing the species thought suitable for zoos, since some species can't easily achieve good welfare in captivity (e.g., forest duikers and cheetahs; Mason 2010). But zoos may approach this problem in different ways, depending on how they prioritize the value of animal welfare. Four alternatives are:

1. Absolute conservation value, no welfare side constraint: When conservation values and the value of good animal welfare are likely to conflict, conservation values should always take priority, even if this means causing or allowing animal suffering, frustration, or restriction on natural behavior.
2. Conservation value primary, absolute welfare side constraint: Conservation value is the priority for zoos, but it should not be pursued where doing so would lead to poor animal welfare (so, for example, species that do not thrive in captivity should not be kept). Depending on what

understanding of animal welfare is adopted, such side constraints will be more or less severe.

3. Conservation value and animal welfare commensurable, with weighing of values: Values must be weighed to bring about the best overall consequences in terms of both values. So zoos should not sacrifice good animal welfare for a small gain in conservation value. But for a large gain in conservation value—for instance, saving a key species with a strong likelihood of successful captive breeding and reintroduction—sacrificing a good deal of animal welfare is ethically permissible if alternatives aren't available.
4. Animal welfare absolute or primary: Here welfare would take priority over any conflicting conservation values.

A zoo that accepted alternative 1 would not appear to be taking animal welfare as seriously as zoo association commitments require. The 2015 WAZA report, *Caring for Wildlife* (Mellor, Hunt, and Gusset 2015, p. 11) seems to support alternative 3, recommending that zoos "evaluate whether the animal welfare implications of management interventions are outweighed by their conservation benefits." But *Caring for Wildlife* seems to propose a version of alternative 2 when values conflict, since "the welfare of the individual animal should not be compromised (Mellor, Hunt, and Gussett 2015, 84)." However, depending on what's meant by "welfare compromise," this may not amount to a very strong side constraint. The species selection criteria outlined in the AZA "Regional Collection Plan Handbook (2012)," for instance, while emphasizing conservation, says of welfare only that member institutions need to consider whether they have sufficient expertise "to meet the species' basic biological needs (i.e., nutritional, medical, social, etc.) as related to maintaining and propagating them in AZA member institutions." This doesn't, for instance, mention *positive* welfare, although this does appear in other more recent documents (e.g., Mellor, Hunt, and Gusset 2015).

On approaches 2 and 3, zoos should consider whether they have sufficient capacity to provide good welfare for animals before acquiring them (IUCN/SSC 2014). Research suggests that even among closely related species some do much better in zoos than others. For instance, among raptors, kestrels seem to do better than sparrowhawks, especially in mortality rates; and among psittacines, macaws do much less well in feather plucking and breeding success than lorikeets (Mason 2010). Given zoos' welfare commitments, and the need to choose between species, these kinds of difference (taking into account any known environmental changes that could help improve welfare) should influence collection decisions.

However, welfare as a selection criterion might still conflict with conservation values. A number of studies have suggested that endangered species are more likely to have poor welfare, since it may be harder to meet their needs. Martin et al. (2014), for instance, note that more threatened species existing only in the wild were larger, and occupied more land, than closely related species found in zoos, and that these features might make them harder to keep in captivity. Mason et al. (2013) hypothesized that anthropogenic changes to the wild—changing habitats, restrictions on ranging and dispersal, shifts in climate, new infectious diseases, pollutants, and changing social structures—pose challenges to animals similar to being taken into captivity. It's possible, then, that some species struggling from anthropogenic influences on their wild habitat may also struggle living in an anthropogenic captive environment, and for similar reasons. This conclusion may at least suggest caution about keeping some species endangered for anthropogenic reasons in zoos. Furthermore, since the world outside the zoo is likely to come under increasing anthropogenic influence, reintroductions from these species in the future are likely to pose problems.

To improve welfare and meet conservation goals, some recent research suggests that zoos could include more indigenous and native animals in their collections (Maple and Perdue 2013). The Phoenix Zoo, for example, has been breeding and reintroducing into Arizona species native to the Southwest, including Chiricahua leopard frogs and desert pupfish (see Allard and Wells, this volume). Such programs of breeding locally threatened fish, amphibians, reptiles, and invertebrates seem ideal for zoo collections: they focus on threatened species needing conservation; they are part of current and successful reintroduction programs; these species may take up less space than large mammals; and they may raise fewer welfare problems because they are already in an appropriate climate and need less travel to be reintroduced. Research suggests that visitors do want to see local animals. A recent study by Roe, McConney, and Mansfield (2014), for instance, found that 70 percent of visitors rated seeing endemic or local species as a high or very high priority, as opposed to fewer than half of zoo officials. One might hypothesize (as Roe, McConney, and Mansfield do) that exhibits of regional wildlife could help both in developing understanding of local species and in suggesting how visitors might protect local environments.

CONCLUSION

This chapter has explored how zoos committed to the values of conservation and animal welfare might strategically develop the composition of their collections. We suggested that:

- Zoos could hold a higher proportion of IUCN-listed threatened species to better meet their commitment to conservation, although these species may pose particular welfare challenges in captivity.
- Zoos should consider expanding their collections to include less space-intensive species, particularly amphibians, reptiles, invertebrates, and some fish. These species may also raise fewer welfare concerns.
- Evidence suggests that holding more small and nonmammalian species would not discourage visitors, but there is also evidence that holding some charismatic mammals would attract visitors and promote zoos' conservation mission.
- Reintroductions from zoos have so far had limited success, and climate change will make reintroductions even more difficult. This may weigh against assurance populations and suggest that species should be kept for reintroductions only where plausible in a climate-changed world.
- Where species cannot have good welfare in zoos, this should normally count against their inclusion in zoo collections. Animal welfare concerns should include both positive and negative welfare, but there is debate about the relative significance of positive welfare and whether to include opportunities for natural behavior as a consideration in its own right.
- Zoos should consider expanding their native and endemic collections—the Arizona-Sonora Desert Museum in Tucson (see Ivanyi and Colodner, this volume) is a good example of this.

So, to answer the question in our chapter title: given their own value commitments, zoos may do best to expand their collections of less space-intensive, local, threatened, invertebrate populations, especially where animals lack subjective welfare (such as butterflies) or where their welfare needs are relatively easy to fulfill. But they could hold, in addition, a few charismatic "flagship" mammals, if they can thrive and have good positive welfare in captivity.

CHAPTER FIFTEEN

Why Zoos Have Animals: Exploring the Complex Pathway from Experiencing Animals to Pro-environmental Behaviors

Alejandro Grajal, Jerry F. Luebke, and Lisa-Anne DeGregoria Kelly

Modern zoos and aquariums are important venues for family fun, science learning, and developing personal connections with animals and nature. Zoos and aquariums are extremely popular, and their visitors reflect a wide cultural and socioeconomic diversity. They play key roles as civic and cultural institutions in their metropolitan areas, contribute to economic activity and jobs, advance science and education, and are prominent in global conservation. It is clear that biodiversity conservation remains a high calling for zoos and aquariums, and this is regularly emphasized in regional and international position papers, reports, and policies, such as the 2015 WAZA Conservation Strategy (see Barongi, this volume). In the face of a growing global biodiversity crisis, this renewed emphasis on conservation is a timely reminder that zoos around the world have a key part in the future of the planet.

At the same time, and as several of the other chapters in this volume demonstrate, zoos and aquariums in the twenty-first century are at a critical crossroads and are confronted with tectonic shifts that can be summarized in two emerging, intertwined challenges. The first we term the "cultural shift challenge": zoos and aquariums are tasked with meeting the needs of their audiences in the face of rapid worldwide socioeconomic and demographic changes and a personal detachment from nature owing to increasing urbanization. The second challenge we term the "relevancy challenge": the original Victorian premise that zoos presented the "wonders of the world" to an audience with limited access to information looks perilously outdated. This premise has been permanently changed by a digital media revolution that is rapidly widening public access to information and changing perceptions about animal welfare.

Given these challenges, and the industry's view that biodiversity conservation should remain a high priority for zoos and aquariums, the role of zoos in conservation can take many different iterations. Here we argue that the opportunity to experience zoo and aquarium animals forms a basis for a complex collection of affective, cognitive, and social responses that may ultimately influence visitors' pro-environmental actions that can support biodiversity conservation efforts. Personal engagement in pro-environmental behaviors is an important component of the United Nations' objectives known as the "Aichi Biodiversity Targets" (CBD 2010). Target number 1 of this global strategy points out that personal decisions and personal support for public policies are critical for biodiversity conservation. Zoos and aquariums are significant players in direct field conservation efforts to save species and habitats. Compared with the expenditures on field or ex situ conservation efforts, however, zoos and aquariums spend more of their budgets and resources on managing the visitor experience. The visitor experience entails an engaging living animal collection and exhibits that build a connection between people and animals. While a popular assumption is that animals and exhibits serve as recreation and entertainment, we contend that this connection between people and animals is one of the principal conservation assets of zoos and aquariums. Surprisingly, this strength in engaging diverse and large audiences in pro-environmental actions has been consistently downplayed, not only by avowed zoo critics (e.g., Zimmermann 2015), but also by zoos themselves, which tend to underestimate their social value compared with other field or ex situ conservation strategies.

We bring together emerging evidence that links perceptions of animal welfare and the zoo experience while reaffirming the effectiveness of zoos and aquariums in building connections with animals and influencing visitors' pro-environmental behaviors. We call for a bolder approach to engage zoo and aquarium visitors. Although there has been discussion about ethical dilemmas related to animals in zoos, we hope to expand this discourse to bring to the fore the truly existential dilemma of our time: how to engage humans in biodiversity conservation (see also Clayton and Le Nguyen, this volume; Norton, this volume; Maple and Segura, this volume). Particularly, How can zoos and aquariums become agents of social change and innovation for the future of biodiversity on our planet?

THE "CULTURAL SHIFT CHALLENGE" FOR ZOOS AND AQUARIUMS

For the first time in human history, much of the world population lives in large cities. This increased urbanization diminishes a once common connection between humans and nature, particularly when urban wildlife and parks are not always even recognized as nature (e.g., Pyle 2003). Connections between individual actions and the consequences for animals and habitats are often muddled, depersonalized, and distant. Access to nature and green open spaces is rapidly becoming one of the needs of urban communities, particularly for economically depressed or marginalized segments of society. Furthermore, leisure time is increasingly driven by the virtual experience of television and the Internet (Kaplan 1995; McDonald, Wearing, and Ponting 2009). Two-dimensional digital images of nature create a shifting paradigm in which virtual experiences of nature become the "new normal" and personal expectations and connections with nature are progressively eroded (Kahn et al. 2008).

For a vast portion of the world's humans, zoos and aquariums are rapidly becoming the most accessible and in many cases the only places to experience diverse live animals from around the world. This is particularly true for underserved audiences in urban cores that seldom experience nature in more naturalistic settings such as national parks or reserves (e.g., Floyd 1999). The uniqueness of authentic personal experiences with real animals may explain why zoos and aquariums enjoy widespread popularity. In North America, accredited zoos and aquariums receive about 180 million visits annually (AZA 2014c). Worldwide, visits to accredited zoos and aquariums are estimated at 700 million annually (Gusset and Dick 2011). Many zoos and aquariums are in large metropolitan areas and provide opportunities for a social outing with friends, family, or school groups. This high attendance cannot be explained solely by the zoos' entertainment value or marketing prowess, as cynics like to assert. Actually, the popularity of zoos may be better explained by a primeval affective bond that humans have with animals (Briseño-Garzón, Anderson, and Anderson 2007; Myers, Saunders, and Birjulin 2004; Vining 2003; Wilson 1984). This bond is enhanced during the zoo and aquarium visit by encounters with animals that encourage a combination of cognitive, affective, introspective, and social experiences (Clayton and Le Nguyen, this volume; Schwan, Grajal, and Lewalter 2014).

THE "RELEVANCY CHALLENGE" FOR ZOOS AND AQUARIUMS

Modern zoos were originally created based on a Victorian model that presented the "wonders of the world" to an audience with very limited access to information about animals and nature. This model has been progressively refined, based on the underlying assumption that fulfilling a deficit of knowledge about animals and nature would nurture education goals and connect visitors with the natural world. Today this model looks perilously outdated, as it has been permanently changed as global communication capabilities such as the Internet and social media are rapidly widening public access to information. Associated with this trend is the significant change in public perceptions about animal rights, particularly the viewpoint of animals as individual sentient beings, which contrasts directly with the population or ecosystem focus of traditional conservation efforts (Palmer, Kasperbauer, and Sandøe, this volume). Empathy for individual animals has always been part of the human emotional repertoire, but the emergence of social media and expansive campaigns by animal rights groups have created a "perfect storm" that severely questions the relevancy of zoos and aquariums (e.g., Marino, Bradshaw, and Malamud 2009). Consequently, with large amounts of information readily accessible, articulating the role of a living animal collection becomes increasingly crucial.

The relevancy challenge also encompasses the need for diverse audiences to participate in solving large global problems such as biodiversity conservation. Individuals with disabilities and minorities, for example, are underrepresented in science and engineering higher education and careers (NSF 2015). Zoos and aquariums must provide opportunities to engage all audiences in saving biodiversity. The relevance of zoos and aquariums depends on their engaging the public in conservation and fighting the global extinction crisis while remaining pertinent in this new informational landscape.

SAVING BIODIVERSITY

Biodiversity conservation remains a high calling for zoos and aquariums. Although this was already a strong point in the Conservation Strategy of the World Association of Zoos and Aquariums (WAZA 2005), the renewed emphasis at the international, regional, and national levels is a timely reminder for zoos around the world. In fact, zoos should increase their role in field and ex situ conservation in every possible way, as clearly stated in WAZA's

"One Plan Approach" (Traylor-Holzer, Leus, and Byers, this volume). The knowledge and experience of zoos in combining both efforts is significant (IUCN/SSC 2008). Zoos and aquariums are effectively working ex situ with a few critically endangered species. But this strategy reaches only a relatively small number of species (and individuals), and efforts are taxonomically biased and expensive (Conde et al. 2013), particularly compared with the Global Biodiversity Goals or the IUCN Red List (www.iucnredlist.org). This does not mean zoos and aquariums should abandon those efforts, since they have been an important option for species like the California condor or the Mauritius kestrel. Comparatively, the support offered by zoos and aquariums to field conservation efforts is significant (AZA 2014c). For example, AZA estimates that its accredited institutions contribute about $154 million, which by itself is one of largest nongovernmental sources of funding for biodiversity conservation efforts worldwide. The cumulative support by zoos and aquariums cannot be underestimated. But these efforts are dwarfed by the proportion of operational budgets that zoos and aquariums dedicate to maintain their living collections, exhibits, and visitor experiences. It is therefore essential that they maximize the ways managed living animal collections positively affect the visitor experience and ultimately engage visitors in pro-environmental behaviors.

WHY DO ZOOS EXHIBIT ANIMALS?

Given the challenges described above, it is imperative that zoos and aquariums have actionable answers to the question "Why do zoos exhibit animals?" For a long time they answered this question by saying that their live animal collections increase public knowledge and awareness about animals. Studies show that they do create important learning environments (Jensen 2014). And while it is true that zoos create didactic understanding and intellectual awareness (Moss et al. 2015), we also know that awareness and understanding by themselves are not strongly related to inspiring pro-environmental behaviors (Kagan 1984; Hungerford and Volk 1990; Moss, Jensen, and Gusset 2016). Yet many zoos continue to sell their role short, with vague messages and facts about animal biology or simple informational (and often gloomy) stories about conservation problems. Modern zoos cannot be satisfied with simply informing visitors about species in their collections or about the extinction crisis. It is critical that they actively motivate pro-environmental behaviors by their constituencies and that such behaviors be relevant to diverse constituencies. Engaging in personal conservation behaviors is an important component of support for public policies. Individuals who engage in progressively assertive environmental actions increase their skills in communication and activism (Dietz et al. 2009).

Zoos and aquariums may be oversensitive to the perception of environmental advocacy (Spring, this volume). Furthermore, until recently zoos and aquariums felt hesitant to play a more assertive role in public engagement in conservation, mainly because of the disappointing lack of strong empirical evidence of a relation between education and conservation outcomes. This lack of evidence has remained one of the main institutional barriers for significant investment in effective educational strategies (Ardoin and Heimlich 2013). But recent findings provide tantalizing indications that zoos can change visitors' behavior. For example, a recent large multi-institutional survey showed that many visitors are already concerned about environmental problems such as climate change (Clayton et al. 2014). We also know that animals elicit positive emotional responses in most zoo visitors (Myers, Saunders, and Birjulin 2004; Luebke et al. 2016).

Although zoos and aquariums are committed to pursuing public engagement in conservation, observable behavior changes cannot be the only indicator of the effectiveness of education and communication efforts. Zoo visitors come with all kinds of preconceptions and experiences, so their responses to the zoo visit can vary. Positive affective encounters with zoo animals can lead to visitor changes that may not be self-evident, such as changes in identity, caring attitudes and values toward animals, or more confident communication skills about conservation. The relationship between affective connections with zoo animals and changes in visitors' knowledge, attitudes, skills, and behaviors is an area with an urgent need for more research and evaluation.

Here we present a hypothesized model of the zoo and aquarium visitor experience. The model illustrates that engagement in pro-environmental behaviors is related to strong and complex relationships among visitors' predispositions, animal experiences, affective connections, social exchanges, and the designed zoo setting. This model is based on recent findings along several lines of research, and the relationships are graphically presented in a simplified diagram (fig. 15.1). This model leads to recommendations of education strategies to influence a positive relationship between the zoo and aquarium visitor experience and pro-environmental behaviors.

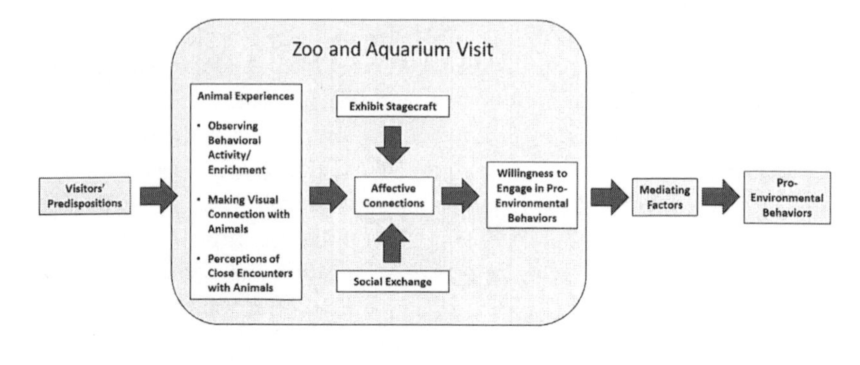

FIGURE 15.1. The zoo and aquarium visitor experience: our hypothesized model illustrates that engagement in pro-environmental behaviors is related to strong and complex relations among visitors' predispositions, animal experiences, affective connections, social exchanges, and the designed zoo setting. The model leads to recommendations for specific education strategies to influence a positive relation between zoo and aquarium visitor experience and pro-environmental behaviors.

ANIMAL EXPERIENCES AND AFFECTIVE CONNECTIONS

Zoo and aquarium visitors arrive with various interests and predispositions, but most already are interested in animals and predisposed toward conservation actions (Luebke and Matiasek 2013). Zoos and aquariums afford visitors the unique opportunity to observe or interact with a diversity of animals they otherwise would not regularly see. This opportunity can be vital in influencing positive feelings toward animals and ultimately inducing visitors' pro-environmental behaviors. Recent studies demonstrated a strong relationship between animal activity and visitors' affective and cognitive responses (Luebke et al. 2016; Powell and Bullock 2014). Close encounters with zoo animals, and particularly with observable active animal behaviors, directly amplify visitors' positive affective responses toward animals. These findings reinforce the importance of displaying animals in positive husbandry environments that not only support an animal's diverse behavioral repertoire but also provide visitors with a personal perception of positive animal welfare (e.g., Kreger and Mench 1995; Melfi, McCormick, and Gibbs 2004).

Animals in their designed exhibits are still—by far—the strongest motivators to engage visitors in learning. This point, while obvious, is important: public perceptions of poor animal welfare, such as animals pacing, unnaturalistic or poor exhibit spaces, and perceptions of animal "sadness" or "loneliness" can lead to a negative affective state in which a potentially positive environmental message is erased. So experiencing realistic animal behaviors during a zoo visit can be critical to creating a positive affective relationship with animals. The emerging evidence of a possible causal link between positive affect and learning in the free choice learning environment of a zoo or aquarium is critical in achieving conservation education and animal welfare outcomes (Luebke et al. 2016).

EXHIBIT STAGECRAFT

The relationship between animals' activity and visitors' affective responses is also influenced by the combined atmosphere, arrangement, and perceived authenticity of animal exhibits and the zoo experience. Schwan, Grajal, and Lewalter (2014) call this the "stagecraft" of the visit. In this free choice setting (Falk and Dierking 2000), visitors arrive with "mixed motives" that range from simple enjoyment to uplifting spiritual reflections regarding their connection to the natural world. So zoo and aquarium exhibits are regularly designed with the objective of stimulating feelings of surprise, suspense, discovery, and enjoyment. While curators and exhibit designers pay particular attention to storytelling and creating an atmosphere that uses color, lighting, staging, or animal presentation, the learning and knowledge acquisition responses of exhibit design factors have been largely ignored by researchers. Naturalistic displays are regularly used to present animals in detailed re-creations of original environments, which often immerse visitors in a staged scenario with high authentic appeal (Mortensen 2010). Well-designed zoo exhibits have great attraction power (Peart and Kool 1988), engage family visitor groups in learning interactions (Ash 2004; Reiss and Tunnicliffe 2011), and in some cases increase visitors' knowledge (Peart and Kool 1988). The creation of authentic, multisensory zoo experiences with living organisms clearly provides immense opportunities for learning and hands-on experimentation. Yet empirical research has just begun to explore how design principles affect learning. More research is needed to identify and incorporate learning components of exhibit stagecraft.

SOCIAL EXCHANGE

Museums and zoos are highly social settings. Most people plan their visits primarily as social events (such as spending time with family) rather than as learning opportunities (e.g., Clayton, Fraser, and Saunders 2009). Thus the zoo is not just a place where visitors explore nature but also a place that promotes and shapes social interactions. Within this social setting, the zoo and aquarium visit combines science learning, enjoyment, and family fun. Visitors also report enjoying the opportunity to discuss their relationship with animals and nature among their family or social group and with zoo educators (Luebke and Matiasek 2013). The conversational and social context of the zoo and aquarium visit may enhance learning even when visitors do not actively seek to learn or even recognize that they are learning (Clayton, Fraser, and Saunders 2009). It is common for members of the social group to interpret animal observation for others, such as adults explaining to children or vice versa. Explaining and interpreting lead to higher retention of scientific concepts, appreciation of new knowledge, and longer retention of facts and experiences for adults (Briseño-Garzón, Anderson, and Anderson 2007). Furthermore, these enriching social interactions can affect shared values or understandings (David and Bar-Tal 2009) and emotional experiences that are recalled later (Thomas, McGarty, and Mavor 2009). Thus social interactions can lead to a sense of collective identity, which in turn can influence or even modify group-relevant social norms (Steg and de Groot 2012). Social norms and the favorable conditions of positive social discourse and interpretation at zoos and aquariums can be important in motivating pro-environmental behavior (Bamberg and Möser 2007; Osbaldiston and Schott 2012). Finally, the social interactions with zoo and aquarium interpreters (e.g., docents, keepers, and naturalists) can be particularly powerful in helping visitors make connections between the exhibits and complex scientific topics (Weiler and Smith 2009) or even conservation issues and solutions. Zoos and aquariums have a powerful opportunity to leverage visitors' sense of connection to animals and weave together interpretive displays, educational opportunities, and social interactions to engage receptive audiences in pro-environmental conversations.

MEDIATING FACTORS

Despite visitors' affective connections and their willingness to engage in pro-environmental behaviors, other factors may influence or mediate environmental actions. Furthering this line of research, Grajal et al. (2017)

recently uncovered a directional relationship between affective connection with animals and visitors' engagement in pro-environmental behaviors. The study, based on a recent visitor survey on climate change attitudes across fifteen zoos and aquariums in the United States (Luebke et al. 2012), used a moderated-mediation model. Overall this statistical model was significant, and the results suggested that certain mediating factors explain how the sense of affective connection with animals is correlated with pro-environmental behaviors. In particular, the analysis indicated that individuals who reported a stronger connection with animals were more likely to report a greater level of concern for the effects of climate change on animals and people, a greater degree of certainty that climate change is happening, and a stronger sense of their ability to address climate change personally; and in turn they reported more frequent pro-environmental behaviors.

The directionality of this relationship from sense of affective connection with animals to self-reported pro-environmental behaviors suggests that a sense of connection with animals influences participation in behaviors to address climate change, not the other way around. This agrees with recent findings about how emotions mediate pro-environmental behaviors (Carmi, Arnon, and Orion 2015). While these results are focused on behaviors to address climate change, we suspect similar links may exist between the sense of connection to animals and behaviors to address other global biodiversity threats. Therefore strengthening education and communication strategies that leverage an affective connection to animals would bolster citizens' participation in pro-environmental behaviors that support biodiversity conservation efforts.

While this moderated-mediation model was significant, some caveats are necessary. First, as with all surveys, self-reporting, particularly about pro-environmental behaviors, brings some uncertainty about the veracity of the individual answers. The large samples in these studies tend to compensate for the uncertainties of survey answers. Second, it is possible that the pathway among the animal experience, the affective connection with animals, and pro-environmental behavior may not be the only, or even the strongest, one. Other pathways are possible (e.g., those including animal knowledge, spirituality, or learning modes) that affect and covary with other factors in the relationship. We encourage additional research into understanding different pathways between the zoo experience and pro-environmental behaviors. Third, the findings were focused on the individual perspective of the zoo experience, a necessary bias of individual surveys. It is clear that social and communal learning interactions play an important role that can be difficult to measure with survey tools, and more adaptable methods, such as observational protocols, may yield further insight into the social learning effects of the zoo experience. Finally, although we focused on specific climate change pro-environmental behaviors, other behaviors (e.g., Pearson et al. 2014) and other potential dependent variables need to be explored, such as developing a stronger environmental identity, observed consumer choices, or patterns of social influence. More research can illuminate this tantalizing evidence.

WHAT ZOOS AND AQUARIUMS CAN DO

Our hypothesized model of the zoo and aquarium visitor experience leads to recommendations for specific education strategies:

1. Improve visitor outreach and diversity: Remove barriers to visitation and invite a variety of audiences including underrepresented communities.
2. Facilitate animal experiences and address animal welfare concerns: Encourage animals' natural behaviors through behavioral enrichment techniques. Assume that zoo and aquarium visitors are already making an ethical choice by visiting. So emphasize institutional care and research in advancing animal welfare, and the zoo's role in biodiversity conservation.
3. Encourage visitors' participation and engagement: Design exhibits that increase sensorial and emotional connections with animals. Develop effective and assertive education and communication programs that promote social learning and group participation. Abandon bland "awareness" campaigns and embrace evidence-based education and communications strategies promoting personal engagement in pro-environmental behaviors. Abandon simple didactic approaches that explain scientific facts, such as the mechanics of climate change.
4. Empower visitors to engage in pro-environmental behaviors: Provide accessible and practical examples of pro-environmental behaviors that build visitors' confidence or self-efficacy in addressing biodiversity conservation issues.

CONCLUSION: THE ETHICS OF CONNECTING PEOPLE WITH NATURE

A strong campaign against zoos and aquariums has been launched by self-proclaimed animal rights entities. The traction of such campaigns is built on a growing paradigm shift in society about the rights of individual animals and on public sensitivity to animal welfare issues. Zoos and aquariums have developed some of the strongest tools for enhancing and measuring animal welfare in their collections. However, a focus on the ethics of animal rights, while necessary, does not squarely address the ultimate ethical dilemma of how to engage humans in active conservation efforts. Zoos and aquariums, as strong conservation organizations and as true agents of social change, should be at the forefront of this existential dilemma. Here we have shown compelling evidence of the effectiveness of zoos and aquariums in engaging their visitors in conservation action and contributing to social change. To fully achieve their mission, zoos and aquariums should not passively wait for visitors to arrive at their gates. They should actively reach out to those segments of society that have been disenfranchised and underserved by mainstream conservation messages. In an increasingly multicultural world, the social role of zoos is critical in providing diverse points of view on biodiversity conservation, building community support, and generating political capital for biodiversity conservation. Zoos and aquariums should be the first to encourage new voices to become active in conservation efforts by providing greater opportunities and access to authentic animal experiences, science education, culturally relevant engagement, and social innovation. Most zoos and aquariums represent inspiring and accessible portals to nature in large metropolitan areas, where humans are surrounded by the common ills of increased urbanization. Let zoos be active players in advancing effective social change to the relationship between humans and nature.

CHAPTER SIXTEEN

People in the Zoo: A Social Context for Conservation

Susan Clayton and Khoa D. Le Nguyen

INTRODUCTION

Zoos were originally designed to satisfy human interests: curiosity about animals, desire for social interaction, and the hope to signal status by possessing exotic creatures. Over the years, creditable zoos have broadened their focus to include the conservation and preservation of species and habitats while still emphasizing anthropocentric goals such as education and entertainment. There can be an uneasy tension between the desire to attract visitors and the concern for conservation (Mendelson, this volume; Palmer, Kasperbauer, and Sandøe, this volume). But there can also be synergy between the two goals. People's interest in animals and their desire for social interaction can be supported in ways that also further the conservation agenda.

Accredited zoos, operating as conservation organizations, hope the captive animals can increase concern by serving as "ambassadors" exposing threats that endangered species face in the wild. In this view, educating zoo visitors can be an indirect form of conservation (Palmer, Kasperbauer, and Sandøe, this volume). Increasing visitors' concern about endangered species should heighten their motivation to act to protect those species, for example, by supporting relevant policies, donating to conservation organizations, and changing their own behavior. Others, however, have argued that zoos have the potential to decrease sympathy for animals by objectifying them and placing them in cages (Beardsworth and Bryman 2001; Jamieson 1985). The question about zoos' ultimate effect on animals' well-being is one of the core disagreements in the debate over the ethics of the zoo (Maple and Segura, this volume).

For creating conservation concern, the question is whether animals are seen to matter; that is, to have moral significance (see, e.g., Goodpaster 1978). Do zoos encourage a sympathetic perception of animals? Further, is sympathy toward animals connected with concern about conserving species? These questions can be addressed empirically by examining how animal displays affect their human observers.

PEOPLE IN THE ZOO

Zoos are primarily a social context. Although people visit the zoo to see animals, they mostly come with other people. A number of researchers have concluded that the principal reasons for visiting a zoo are social, such as spending time with friends and family and helping one's children learn about nature (Moss, Jensen, and Gusset 2014). Thinking about zoos as a social context suggests that their influence depends at least in part on the social interactions that take place there.

Our research examining the ways visitors respond to animal exhibits has highlighted several important themes. One is simply that people value the animals, enjoy seeing them, and are made happy by the visit. Zoo members are also typically higher in environmental concern than nonmembers (as assessed by questionnaire items asking about attention to, and worry about, environmental issues; Clayton, Fraser, and Burgess 2011; Clayton et al. 2014; Kelly et al. 2014). This doesn't prove that zoos create concern, because zoo members could already have been more concerned (although see Clayton et al. 2014), but it does suggest that zoos don't discourage it.

A second theme is that people create social interactions around the animal exhibits in ways that strengthen social bonds and establish a shared value for nature. That is, the social interactions can encourage people to consider animals to be more significant. Parents, for example, don't simply use the animals to entertain their children: they encourage the children to interact with the animals, such as by waving to them; they may establish a personal bond with a particular animal ("Let's go visit our lion!"); and they may use language signaling that the animals are worthy of care and respect. Social interactions at the zoo may also include conversations about environmental issues and the importance of caring for the natural world (Clayton et al. 2014).

A third theme reflects the ways people seek to connect with the animals in some way that suggests a shared perspective or experience. This can occur through eye contact, waving at the animal, making a comparison between the animal and humans, and even imitating the animal or speaking for the animal (for example, "Why don't these people feed me?"). Although people's perceptions that they have insight into the animal mind are almost certainly inaccurate, their feeling that there can be shared experience—what Pekarik (2004) referred to as "transcending the species barrier"—provides a foundation for greater empathy and concern toward the animal.

PERCEPTIONS OF ANIMALS

The opportunity for connection matters (Grajal, Luebke, and Kelly, this volume). Nature and environmental concerns are distant for many people, but encounters with specific animals can make remote issues seem more personal. Although an ability to take the animal's perspective does not translate directly into concern about the ecological threats a species faces, let alone a tendency to take ameliorative action, it may lower the threshold for such a response. In general, perceptions of similarity increase empathy and a willingness to help (Batson et al. 1995; Ovels, Horberg, and Keltner 2010). Positive attitudes toward animals seem to be correlated with a perception that they are phylogenetically or behaviorally similar to humans (Allen et al. 2002; Harrison and Hall 2010; Serpell 2004; Westbury and Neumann 2008). Self-reports and physiological data indicate that perceived similarity or phylogenetic relatedness between human and animals is associated with a higher empathic response (Plous 1993; Westbury and Newmann 2008). In survey research at the zoo, Clayton, Fraser, and Saunders (2009) found that a perception that animals were similar to humans was significantly correlated with the desire to help care for them.

Perceiving animals as similar to humans is likely to give rise to anthropomorphism, the tendency to attribute humanlike mental states to animals and to interpret animal behavior as parallel to human behavior (Waytz, Cacioppo, and Eply 2010). People who score higher on a measure of anthropomorphism are more likely to evaluate nonhuman entities in moral terms (Waytz, Cacioppo, and Eply 2010). Anthropomorphism and enhanced perceived human-animal similarity generally lead to positive attitudes and moral inclusion of animals (Bastian et al. 2012; Butterfield, Hill, and Lord 2012; Costello and Hodson 2010). Indeed, Kahn (1999) suggests that anthropomorphism is a way of thinking about other species that allows people to recognize the intrinsic value and moral standing of the natural world and thus to feel more concern about its protection and conservation. (There is a potential downside to anthropomorphism, as discussed below.)

ENCOURAGING A CONNECTION WITH ANIMALS

What can zoos do to increase visitors' feeling of connection to the animals they observe? Zoos seem to be increasingly aware of the importance of fostering a sense of shared identity with animals among their visitors. Some zoos have adopt-an-animal programs making the animal "part of the family." Some exhibits encourage people to recognize the similarities between themselves and other animals, for example, by describing shared physical traits, motivations, and behavioral patterns. How effective is this focus?

We conducted a small study to examine the consequences of asking people to view animals as similar to or different from humans. Taking advantage of a small captive monkey colony used for behavioral research on a university campus, we experimentally manipulated human participants' focus on their own similarity to or difference from the animals as they observed the animal enclosure. Fifty-eight undergraduate students viewed a colony of three capuchin monkeys (*Cebus apella*) housed in a basement enclosure. The enclosure, approximately 3.5 meters wide by 3.5 meters long by 3 meters high, included many toys and climbing opportunities for the monkeys, and observers watched from behind a one-way window 2 meters wide by 1.2 meters tall. The monkeys could not directly observe the participants, but they could sense movement behind the glass and frequently approached the window and tried to see through it. They often shaded their eyes with their hands, which tended to evoke an appreciative "Awww!" from viewers.

Participants were told to observe the monkeys for about five minutes, thinking about the ways the monkeys were similar to ($n = 22$) or different from ($n = 21$) humans, or they were given no instructions ($n = 15$). They were told they would be asked to record their observations. After five minutes, they wrote their comments in response to a simple open-ended prompt.

We hypothesized that focusing on the similarities between monkeys and humans should increase empathy and lead participants to make different observations of the monkeys compared with focusing on differences. The experimental condition in which participants were given no instructions was included to obtain an unprompted reaction to the monkeys.

Participants' written comments about the monkeys were coded for the presence of certain predetermined content categories, as shown in table 16.1.

TABLE 16.1. Percentage of participants commenting on particular attributes of the monkeys in the different conditions

	Overall	*Similar*	*Different*	*Control*
Feelings	67	81	48	77
Interactions among monkeys	60	71	43	69
Statements of similarity	56	86	10	85
Statements of difference	42	5	76	46
Physical characteristics	44	3	57	33
Anthropomorphizing	27	38	14	31
Face	14	29	5	8

Note: Distinctively high percentages are in boldface.

Substantial differences were associated with the experimental manipulation. Observers who were focusing on similarities were more likely to comment on the monkeys' feelings, their interactions with each other, and their faces and were most likely to make specific statements about similarity as well as to explicitly anthropomorphize them. Participants focusing on difference were more likely to describe physical characteristics and to make explicit statements of difference. Those in the control condition showed a pattern of comments much like those in the "similar" condition except they did not often comment on the monkeys' faces, and they also described ways the monkeys were different from humans.

These results, although based on a very small sample, show that encouraging people to think about ways they are similar to animals leads them to focus on different things, with possible consequences for a feeling of connection. One implication is that zoo signage that emphasizes similarity or difference can also influence the way people observe the animals and the exhibit.

FROM CONNECTION TO CONCERN

Empathy for animals in itself is unlikely to lead people to support conservation or to engage in more sustainable behavior unless they also recognize that there is a problem to be solved. Realizing that another is in need of help is crucial to the activation and intensity of empathic concern, which subsequently predicts helping behavior (Batson and Shaw 1991). Most people are aware that some species are endangered, and it is common for zoos to have displays about the need to conserve species as well as to indicate whether an animal is endangered. But to be motivated to take action, people also need to recognize that humans are implicated in the threat.

In our study we were also interested in levels of environmental concern among our participants. We expected that manipulating similarity focus versus difference focus could change the level of concern but that the effect would depend on the availability of information about threats to species in the wild. People who view the monkeys as similar and who are informed about the role of human behavior in threats to endangered species should show more interest in taking action to protect animals in the wild than people who view the monkeys as different, are uninformed about the role of human behavior in threatening ecosystem, or both, so we also manipulated the information provided. In a "short" condition, an information sheet described the loss of animal species and the mission of zoos to support conservation in the wild. Participants in a "long" condition received an additional paragraph about the role of humans in depleting animal habitat, which also stressed that individuals could help conserve habitat, specifically by reducing food waste and using less paper. As dependent variables, we measured general environmental concern (five questions, such as "How concerned are you about the loss of animal species due to environmental degradation?"), perceived similarity to the monkeys (whether the monkeys were different from or similar to humans), and intention to engage in more sustainable behavior (two items, including "How likely are you to make purchasing decisions based on their impact on wild animal species?", all on seven-point scales). Finally, we measured participants' environmental identity (EID; Clayton 2003) to assess their preexisting feelings of connection to the natural world.

As expected, the interaction between information and focus conditions had a significant effect on perceived similarity and approached a significant effect on behavior. When EID was included as a covariate to statistically control for preexisting differences while assessing the results of the experimental manipulation, the effect on behavior became significant and the effect on perceived similarity remained significant. The pattern of results for environmental concern was similar, although the effect was not statistically significant (see table 16.2).

In general, the information sheet had little effect on participants who were focusing on similarities between humans and the monkeys; they were already sensitized to the issue. In the control condition, however, participants who read about human involvement rated monkeys as more similar

TABLE 16.2. Mean concern and intention toward sustainable behavior in the experimental groups

	Concern	*Behavior*	*Perceived similarity*
Similar			
Short form	5.5	4.6	5
Long form	5.6	4.8	5.2
Different			
Short form	5.8	5.5	6
Long form	5.6	4.8	4.5
Control			
Short form	4.8	3.5	4
Long form	6.1	5.2	6

Note: Scores are based on a seven- point scale.

and reported that they were more likely to engage in sustainable behavior than those who were not informed about human responsibility. Meanwhile, when they read about human involvement, participants who were focusing on differences saw the monkeys as slightly less similar and were less likely to plan to engage in sustainable behavior. This may reflect a motivated denial: if humans are having a negative impact on something we feel relatively unconnected to, we look for reasons not to care and not to have to change our behavior.

IMPLICATIONS FOR ZOOS

Information alone is rarely enough to promote pro-environmental behavior (Schultz and Kaiser 2012), and this was true in our study. By itself the information provided (short versus long) had no effect: simply giving our participants more information about human impact on endangered species made no difference to environmental concern and behavioral intentions. Although knowledge is important, to make a difference it needs to be combined with motivation. Feeling a connection to an animal based on an emotionally significant zoo visit, may provide this motivation.

However, simply observing the monkeys (in the control condition) did not lead to the highest level of concern or behavioral intent in the absence of information about human involvement. Participants who were given no directions about focusing on similarity or difference, and who received the short information form, showed the lowest concern and the least tendency toward supportive behavior. It seems likely that these participants were simply not as engaged by the experience as were the others. Significantly for zoos, either looking carefully at the monkeys (in the "similar" or "different" condition) or receiving information about human involvement was necessary to create concern and a tendency toward supportive behavior.

The results of this study remind us of the potential for zoos to include conservation education in the entertainment experience they provide for their visitors. Although zoo visitors do not generally record their observations of animal exhibits, they do have conversations about the animals. Visitor groups may use informational signs as the foundation for specific types of conversations about the animals—indeed, zoos could provide prompts that encourage such conversations. It seems likely that carefully constructed sign and exhibit design could encourage zoogoers to think about the ways they are similar to animals and to engage other members of their group in this discussion. Based on the results of this study, it seems that if zoos want to encourage people to feel more concern about endangered animals they should continue emphasizing the similarity between humans and the animals being exhibited. In combination with the feeling of connection this engenders, the broader conservation messages found at the zoo can then lead the zoogoers to think about the ways they can change their own behavior to support conservation initiatives.

A COMPLEX TASK

Providing the appropriate amount of information about threats to species can be a difficult line to walk, because zoogoers may react negatively to signs that appear to assign blame. Anthropomorphism can also lead away from a more ecological perspective, which is less about identifying with and caring about phylogenetically similar organisms than about understanding how populations and communities evolve over time and space and about the interactions among the abiotic and biotic elements of the system. Emphasizing similarities between animals and humans could also cause problems if it decreases an accurate understanding of wild animals and their needs. Indeed, zoo visitors sometimes misinterpret the well-being of the animals on display by assuming the animals would have the same desires as the visitors. Finally, empathy toward primates may be relatively easier to create; research is needed to see if these findings also apply to other species. But the right combination of heightened connection to animals and heightened awareness of conservation needs may encourage greater concern about species protection among zoo visitors.

CHAPTER SEVENTEEN

From Sad Zoo to Happy Zoo: The Changing Animal Welfare and Conservation Priorities of the Seoul Zoo in South Korea

Anne S. Clay

In 2013 the Seoul Grand Park Zoo in South Korea released an Indo-Pacific bottlenose dolphin named Jedol into the ocean near Jeju Island—the first time an Asian zoo had reintroduced a captive dolphin to the wild. Illegally caught and sold to the zoo in 2009, Jedol had spent four years performing for visitors before his liberation—his release the result of an intense public campaign by animal welfare activists, scientists, and local politicians. Among those assisting in Jedol's rehabilitation and release was the well-known dolphin activist Ric O'Barry, founder of the Dolphin Project (and prominently featured in the 2009 Academy Award-winning film *The Cove*). The interest surrounding the release was unprecedented; indeed, no animal in South Korea's recent history had provoked as much media attention as Jedol (Jang 2014). The zoo's annual journal praised his release as "an exemplary representation of the cooperation between society and a citizens' committee comprised of academics, civic organizations, and government officials" (Seoul Grand Park 2013b, 78).

Jedol's story provides a window into the dynamic between conservation and animal welfare at the Seoul Zoo. Although the return of one dolphin might be considered only an act of animal welfare, zoo officials insisted it was in fact part of the institution's broader conservation agenda (No 2015). There is some evidence for this claim. Currently, the only existing pod of Indo-Pacific bottlenose dolphins in South Korean waters lives around Jeju Island. The 114 individuals within that pod are predicted to decline to 20 by the year 2050. Every individual therefore "counts" in conservation terms (at the level of the population). Furthermore, mass media coverage not only created awareness of the conservation status of the Jeju dolphin population, it also helped enforce antipoaching efforts. For example, Jedol's captors were punished, and the illegal dolphin trade significantly decreased (Jang 2014). Finally, the animal's release also stoked greater awareness of animal welfare in the country and spurred a broader conversation about the use of marine wildlife in shows (Jang 2014; Seoul Grand Park 2013b).

For the Seoul Zoo, then, Jedol's release clearly signified far more than the welfare of one animal: it represented the kind of collaborative social, academic, and scientific integration that the zoo aspired to create in its other conservation endeavors. It may also have signaled something of an evolution in Korean environmental ethics. As a piece in the zoo's annual yearbook put it, "Jedol's release stemmed from a reexamination and a new establishment of the relationship between people and animals; human beings and nature" (Seoul Grand Park 2013b, 83).

The zoo's emerging mission to create a harmonious relationship between society and the natural world through collaboration in conservation and animal welfare, however, remains a challenge. On the one hand, Jedol's case illustrates how welfare and conservation policies can work together to focus public attention on environmental problems such as species decline. On the other hand, successful conservation projects increasingly demand complex collaborations and significant resources. It seems clear that Jedol's release occurred only because a critical mass of animal welfare and academic institutions actively provided the necessary resources, expert opinion, and media attention that the Seoul Zoo could not provide on its own.

Ideally, the consistent deployment of such collaborations would aid the Seoul Zoo in expanding its research and conservation projects. Unfortunately, the exclusionary institutional culture in South Korea, along with the Seoul Zoo's status as a subdivision of the leisure-oriented Seoul Grand Park, has made it difficult for the zoo to establish these collaborations on a larger and more permanent basis.

In this chapter I examine how the Seoul Zoo has attempted to move away from its origins as an entertainment site to become a modern conservation institution by developing partnerships with outside organizations and institutions. Jedol's case not only demonstrates the Seoul Zoo's changing character and hopes for the future, it also emphasizes the difficulties the zoo has faced while navigating the tricky relation between animal welfare and conservation in its practices and policies. The Seoul Zoo's most successful collaborations have historically been with local animal welfare organizations that encourage the institution to conduct more ecological research on animals in order to improve their quality of life. Yet as the leaders of the zoo have touted the zoo's potential in animal conservation and research, limited resources have frustrated their attempts to create significant collaborations and partnerships that would help them realize this more ambitious vision. As a result, the Seoul Zoo has evolved into an institution that presents some of its animal welfare projects as "conservation," blurring the important distinction between the two motivations and programs.

The discussion that follows is an attempt to explore this changing dynamic between animal welfare and conservation based on my analysis of key zoo-related documents and interviews with Seoul Zoo officials, Korean academics, and conservation experts. Korean zoos are not nearly as well documented as their Western counterparts, so I hope this chapter will provide a glimpse into a distinctive cultural and institutional context that offers interesting comparisons with the largely US-centered focus of this volume. At the same time, even though the Seoul Zoo conservation story is in several respects a uniquely Korean one, we will see that it shares many features with other modern zoos seeking to balance a diverse set of organizational and public priorities (entertainment, animal welfare, conservation, and research). And like many of these other zoos, it has struggled to break free of its own institutional history (Henson, this volume).

SEOUL GRAND PARK: A BRIEF OVERVIEW

Originally created in Seoul's Changgyeong Palace in 1909 for the amusement of the Japanese occupiers, in the past thirty years Korea's first zoo has sought to become a modern zoological park that offers entertainment, educates the public, and conducts advanced research. In 1984 the zoo moved to its current location in the mountains of Makgyedong, Gwacheon, and it has since become of part of Seoul Grand Park, a large leisure facility that also has an amusement park, botanical garden, and natural history museum (Choi 2013a). Extending over fifty-eight hectares and containing 1,810 individual animals representing 132 species, Seoul Grand Park is the only zoo in South Korea to have its own research laboratory (Seoul Grand Park 2013a).

More than anything, the construction advisory committee directing the new zoo project wanted to rid the zoo of its historical Changgyeong Park image. The new venue was to be spacious and pasturelike—a vast improvement over the cramped quarters at the zoo's previous location. The planners sought to blend modern buildings and facilities with the geographical features of the surrounding landscape. Such facilities were to be one-story buildings evenly spaced inside the zoo and concealed within the existing environment. Wide, winding paths would avoid crowding and provide a pleasant and safe way for visitors to move around the zoo while making plenty of discoveries along the way (Seoul Grand Park 1996). Certain parts of the zoo were designed based on zoogeography. Other areas grouped animals from similar taxonomic groups for convenience. For some exhibits, Carl Hagenbeck's innovative design of natural-looking exhibits with moats and ditches instead of bars was used to allow visitors a closer, more immersive interaction with animals. However, as with many Western zoos that adapted this method, the design of these natural-looking illusions at the Seoul Zoo was often merely for the visitors' aesthetic satisfaction rather than the welfare of the animals (Seoul Grand Park 1996; Hanson 2002).

The challenges faced by the Seoul Zoo are again similar to those of other modern zoos in the United States, which also started out as entertainment venues (Henson, this volume). Created as a place for leisure, the Seoul Zoo only later made conservation a core priority. Also, as in the US case, the animal rights movement played an important role in improving animal care in South Korea. Like other modern zoos, Seoul Grand Park faces a complex set of trade-offs among conflicting priorities, including conservation, animal welfare, and public education (see also Palmer, Kasperbauer, and Sandøe, this volume). It is not always clear, however, how such trade-offs are to be made responsibly and consistently in practice. For example, although zoo associations, such as the World Association of Zoos and Aquariums (WAZA; see, e.g., Barongi, this volume), emphasize animal welfare, they do not specify exactly how the aim at animal welfare fits together with conservation (Palmer, Kasperbauer, and Sandøe, this volume).

However, unlike zoos in the United States, South Korea's rapid modernization caused the Seoul Zoo to follow a more compressed path from entertainment venue to modern zoological institution. Modeling itself on the best American zoos of the 1970s, it found itself trying to catch up to changing international standards. With democratization, which occurred in the 1980s, the growth of animal rights movements put new pressure on zoo administrators. The suddenness of this transition, combined with a lack of resources for the common development of animal welfare and conservation research, resulted in an often ambiguous and undifferentiated understanding of conservation and animal welfare. In fact the two concepts frequently were (and remain) conflated in many areas of promotion and practice at the Seoul Zoo, even though welfare and conservation are ultimately grounded in different value systems that often specify divergent policies and interventions.

ANIMAL WELFARE, ANIMAL RIGHTS, AND CONSERVATION

Animal welfare concerns humane treatment and an animal's physical and mental well-being (Hutchins 2007, 94). According to one influential argument, sentience—an animal's ability to experience pleasure and pain—is the ground of moral standing or considerability (Singer 1975). As a result, animal welfare advocates are often "looking to balance overall harms and benefits rather than to allow individual interests to 'trump' the good of the many" (Minteer 2013, 79). In contrast, the ethically more stringent animal rights perspective argues that "sentient beings have an intrinsic and inviolate right to life, liberty, and bodily integrity" (Hutchins 2007, 93). Proponents of animal rights, in other words, generally take the animal welfare viewpoint further by ascribing the moral equivalent of personhood to sentient animals, especially mammals, which are thought to be capable of more complex cognitive experiences (Minteer 2013, 79). Although in some cases some animal welfare proponents might reluctantly accept the human use of animals as long as their pain and suffering is minimized and the net benefits of the activity are significant, animal rights proponents believe exploiting animals (for food, entertainment, research, etc.) violates their fundamental rights to liberty and autonomy. They reject the very concept of the zoo—which places animals in captivity primarily, they argue, for human benefit—as fundamentally immoral (see also Palmer, Kasperbauer, and Sandøe, this volume).

Conservation is different. It entails "the securing of long-term populations of species in natural ecosystems or habitats wherever possible" (WAZA 2005, 9). Unlike the animal rights and welfare views, which ascribe intrinsic value to the interests and well-being of individual animals, the conservation viewpoint prioritizes populations, habitats, and species over individuals—as well as more ecological notions such as ecosystem health and integrity. The upshot is that, although the animal welfare/rights and the conservation agendas share a concern for animals' well-being, they are distinctly different concepts, rooted in different concerns. Although in some cases these agendas overlap (both animal welfare/rights and conservation advocates condemn the poaching of endangered species, for example), they can also diverge, in some case dramatically (see the discussion of Marius the giraffe in Greene, this volume).

ANIMAL WELFARE AT THE SEOUL ZOO, PAST AND PRESENT

The South Korean activists who seek to promote animal welfare often ignore the fine philosophical distinctions between animal rights, welfare, and conservation. For example, rather than attempting the radical approach of shutting down zoos because they are thought to violate animals' fundamental rights, current South Korean animal rights organizations seek to improve the overall quality of life of zoo animals, an agenda that puts them more in line with traditional animal welfare (and not rights) philosophy.

The entertainment purpose of the Seoul Zoo initially presented the biggest obstacle to maintaining the psychological and physical well-being of its animals. Nevertheless, from its founding in 1984, the Seoul Grand Park Zoo made animal welfare an important priority in its exhibit designs. Under the motto "animal welfare first," the institution sought to accommodate its animal tenants with enough trees and rocks for hideouts, appropriate lighting, ventilation, and enrichment (Seoul Grand Park 1996; Cho et al. 2009). However, without sufficient funding, the animal facilities deteriorated over time, and the "prisonlike" confinement of the animals became the center of a debate over the treatment of zoo animals in South Korea (Cho et al. 2009).

In 2001 Dr. Jaecheon Choi, current director of South Korea's Biodiversity Foundation, highlighted the sad state of the Seoul Zoo by publishing an article titled "I Hate Going to the Zoo" in the *Hanguk Ilbo* newspaper. Returning to Korea after living for several years in the United States, Choi took his young son to the Seoul Zoo. "It only took two hours at the zoo for my son—my son, who loved going to the zoo so much—to turn to me and ask me to go home because he felt sorry for the animals" (Choi 2014). Saddened by how the zoo in Korea had made his son feel, Choi realized this experience had opened his own eyes to the desolate plight of the zoo animals. "In this young child's eyes there was a difference with the zoos he visited in the United States. What I mean is, he could see that the animals did not seem happy. But as an adult, even though I saw a difference, I just thought circumstances can be this way" (Choi 2014).

Choi's article stimulated public discussion about conditions at the zoo. A group of anxious, animal-loving Korean citizens created Haho, an "environmental movement alliance for the protection of wild animals and the advancement of animal welfare" (Haho 2004). In 2002 Haho published *Sad Zoo*, the first report on zoo animal welfare in South Korea, which criticized the Seoul Zoo's lack of enrichment and unnatural exhibit settings that often caused physical and psychological injury to the zoo animals (Haho 2004). A follow-up report, *Sad Zoo 2* (Haho 2004) highlighted changes since 2001 but indicated that little had improved in three years (Haho 2004; Cho 2007).

Nevertheless, the *Sad Zoo* reports resulted in some positive changes. In 2002 Seoul Grand Park created a plan for the zoo to become an "ecological zoo" with exhibits that more closely reflected the natural habitats of the animals on display. The plan also placed greater emphasis on wild animal species conservation, education, and research (Seoul Grand Park 2002). Associations such as the Korean Animal Rights Advocates have encouraged zoos to participate more in research on animals' natural behavior and ecosystems to help improve enclosure environments at the zoo. However, an emphasis on research and conservation may cause more problems when research begins conflicting with animal welfare.

In 2003 the Seoul Zoo began establishing enrichment programs for its animals, and in 2005 the Animal Breeding Department became the Animal Welfare Department, indicating a movement toward animal well-being and away from the practical use of zoo animals (Cho et al. 2009; Kang et al. 2013). In a break with its previous practice, the zoo began keeping families of chimpanzees together, stimulating their natural ability to teach and learn. Likewise, the zoo also found a way to display all the lions together, sparing individuals from spending their days in cramped cages (fig. 17.1). To help stimulate giraffes' natural behavior of feeding from high places, the zoo created a new feeding system that used pulleys to lift up the troughs (fig. 17.2).

Other new enrichment programs stimulated the natural behavior of the

FIGURE 17.1. Trees were planted for shade in the lions' enclosure. Photograph by Anne Safiya Clay.

FIGURE 17.2. Enrichment programs at the Seoul Zoo first began in 2003. Photograph by Anne Safiya Clay.

great apes (Haho 2004; Seoul Grand Park 2012; Kang et al. 2013). In 2009, declared the "Year of the Gorilla" by the World Association of Zoos and Aquariums, the zoo launched its "Happy Gorilla Project" to welcome its new tenant, a male lowland gorilla named Ujiji. The zoo redesigned the exhibit, replaced the outside enclosure's concrete floor with grass, planted trees, and provided more hiding places (Yang 2014a; Seoul Grand Park 2009; see also Lukas and Stoinski, this volume).

The 2009 Happy Gorilla Project was the start of a series of radical improvements in animal welfare at the Seoul Zoo. The practice of allowing visitors to touch, hold, and be photographed alongside certain zoo animals outside their enclosures was terminated in 2012. At the same time, the zoo began a program to improve the nutrients in the animals' diets. In 2013 visitors were no longer allowed to feed wild zoo animals for fear of transferring bacterial diseases and disrupting the animals' strict diets. In that same year, the Seoul Zoo began to apply positive reinforcement training to strengthen the bond between zookeepers and animals and to relieve the animals' stress during medical procedures. In addition, the zoo replaced their Artificial Nursing Center (where zookeepers raised babies rejected by their mothers) with the Conservation Education Center. Previously, visitors were given opportunities to pet and fondle these baby animals, which were even sometimes sent to local orphanages for the children to enjoy. Realizing the stress such situations put on these young animals, the zoo decided to close down the nursing center, and zookeepers began rearing rejected infants where the mother could hear, smell, or see her offspring (No 2015; Seoul Grand Park 2013b).

There were other developments. By 2014 the zoo instituted regular medical checkups for all animals (No 2015; Kang et al. 2013; Seoul Grand Park 2013b). It also ended its flamingo show, in which a costumed zoo employee chased flamingos to the music of a waltz. And for the past several years the zoo has made efforts to educate people about the ecology of dolphins through their dolphin show rather than having the dolphins perform solely for entertainment (No 2015).

Overall, then, the Seoul Zoo's journey from "sad zoo" to "happy zoo" reflects a significant effort by the institution to collaborate with citizens on these issues (Yang 2014a). At the same time, animal welfare can never be the most important priority of any zoo, since zoos by their very nature sacrifice the interests of individual animals (which presumably would be better off in the wild) for some greater purpose, such as public education, research, or conservation. Like other zoo administrators, the leaders of Seoul Grand Park seek to balance these conflicting priorities as they increase their involvement in conservation and research projects.

Many of those who work at the Seoul Zoo share curator Hyojin Yang's sentiment that "the animals' happiness and comfort is most important [for a successful zoo]" (Yang 2014a). Yang believes that if the animal is happy, people will sense this and be more responsive to the zoo's larger environmental message (species conservation, ecological protection). Proper welfare not only establishes deep connections between the zoo and its visitors, it also allows animals to display their natural behaviors in a captive setting (see Maple and Segura, this volume). This sentiment is also in keeping with the vision of the World Association of Zoos and Aquariums (WAZA), which specifically states: "Whilst conservation of wildlife is the core purpose of modern zoos and aquariums, animal welfare is our core activity" (Gusset and Dick 2015). The implication is that any actions zoos take for conservation must also be compatible with welfare standards.

THE EVOLUTION OF CONSERVATION AT THE SEOUL ZOO

From its founding in 1984, the Seoul Zoo had called for a "scientific and efficient administration" in addition to a "safe and pleasant visit" (Seoul Grand Park 1996). However, a real emphasis on scientific research and conservation was slow to develop. In 1999 Seoul Grand Park established the Korea Wildlife Conservation Center, an extension of Seoul Grand Park dedicated exclusively to the conservation of native endangered species. The following year the South Korean Ministry of the Environment designated Seoul Grand Park as the first environmental protection institution for Korean endangered species and the zoo launched its Eco-Zoo project, an attempt to make conservation of Korea's indigenous species the zoo's ultimate focus (Seoul Grand Park 2001).

Initially the Ministry of the Environment designated a total of ten native species as the objects of conservation facilities at Seoul Grand Park. By 2014 that number had increased to twenty-one. The Ministry of the Environment has praised the Seoul Zoo for steadily promoting conservation endeavors by aiding in the propagation and rehabilitation of endangered species, and the zoo has been named a leader in terms of research resources. Current Seoul Zoo director Jeongrae No told me he believes this focus on conservation, which includes breeding and propagation (especially of rare or indigenous species) as well as education, is the most important element of a successful zoo. During his three years as head of the Seoul Zoo, he has done much to emphasize conservation in the institution's vision, particularly in terms of education. In addition to being the only domestic zoo with a research laboratory, it is also the only zoological institution to be involved in environmental protection (No 2015).

However, my discussions with other individuals at the Seoul Zoo, academic institutions, and conservation organizations suggest that, although the Seoul Zoo is South Korea's leading zoological institution, it is severely limited by a lack of resources and funds. Rather than leading in the conservation of twenty-one species, the zoo seems to be especially invested in only three: the Asiatic black bear, the red fox, and the Asian leopard cat.

The Asiatic black bear and red fox projects are two of the ongoing projects at the Seoul Zoo where the institution actively works with an outside conservation organization. Owing to poaching, habitat loss, and secondary poisoning from rodenticides, the red fox became extinct in South Korea in the late 1980s. In an effort to restore the species, a facility in Sobaek Mountain National Park has been working with the Seoul Zoo to breed and release red foxes brought from North Korea and China (Lee et al. 2013). Seoul Zoo's lab coordinator Gyeongyeon Eo mentions that the zoo has bred six red foxes and supplied them to the Sobaek Mountain red fox restoration facility (Eo 2015). However, director Seongyong Han says that although the Seoul Zoo would like to participate in the breeding and reintroduction of endangered species, many of these conservation organizations regard the zoo as merely a source for an emergency stock of individuals, and as a result the zoo has had no opportunity to participate in the conservation research related to these projects (Han 2015).

The zoo, however, has had an essential role in the conservation of Asiatic black bears in South Korea. Although the bears finally became a locally protected species in 1982, by the late twentieth century the poaching of bears for traditional medicine had caused the population to dwindle to a little over five individuals living in the Jiri and Odae Mountains (KNPS 2014). In an attempt to save the species, Professor Hang Yi organized an Asiatic black bear population and habitat viability assessment workshop at the Seoul Zoo in 2001 in collaboration with the International Union for Conservation of Nature (IUCN). The workshop evaluated four major problems that could ultimately be solved only through close cooperation with the Seoul Zoo: successful restoration of a minimum viable population of Asiatic black bears required more information on wild populations; the ability to obtain purebred bears; increasing the population of bears; and extensive information about breeding captive bears (Lee et al. 2001).

The Seoul Zoo used diplomatic relations to exchange animals with Russia, China, and even North Korea to receive individuals of the native subspecies of Asiatic black bear. Bear cubs are initially bred at the zoo and then sent to Jiri Mountain National Park, where they are habituated and released. Because a sustainable and genetically viable population of bears in that area requires continual addition of individuals, the project depends on the Seoul Zoo for bears. Since 2001 the zoo has sent fifteen bears to Jiri Mountain, a little under half of the current population (Eo 2015).

Within the Seoul Zoo's current conservation programs, the breeding and propagation of the critically endangered Amur leopard cat is the institution's only exclusive project. It's a program, however, that has drawn some concern outside the zoo walls. In his interview with me, Seoul National University professor Hang Yi criticized the zoo for its disorganized releases of this animal. He claims that the zoo neglects to acclimatize the animals before they are released and that the program has had no significant ecological effect. In his opinion the zoo performs such actions only to gain favor with the public (Yi 2015). When describing the process of releasing Amur leopard cats into the wild, Seoul Zoo lab coordinator Eo mentions that the captive-bred offspring of rescued wild leopard cats go through vaccinations and deworming before being experimentally released into the wild to see how well they adapt to their environment. Out of the five animals released, only two leopard cats are still living (Eo 2015).

Like Professor Yi, The Korean Otter Research Center director Seongyong Han is also critical of the zoo's reintroduction project, but for different reasons. By comparing the Amur leopard cat project with the Asiatic black bear restoration mission, he points out that the Seoul Zoo is sorely lacking in the resources and equipment needed to manage successful reintroductions. Whereas the Asiatic black bear project received much financial support and extensive input from outside experts, the Seoul Zoo does not have the expertise and manpower to do necessary tasks such as frequent monitoring of the leopard cats' telemetry transmissions, which greatly limits data collection. It's a conclusion supported by Director No, who told me that the Seoul Zoo has not been able to attain as much of a conservation role as it would like because of a shortage of funds and resources for research and species reintroductions. This lack of resources, furthermore, works against the development of greater scientific and conservation capacity at the Seoul Zoo. According to its critics, the zoo fails to conduct its conservation programs at a high level and so discourages collaboration with other institutions (Han 2015).

What the Seoul Zoo lacks in its conservation initiatives, however, it attempts to make up for in its education programs. Director No insists that the zoo must send a strong conservation message to its visitors: "The message we want to give is not for people to look at these animals because they are impressive, but for people to practice conservation as a result of seeing these animals" (No 2015). No has pushed for the Seoul Zoo's education programs to include elements of ecology and conservation. This curriculum has expanded to thirteen ongoing programs in addition to over twenty-five courses in the zoo's database for kindergarten, elementary, and middle-school children. Inyeong Yeom, the head of the Seoul Zoo's education team, believes educational programs are vital to helping ordinary people understand the zoo's research. "While practicing conservation, zoos must tell their visitors what kind of research they do in a way that is fun and comprehensive" (Yeom 2014).

Juhui Bae, the zookeeper in charge of the center, says that the establishment of the Seoul Zoo's Conservation Education Center, where visitors can learn about the conservation of native Korean species, shows how much the Seoul Zoo's consciousness regarding its role in education has changed. The zoo's goal is now to instill this awareness of environmental issues in its visitors. However, Bae laments that not many people seem interested in the new education programs at the center. In fact, most of those who attend the lectures all the way through already have a significant interest in environmental conservation. Yet Bae does not give up hope, telling me that she believes reaching even one person is important in order to bring change (Bae 2015).

Rather than preaching to the choir at a specialized facility, such as the Conservation Education Center, it may be more effective to reach visitors by creating more habitat-based exhibits that can demonstrate how animals live in their natural state. Recently the tiger exhibit was completely remodeled according to the conservation master plan for the EcoZoo concept, which hoped that "the opportunity to allow visitors and animals to study each other nose to nose, with nothing in between but a pane of glass, will make the conservation message memorable and believable" (Seoul Grand Park 2001). However, the remodeling of exhibits is constrained by the byzantine regulatory structure that governs the leisure-oriented Seoul Grand Park.

Many exhibits at the zoo still do not accurately represent habitats and therefore are not suitable to educate visitors on their own. The baboon exhibit, for instance, contains generic representations of African huts as shelter for the baboons. Similar to the architecture of Western zoos during the colonial era, this exhibit is more a nonspecific representation of the people of Africa than of the animals from that continent. Moreover, it could also be said that the more naturalistic tiger exhibit does not offer an accurate representation of the Siberian tiger's ecosystem. Using native Korean plants and stones, this particular exhibit is designed to look like a majestic representation of the Korean mountains. It is an exaggeration of nature meant to be an idealistic portrayal of the animal as the "Korean" tiger, which used to roam throughout the peninsula. Yet, although the representation may be inaccurate, the exhibit's appeal to the Korean people may be enough to get them to care about the species' fate as well as the importance of protecting other indigenous species from extinction.

CONCLUSION: JEDOL'S UNCERTAIN LEGACY

In many ways the harmonizing of welfare and conservation at the Seoul Zoo fits into the emerging framework of "compassionate conservation" (see, for example, http://compassionateconservation.net/). This approach "builds on an agenda that calls for 'doing science while respecting animals' and for protecting animals because they are intrinsically valuable, and do not only have instrumental value because of what they can do for us" (Bekoff 2015). According to Marc Bekoff, an animal activist and also an emeritus professor of ecology at the University of Colorado, compassionate conservation advances an innovative view of conservation in which empathy plays an important role in decision making. In his view, it "allows for—but does not dictate—outcomes in which the interests of others supersede those of humans" (Bekoff 2015). The Seoul Zoo's decision to release their bottlenose dolphin, Jedol, back to the ocean around Jeju could therefore be interpreted as an act of compassionate conservation.

Yet Jedol's story importantly also symbolizes the zoo's vision for its own future. This single act of animal welfare helped bring awareness not only of the plight of dolphins in zoos, but also of their endangered state in the wild. In particular, the efforts put into freeing Jedol represent the type of successful collaboration the Seoul Zoo continues to replicate. It demonstrated the value of engaging many social actors in a symbolic but also tangible conservation project that enjoyed considerable support among the public.

But the Jedol case is finally also a reminder of the challenge of balancing the zoo's animal welfare mission with its conservation ambitions. Although the zoo's heightened focus on animal welfare was an important corrective to some of its historical practices of animal exhibition and care and taps into strong popular sentiment, a programmatic and public overemphasis on individual animals' well-being by and at the zoo may ultimately present more barriers than bridges if the institution seeks to focus more on (and becomes more able to support) animal research and species conservation, especially when these latter pursuits conflict with strong animal welfare norms (Minteer 2013; Palmer, Kasperbauer, and Sandøe, this volume). As with other zoos around the globe, hard choices will likely await the institution as it tries to develop a conservation profile that is deliberately distinct from its emergent and obviously very popular animal welfare agenda. The compassionate conservation ideal, however, suggests a possible path forward.

Indeed, Jedol's release clearly touched the hearts of many Koreans including, not least, the scientists and conservationists involved in his return to the sea. On the shore of Jeju Island, the dolphin research team erected a large stone to commemorate the animal's release. Its short inscription perhaps conveys better than anything else the zoo's hopes for creating an inspiring and unified vision of animal welfare and conservation in this century: "Jedol's dream was the ocean."

CHAPTER EIGHTEEN

Wildlife Wellness: A New Ethical Frontier for Zoos and Aquariums

Terry L. Maple and Valerie D. Segura

In the decades since *Ethics on the Ark* was published, many zoos have adopted practices that improved the welfare of the animals in their care. In this chapter we will discuss some of the history leading up to the call for the Ethics on the Ark meeting and the near-perfect storm of events that set in motion a shift toward science-based management practices at Zoo Atlanta, transforming it from a traditional zoo to an empirical zoo (Maple 2008). We will also argue that zoos will be successful if they implement and test the psychological construct of "wellness" for each individual animal living naturally in human care. An extension of animal welfare, wellness is distinguished from welfare by its aspirational, optimal goal of complete and total psychological wellness. This goal unleashes our creative juices to design and develop zoo environments that encourage activity, affinity, reproduction, and resilience.

WHAT WE LEARNED IN ATLANTA

As a college professor teaching environmental psychology and animal behavior in the School of Psychology at Georgia Tech, one of us, Terry L. Maple (TLM), found himself in the midst of a highly publicized management crisis in 1984 when Mayor Andrew Young asked him to intervene as the new interim director of Atlanta's struggling metropolitan zoo. As the reform director, TLM was hired to "stop the bleeding" despite having only nine months of administrative experience as the general curator of the Audubon Zoo while on academic leave from Georgia Tech. He may have been the least experienced zoo director in the nation, but as one who closely observed zoos during a decade of research on zoo animal behavior, he was confident he knew what needed to be done. The key to his success as director in the early days and for the entire duration of his eighteen years of service was his tenured professorship at Georgia Tech. In 1984 the zoo had reached a turning point, spinning out of control after years of low budgets and low expectations. Efforts by private citizens organized into a zoological society failed to reverse the downward spiral. Once the Humane Society of the United States (HSUS) identified the Atlanta Zoo as one of the nation's ten worst zoos, even local government recognized that it was time to right the ship.

For several years, local and national media had been pounding Atlanta for the dead and missing children mystery, followed closely in time by the zoo's own version of this tragedy when investigative reports began to turn up dead and missing zoo animals. One editorial in the *New York Times* opined that how a community treats its animals says something about the human beings who run it (Maple and Archibald 1993). The most painful case was revealed when the twelve-year-old elephant Twinkles was discovered by reporters buried in a shallow grave in North Carolina, hastily abandoned by the traveling circus that had removed her from the zoo with the alleged complicity of zoo employees. Following this event, every zoo in America wanted their animals that were on loan to the Atlanta Zoo immediately returned. Public confidence in the zoo and the trust of its peer institutions was in free fall, and the hysteria could not be contained. The silver lining in this scandal was the compelling public demand that was gaining momentum. Our community was prepared for a protracted struggle, knowing this might be the last and certainly was the best chance to deliver to Atlanta citizens a quality zoo like the one TLM enjoyed in his hometown of San Diego. In 1984 the stage was set for significant social change at the zoo (Desiderio 2000).

As local government and business leaders began to coalesce around a plan for privatizing the zoo, our small zoo team was tasked with putting together the details of how a reputable zoo should operate. The first draft of a new model was ambitious and unrealistic, but the city eventually agreed to provide sufficient bridge funding to elevate operations to a much higher level of achievement. The team envisioned a new zoo with significantly higher operating standards and better management and husbandry practices than before. Indeed, we wanted to be leaders in ethical animal care even before the zoo profession had started discussing fundamental innovations and applications of animal welfare. The intent was to create a zoo that would be a model for all others, since leaders of our institutional peers were aware that the Atlanta Zoo was not alone in its deficiencies. Once a nonprofit board of directors was assembled in 1985, TLM was elevated to the position of president and chief executive officer. The new board leaders envisioned a zoo that reflected superior business practices with an empowered leader at the helm. This change of direction was remarkable in another way. At the center of the civil rights movement, the local government's best practice of racial and gender balance was implemented in the composition of the new zoo board. Four of the nine members were prominent African American business leaders, and four were high-achieving women. It was an outstanding group of citizens who demonstrated their commitment to the cause with generous giving, expansive community networking, meticulous management, and strategic leadership. Their collective business acumen generated an incredible marketing effort as our new brand was promoted throughout the nation. For its part, city and county government, through the auspices of the Atlanta-Fulton County Recreation Authority (enacted in 1961 to build stadiums and arenas in Atlanta), issued $16 million in revenue bonds to jump-start the zoo's transformation from a prison to a paradise for animals. Once the pariah of America's national zoo association, the Association of Zoos and Aquariums, the newly constituted Zoo Atlanta reentered the fraternity of zoos in 1987 as a fully accredited member of the AZA. No zoo ever worked harder to get back into the good graces of its peer institutions. Local pundits proclaimed far and wide our stunning achievements in making a silk purse out of a sow's ear. At that moment, after years of mediocrity and despair, Zoo Atlanta was acknowledged as the city's most endearing phoenix story.

A ZOOLOGICAL TABULA RASA

One of the most significant achievements of Atlanta's zoo renaissance was the emergence of new ideas based on psychological science. Zoo Atlanta, as it was rebranded, almost overnight became the nation's best example of an authentic empirical zoo, that is, a zoo that encouraged scientific collaborators to assist zoo managers in making evidence-based decisions (Maple 2008). In 1985 the doors were opened and researchers were invited to study everything, even the most basic assumptions underlying animal behavior and husbandry at the zoo. This was achievable given our proximity to major institutions of higher learning: Emory University, Georgia Institute of

Technology, Georgia State University, and three veterinary schools within an hour's drive. Efforts were made to ensure that Zoo Atlanta had strong partnerships with each of these institutions. It was fortunate that TLM was encouraged by administrators at Georgia Tech to continue his research on animal behavior and environmental psychology while serving as the zoo's new chief executive. The multidisciplinary research group, Georgia Tech Laboratory for Animal Behavior (TECHLab), comprised faculty collaborators and highly motivated graduate students, many of whom have since become leaders in the zoo profession. One of the key trends in the past twenty years is the number of highly educated young leaders who have entered the profession throughout the nation. It is not unusual now for a scientist to be selected to lead a zoo, for example, Dr. Dwight Lawson, CEO at the Oklahoma City Zoo, Dr. Kyle Burks, CEO at the Sacramento Zoo, Dr. Chris Kuhar, CEO at the Cleveland Zoo, and the recently retired Dr. Jackie Ogden, CEO at Disney's Animal Kingdom. All these leaders were educated at Georgia Tech and worked at Zoo Atlanta. Their stories are told in a new book of essays edited by three of TLM's former students (Hoff, Bloomsmith, and Zucker 2014).

As agents of change, we seized the opportunity to think big about the zoo's future with the strong support of government and local business leaders. As a community, Atlanta was determined to find a way to end the drama and repair the city's damaged reputation. Our unity of purpose encouraged risk taking at all levels of the organization, and this is one reason so many of the zoo's employees and student collaborators became leaders of other zoos: they emerged from the Atlanta experience confident in their ability to confront and overcome adversity.

TALKING TO ADVERSARIES

The ethics conference in Atlanta was the first time zoo professionals sat across a table and debated animal rights/welfare critics. We were careful not to invite extremists who were irrational or unreasonable. In addition, we required that they be at least somewhat scholarly. They had to be able to argue with us in a civil manner and write a cogent paper for publication. We didn't want our intellectual adversaries to organize a demonstration or generate their own self-serving media. We were seeking an honest, objective debate on the issues to reach common ground where we could. When the Ethics on the Ark meeting was first proposed, many of our more cautious peers regarded it as a risky project. AZA's director of conservation and science, Dr. Michael Hutchins, played a key leadership role in winning the AZA board's support for this meeting. Several of the proposed speakers were on record as harsh critics of zoos, but every presentation was constructive and engaging. The professors among us were the reason we could not be intimidated by our most severe critics: those with a background in academia were well versed in holding constructive critical debates with detractors. Our zeal to debate them was based on our shared vision that zoos could be much better if we embraced constructive criticism. Ethics on the Ark was essentially a scientific exercise, and it broke new ground for the zoo profession. Before this meeting, ethics was a topic explored by philosophers, not zoo biologists. By involving experts outside the zoo profession, we elevated the conversation to include broader issues of animal rights and conservation. This meeting was another testimony to Atlanta's civil rights origins, as Martin Luther King Jr. repeatedly advised his supporters never to walk away from an opportunity to talk with adversaries. Unfortunately, with few exceptions there still exists a general reluctance within the zoo community to debate responsible critics. A concerted effort to welcome diverse views and debate from within and outside the zoo community is certainly needed. The best venue for these conversations is our national conference (AZA). We should actively recruit proponents of controversial viewpoints and offer our members a chance to hear both sides. Based on recent events, HSUS should be the first organization invited to engage in a constructive conversation.

Another step in the direction of constructive criticism was the publication of David Hancocks's book *A Different Nature* (2002). An experienced innovator as director of Seattle's Woodland Park Zoo and the Arizona-Sonora Desert Museum, Hancocks was one of the keynote speakers at the Atlanta ethics conference. His bold criticism stung and even alienated many zoo leaders. The fallout meant that any valid point Hancocks may have raised was disregarded simply because of his unapologetic tone. This was a missed opportunity because as a community of zoo professionals we failed to follow up on many of the issues and ideas he introduced in his stimulating book. Since its publication, he has not been invited to speak at the AZA conference. It is not too late to debate Hancocks and other responsible critics on the pace and direction of social change as zoos must continue to evaluate progress and evolve to meet, and we hope exceed, the public's growing expectations. By exceeding expectations we can inoculate ourselves against scandals like the Atlanta Zoo crisis and, more important, we can fulfill our ethical obligation to deliver the best practices and highest standards for the wildlife in our care.

WELLNESS AS WELFARE

The ongoing debate about our collective treatment of elephants is a powerful example of the challenge of exhibiting megafauna with special needs. It has become clear that radical revision of elephant exhibition and management is absolutely necessary. Success with these massive and highly intelligent creatures will require innovations far beyond the minimum standards required by current regulations (Maple, Bloomsmith, and Martin 2008). In revisiting the progress made since *Ethics on the Ark* was published, we are once again prepared to issue a challenge to the zoological community. If for the most part satisfactory welfare for animals has been achieved, it is time to look for new and better ways to build on that success and significantly raise the bar on our practices and standards.

Derived from visionaries in the human potential movement and embodied in the humanistic psychology of Abraham Maslow (1962), the achievement of wellness leads to physical, mental, and social fitness. In both animals and people, fitness is best managed by "personal trainers" skilled in the psychology of conditioning and learning. Training has become an essential technique for enriching animals' lives in zoos and aquariums, so it is worthwhile to do it in a way that is maximally effective. Today zoo animals need not be chemically immobilized to draw blood, measure blood pressure, conduct an ultrasound examination, or record body temperature. These tasks can be accomplished through safe, low-stress positive reinforcement (Maple and Perdue 2013). Despite these safer animal interventions, there are further advances to be made. As we mentioned, trainers' most commonly used tool is based on operant conditioning, but that alone is not their only one. The most effective option is to enlarge the toolbox to include the entire science of behavior, known as behavior analysis, of which operant conditioning is just one component. *Applied* behavior analysis, employing the science to solve problems of social significance, has the potential to revolutionize behavioral husbandry management and training in zoos and aquariums. However, at present many zoo and aquarium trainers learn on the job from other trainers and operate at a rudimentary level of understanding rather than a sophisticated one. At the same time, we must acknowledge that traditional trainers using Skinnerian operant conditioning are largely responsible for much of the progress zoos and aquariums have achieved in animal welfare. The purpose of asking zoo and aquarium professionals to include behavior analysis as one component of the animal wellness movement is to ensure that research and analysis are the foundation for any decisions related to welfare, enrichment, or wellness.

Adopting the science of behavior analysis, with its emphasis on control and prediction from a natural science perspective, is an extension and advancement of current behavioral husbandry practices. We should reunite psychologists who have expertise in behavior analysis with zoo practitioners (keepers, curators, and veterinarians) to improve wildlife wellness through applied research and practice. The goal of a behavioral wellness program is to prepare environments, opportunities, and schedules that encourage animals to engage in behaviors that indicate thriving (e.g., wellness). A major challenge to the construct of animal wellness is to define thriving for every zoo and aquarium species. At present, with a few exceptions we know only what the absence of wellness looks like. Because wellness is an aspirational construct, it has no upper limits. Since human beings cannot be too well, why would we limit the wellness of zoo and aquarium animals? (See, e.g., Maple and Bocian 2013; Maple and Perdue 2013.)

Once we define wellness, we must find a way to activate the animals by the personal involvement of the animal care staff. A commitment to wildlife wellness programs will let zoo and aquarium visitors observe and appreciate a population of active animals exhibiting natural behaviors in natural social groups. It is possible that when visitors see animals living well they may be better prepared to listen to our conservation messages (see Palmer, Kasperbauer and Sandøe, this volume, for an extended conversation of the relation between animal welfare and conservation efforts in zoos). Of course, in our efforts to provide wellness for animals, we must share our commitment with the public. If we are making great leaps of reform it may be obvious to visitors, but if it is not, we should make the commitment to animal wellness explicit, just as zoos have done with conservation efforts in recent years. One thing is certain: if visitors leave the zoo feeling sorry for the animals, we have failed in our mission to educate and inspire them. A commitment to wellness is good for animals, but it is also changing the workplace climate for frontline caregivers, curators, veterinarians, educators, and volunteers. Although concern with animal welfare has a long history, attention to wellness is just getting started.

ENVIRONMENTAL PSYCHOLOGY AT WORK, THEN AND NOW

In graduate school at the University of California, Davis, TLM was introduced to the ideas of the iconic environmental psychologist Robert Sommer. Professor Sommer's juxtaposition of hard and soft architecture enabled a new vision for naturalistic facilities. Twenty-five years after the publication of his book *Tight Spaces* (1974), Sommer concluded that while airports, mental hospitals, and prisons had not changed, zoos were the exception. Creativity in landscape and zoo design and a steady stream of animal behavior research contributed to the improvements Sommer advocated. Hard zoo architecture was discredited, and exhibit substrates were softened and reformulated. Zoo designers also benefited from the published observations and photodocumentation of wild animal behavior by anthropologists (Lindburg 1988) and field biologists (Harcourt 1987). Zoo professionals who seek innovation in exhibit design should begin by reading a comprehensive textbook on environmental psychology. The basics in this field apply equally to people and to wildlife (Sundstrom et al. 1996).

Environmental psychologists in North America were aware of the findings of European zoo biologists, especially the work of Dr. Heini Hediger, a professor of ethology at the University of Zurich and director of three Swiss zoos from 1938 until 1973. Sommer's research on human use of space was influenced by Hediger's observations of territoriality and by the concepts of flight and critical distance in birds and mammals in the zoo. Hediger demonstrated that a zoo could operate with an empirical foundation, and he inspired many zoo professionals to experiment and innovate with new exhibit ideas. His most influential zoo exhibit was the curvilinear Africa House at the Zurich Zoo. Hediger (1970) repeatedly expressed his disdain for cubic designs, regarding them as essentially "unbiological." He saw nature as the proper model for zoo exhibits.

The ideas of Hediger and Sommer strongly influenced the creative concepts that characterized Atlanta's zoo renaissance. College of Architecture colleagues at Georgia Tech were teaching and research partners who helped students and staff to practice the science of postoccupancy evaluation. Together we published many papers that demonstrated how innovative exhibits affected animal welfare (Maple and Finlay 1986; Wineman and Choi 1991; Chang, Forthman, and Maple 1999; Stoinski, Hoff, and Maple 2001). For two decades, architects and psychologists used the zoo as a teaching laboratory for environmental design, and many doctoral dissertations were later published (e.g., Finlay, James, and Maple 1988). At the time, no zoo had opened its doors wider to research than Zoo Atlanta, and we carefully evaluated everything we did to improve the zoo. Taken together, these publications are a blueprint for reform. Another advantage of the Georgia Tech partnership was the opportunity for employees to pursue degrees with Tech colleagues such as philosopher Bryan Norton (a contributor to this volume).

Deputy zoo director Jeff Swanagan, zoo biologist Gail Bruner, and curator of birds John Fowler received master's degrees in public policy under Norton's supervision (e.g., Swanagan 1992).

Zoos that developed exhibits with features to accommodate appropriate naturalistic social groups were well insulated against the suggestion that they were not committed to individual animal welfare. The shift in exhibit design can be characterized as welfare-oriented. Individuals and populations were well served by this landscape immersion approach, an innovation that accommodated the growing interest in zoo animal welfare. The two chapters TLM wrote or coauthored for the original *Ethics on the Ark* volume discussed Zoo Atlanta's commitment to provide facilities that catered to each species' special needs. The lowland gorilla exhibit we designed and built had five contiguous habitats for groups of gorillas. It was the first time a zoo dared to exhibit a population rather than just a group of gorillas. The Ford African Rain Forest Exhibit was immediately successful as gorillas began to reproduce and raise their young in social groups accessible to an adoring public. An animal welfare highlight of this exhibit was the successful resocialization of the male Willie B., who had lived twenty-seven years isolated from his own kind before he successfully sired offspring at the zoo. At the time of his death at age forty-one he was the oldest gorilla sire in North America. His stern visage adorns the cover of *Ethics on the Ark*, a testimony to his iconic stature. It is the success story of Willie B. and Zoo Atlanta that has inspired our continued effort to build on the concept of welfare-oriented exhibits, toward the aspirational goal of wellness-oriented exhibit design.

In March 2016 the Jacksonville Zoo and Gardens hosted an interactive Elephant Wellness Workshop, with a focus on "wellness-inspired design." Nearly an entire day of the four-day workshop was devoted to panels of leading zoo designers discussing what it meant to design for wellness, incorporate technology and innovation, and explore designers' visions for wellness. Designers and architects engaged in active discussion with attendees and were challenged to propose ideas for exhibits that would support welfare-oriented goals (e.g., general animal husbandry needs, appropriate substrate, temperature regulation) as well as wellness-oriented goals (e.g., providing opportunities for problem solving, encouraging pro-social interaction with conspecifics, evoking novel and exploratory behavior, and incorporating choice and control over the environment). The meeting was so successful that one European zoo director decided to subject all future exhibits to a wellness-inspired design process.

THE JACKSONVILLE MODEL

TLM and Georgia Tech professor Jack Marr successfully collaborated on studies of animal behavior and behavior analysis for thirty-five years (Lukas, Marr, and Maple 1998; Bashaw et al. 2003; Bloomsmith, Marr, and Maple 2007; Clay et al. 2009; Martin et al. 2011). Atlanta was one of the few destinations where graduate students could obtain this combination of research training while concentrating on applications of psychology that directly benefited animals in the zoo. Given our success, we were surprised when our departmental colleagues elected not to continue the animal program in Tech's School of Psychology after we retired. Once retired, TLM endeavored to reinvent himself as a consultant to zoos. In 2011 he implemented a wildlife wellness initiative at the San Francisco Zoo, which continues to this day. In 2014, funded by a private donor, TLM organized another wellness program at the Jacksonville Zoo and Gardens. The coauthor of this chapter, Valerie Segura, the zoo's first "applied animal behavior analyst," was hired to manage this program and build on the successes achieved in Atlanta and San Francisco. A board-certified behavior analyst (BCBA) educated and certified in California, Segura is also coordinator of the model program we are building in Jacksonville and beyond. We are attempting to spread the integration of zoo biology and behavior analytic research and practice throughout the state of Florida to address animal welfare and wellness in zoos. There is an urgent concern among world aquariums that have been pressured by citizens and governments to upgrade the welfare of performing animals. Some governments have banned performing marine mammals altogether. Most recently, SeaWorld has announced that it will no longer breed captive orcas, citing a dramatic shift in public support for the program. This is no doubt a result of the firestorm that followed the movie *Blackfish*. Regardless of whether the film portrayed the situation truthfully, those of us that believe in breeding endangered animals need to be vigilant and to be prepared to *demonstrate* that most managed megafauna can live well and thrive in captive settings (for a broader discussion of these issues see Maple 2016).

Zoo Atlanta, revitalized and redesigned, demonstrated to all that empirical zoos with a strong commitment to wellness and welfare can stay ahead of their critics. Despite this example, many zoos still hesitate to hire dedicated scientists. Lacking dedicated scientists on staff, zoos generally welcome collaboration with scientific partners (see Mendelson, this volume, for a larger discussion on problems relating to insufficient investments in scientific conservation efforts for understudied low-profile amphibians). The zoos of the future, zoos that want to protect themselves from animal rights extremists and lobbyists by making genuine investments in the science and practice of animal wellness, will need to change their position or risk losing hard-won public support. Research is the foundation of animal welfare and should be facilitated in every zoo through formal partnerships with scientists in colleges and universities (Maple and Lindburg 2008). We recognize that zoos cannot change immediately, but the pace of change ought to quicken. It was a decade after the publication of *Ethics on the Ark* that the Detroit Zoo convened its important zoo animal welfare conference. A decade after Detroit, the World Association of Zoos and Aquariums (WAZA) (Mellor, Hunt, and Gusset 2015) proposed welfare-based accreditation standards. Further progress will not be made without identifying new objectives for data-driven zoos.

We believe that zoos with the highest standards and best ethical practices will be trusted and supported by their communities. The best of our zoos and aquariums in the AZA (Association of Zoos and Aquariums) will turn this support into ever more creative exhibits that provide superior quality of life for the animals living among us. *Ethics on the Ark* was an important turning point for AZA institutions and should be credited for the emerging consensus that animal welfare is as compelling a priority as conservation. A new animal welfare strategy is driving the collective commitments of world zoos and aquariums (Maple 2015), and this strategy is the result of our shared understanding of the basic biology and behavior of wildlife under our care. The momentum propelling the wellness movement must continue if zoos and aquariums are to remain relevant while continuing to inform and inspire a world of wildlife enthusiasts. Enacting an empirical approach to animal wellness will be costly. It will require upfront investment to reform exhibits and hire dedicated scientists, and none of this can be done without leadership at the executive level in zoos. Despite this, we are compelled to take on the challenge to protect populations in the wild and act to conserve ecosystems that sustain wildlife. At the same time, we must not only prevent suffering among animals, but also use our resources to enable wellness for individual animals. In this way conservation and animal wellness, both in situ and ex situ, are inextricably bound.

Empirical zoos will be more resistant to challenges from severe critics and media that are biased against zoos. We must continue to monitor and evaluate our exhibits and the individual animals within them to ensure that zoo and aquarium animals can thrive while we support the noble cause of wildlife conservation. Zoos themselves cannot simply survive; they must also position themselves to thrive or they may no longer exist. All of these concerns are the highest priorities of a new generation of ethical, empirical arks.

CONCLUSION

1. In the twenty years since publication of *Ethics on the Ark,* accredited zoos in the Association of Zoos and Aquariums are increasingly willing to accept evidence-based standards and practices.
2. Although accredited zoos generally welcome scientific collaboration, they are not well prepared for scientific endeavors that require dedicated personnel, nor are they inclined to pay for scientific services.
3. Ten years elapsed between the publication of *Ethics on the Ark* and the first zoo animal welfare conference hosted by the Detroit Zoo in 2005. Animal welfare was not quickly embraced by individual institutions, although AZA established an Animal Welfare Committee in 2000.
4. Animal welfare standards and practices generally operate at a minimal level to ensure regulatory compliance, though that falls well short of optimal targets. The psychological construct of wellness represents an opportunity to elevate and expand welfare to an optimal, aspirational standard. In this way individual animals are allowed to thrive in their ex situ environments.
5. A detailed strategic document published by the World Association of Zoos and Aquariums in 2015 has elevated an animal welfare strategy to a position of parity with the global zoo conservation strategy.
6. Although today zoos are managed to a much higher standard, public pressure on zoos and aquariums from animal rights and extreme animal welfare organizations has increased throughout the world. Extremists have effectively lobbied to close some zoos and aquatic parks.
7. To continue operating with the support of the general public, accredited zoos and aquariums must ensure that guests are satisfied with the standards, practices, and ethics of zoo exhibits and programs.

CHAPTER NINETEEN

Zoos and Sustainability: Can Zoos Go beyond Ethical Individualism to Protect Resilient Systems?

Bryan G. Norton

Historically zoos evolved from menageries, typically created to amuse the nobility and, one might say, to illustrate the power wealthy persons held over nature (Hancocks 1995; Hoage and Deiss 1996). Early public zoos mainly extended these attitudes more broadly, as animals were usually displayed in cages, not encouraged to act out their natural habits and functions, and presented simply for the entertainment of visitors. The past century or so has seen gradual, if fitful and spotty, progress by zoos and aquariums into an orientation toward conservation and acceptance of a strong responsibility to engage only in best practices in caring for animals in their charge (Barrow, this volume; Henson, this volume; Kisling, this volume).

After summarizing some trends developed in earlier projects (Norton et al. 1995, for example), and after discussing the goals of zoos and their possible contributions to conservation, I describe the historical trajectory of discussions of protecting species as trending toward protection of biological diversity, which became the main thrust of the emerging field of conservation biology. I then briefly develop the opportunities for justifying zoo activities according to three individualistic ethics and question whether zoos can rely on such ethics to support the full range of their actions and policies. Recognizing that these individualistic ethics might be too restricted to justify protecting habitats in situ, I explore possibilities, including basing conservation of ecological systems on our obligation to provide viable habitats for species that will prove useful to future human beings. I end with some remarks and recommendations that will help zoos contribute to the larger pursuit of sustainable development.

1993–95: THE ORIGINAL ETHICS ON THE ARK PROJECT AND SEQUELS

The Ethics on the Ark conference held in Atlanta about twenty years ago (see also Maple and Segura, this volume) explored ethical concerns, and though no agreement was reached on the exact ethical limits to the invasive or painful treatment of individual animals, the conference and subsequent book had at least two salutary consequences. First, building on that dialogue, zoo professionals articulated a consensus that zoos and aquariums have an unquestioned obligation to treat animals in their keeping with the best known care; second, these professionals acknowledged that their own treatment of wild animals in captivity should illustrate, exemplify, and encourage practices to protect the welfare of animals throughout their local communities (Grajal, Luebke, and Kelly, this volume).

The results of the conference and the subsequent book, *Ethics on the Ark* (Norton et al. 1995), provided balanced discussions of all aspects of keeping wild animals in captivity. The book ends with an appendix that summarizes working group discussions and also describes the best attempts of participants to state consensus agreements on the procurement of animals; the reintroduction, training, and welfare of released animals; captive breeding, surplus animals, and population regulation; captive care, maintenance, and welfare; the use of animals in research; and public relations, fund-raising, and disclosure. The reports by working groups stated the main areas of ethical agreement and disagreement and, as far as possible, offered recommendations to guide the work of zoos and aquariums. Even where important disagreements remained, the groups made some recommendations and also posed caveats about areas that remained controversial.

These points of consensus, elaborated and complemented by other conferences and workshops such as a workshop on treatment of wild animals in captivity organized by Andrew Rowan of the Tufts University School of Veterinary Medicine in 1994, have allowed zoos, aquariums, and captive breeding programs to proceed with their work, producing scientific knowledge of species, specimen behaviors, and their reproductive needs while also contributing offspring to augment populations and in some cases to reintroduce species to habitats from which they have been extirpated. Before these meetings and publications, zoo professionals worked toward conservation goals, but they also found it necessary to respond to critics such as Regan (1995) and Loftin (1995) and to activists such as People for the Ethical Treatment of Animals, who argued that zoos' conservation agenda is a mere rationalization for actions that kill or harm some specimens and that zoos' contribution to in situ conservation is insufficient to justify causing pain to individual animals in order to save abstractions such as species and ecosystems (which cannot feel pain). These meetings brought together zoo defenders and zoo critics, and though disagreements among participants were by no means resolved, zoo professionals learned to articulate an ethical framework based on obligations to protect biological diversity by which they could explain and justify their work.

Zoos today are complex institutions with multiple goals and with varied structures and support systems, and though some zoos have been more aggressive in some areas than in others, leaders in the profession of zoo management seek to fulfill as many of these four goals as possible:

1. Entertainment (and the attendant economic goals of creating local tourist activities, encouraging urban development, and creating a sense of community)
2. Zoological research
3. Education
   a. regarding animals' individual health and ethical treatment
   b. regarding the importance of biodiversity and protection of animals in their natural habitats
4. Conservation of biodiversity of species and habitats (see Maple 1995)

Terry Maple and Valerie Segura (this volume) have done a far better job than I could in summarizing the current status and future progress of zoos with respect to points 1, 2, and 3a. I concur with them in acknowledging and lauding significant progress in these areas. Here I'd like to focus on points 3b and 4: zoos' contribution to conservation, especially conservation in the broad sense, regarding the importance of biodiversity protection and in situ conservation efforts. Zoos, based on Maple's evidence and that of others (e.g., Clayton and Le Nguyen, this volume; Grajal, Luebke, and Kelly, this volume), have made considerable progress in contributing to a broad range of important social goals. Nevertheless critics remain adamant, minimizing their progress in protecting the welfare of wild animals in captivity and scoffing at claims of major contributions to conservation.

Before proceeding, I should clarify what I mean by conservation and what we can expect from zoos—in pursuing their four diverse goals—as contributors to the broader societal goal of protecting biological diversity. The term conservation has historically been used to refer to one wing of the original movement led by Gifford Pinchot and John Muir, as Pinchot's idea of wise use of resources came to be labeled conservation and was contrasted with Muir's goal of preservation, which set out to protect some natural areas *from* human interference. More recently, however, the term preservation has been rejected by many contemporary scientists and activists because it became associated with locking up resources and discouraging economic growth.

In the 1970s and 1980s biologists, who became alarmed by threats to species and ecosystems, adopted the label conservation biology for their work, which coincided with a related movement to concentrate efforts on protecting biological diversity. The phrase was shortened to biodiversity, and this new term was taken to refer to the protection of at least three aspects of nature: genetic diversity, species diversity, and cross-habitat diversity. This useful simplification has been criticized on several grounds (Norton 1987), but the triumvirate has nevertheless coalesced as a rough-and-ready characterization of what conservation biologists wish to save.

Given their emphasis on caring for individual specimens, zoos can contribute strongly to the protection of species by providing detailed understanding of natural processes that can guide actions in the wild, and by engaging in careful captive breeding they can help to maintain the genetic integrity of both captive and wild populations by keeping studbooks and paying careful attention to breeding patterns ex situ. So we may conclude that zoos can help protect species—by maintaining and studying specimens of vulnerable species—and can also contribute to better understanding and management of genetic aspects of conservation. However, we might wonder whether zoos, being focused on individual specimens and populations, are well positioned to protect cross-habitat diversity, which is very important to conservation in the long run because it is in these diverse systems that the forces of evolution unfold.

ZOOS AND CONSERVATION

While some of the larger zoos are able to establish strong contributing relationships with in situ efforts to save some of the species they exhibit and breed, including in some cases providing individual animals to augment flagging populations or to establish new breeding colonies in situ, the core of zoos' contribution to conservation efforts must be education. While zoos can augment in situ efforts by caring for and breeding captive animals, their major advantage over other organizations will be in reaching the public: zoo visitors represent a cross section of society and offer many opportunities to educate previously oblivious people and make them supporters of conservation. So zoos and aquariums should maintain a precise focus on public education regarding wildlife and biodiversity.

Of course some critics would argue that more could be done for biodiversity conservation in situ if zoos were closed and the money devoted to their management were applied to on-the-ground efforts to save habitat. Zoo professionals have reasonably responded that since the greatest share of zoo management budgets comes from visits and gate receipts, the amount saved by closing zoos would be minimal because that major source of revenue would disappear. Further, they point out zoos' special educational value in that they reach the general public rather than preaching to the converted.

A major contribution in this direction requires a commitment of resources and staff members, which may seem to compete with support for the maintenance and care of collections, but zoos can magnify their efforts by embracing efforts by groups committed to local conservation projects. For example, the Brookfield Zoo has supported the Chicago Wilderness objectives through the Chicago Zoological Society, and Chicago Wilderness has in turn incubated further groups who contribute to the goals of protecting local and regional biodiversity.

In general, zoos and aquariums have been important elements in the "ecology" of conservation groups and institutions, in that their direct contact with animals—and their ability to intensely study behavior patterns such as reproductive behaviors—can provide essential information that can be used for in situ efforts in the field. Likewise, zoos have led the way in learning how to treat individual animals humanely, and they have discovered and stored important information regarding breeding patterns among captive animals to maximize genetic diversity in captive populations (Tudge 1992; Ryder, this volume). This is essential to avoid genetic swamping by a few inbred animals when members of the captive stock are reintroduced to the wild.

Zoos can also contribute to conservation by educating the public about the importance of healthy habitats and the resilience of systems. Here critics will continue to note that, by emphasizing charismatic animals and by still presenting many animals in cages and small tanks, zoos and aquariums perpetuate a view of animals as mascots rather than as contributors to complex and interactive systems in the wild. This criticism may be inevitable, but zoos can counteract it by constantly emphasizing the importance of habitat in protecting species in situ and in striving to make ex situ housing as natural as possible in function as well as in appearance. So, in addition to contributing through research and management of wild populations, zoos can and should work to situate their contributions regarding the treatment of individual animals within a larger network of groups that, on the whole, work to protect natural systems that provide the natural habitat within which particular species struggle to survive and maintain healthy populations (see Ivanyi and Colodner, this volume)

More generally, zoos have many opportunities, by partnering with school systems and other organizations, to acquaint visitors and interns with wild and exotic species as well as denizens of local habitats, and thus to inspire more interest in animals, animal welfare, and conservation. A committed zoo or aquarium can be the focal point for a whole community's involvement in ongoing conservation efforts. By educating children and others to respect and care for animals, by exhibiting animals that are important in local ecosystems and, especially, species that are endangered in the wild, education efforts of zoos and aquariums can create focal points for communitywide efforts to improve local environments. For example, the Arizona-Sonora Desert Museum in Tucson has adopted the role of exhibiting local species and illustrating how those species are organized in local habitats (Ivanyi and Colodner, this volume). In this way the institution educates the public not just about specific charismatic animals, but also about how local species survive and about threats to their survival (Hancocks 1995).

INDIVIDUALISTIC ETHICS

Zoos are, and should be, specimen-centric in much of their work: they exhibit individuals, they care for individuals, and they sometimes introduce or reintroduce individual specimens; in doing so, researchers and activists have developed a considerable ethic for welfare of animals in their care.[1] Looking at this progress from the viewpoint of a philosopher, one might say that zoos have made great ethical strides in *individual animal care*. These improvements are supported by extensions of traditional ethics, which have generally involved ethical concerns regarding human individuals. Traditional human ethics come in three varieties, all referencing obligations based on effects on individuals:

- Deontology/rights theory, which attributes important individual rights to all human individuals by virtue of their humanity
- Utilitarianism, which advocates maximizing the aggregate happiness of the human individuals affected by a decision

- Virtue ethics, which encourages individuals to live a life respectful of others, based on an obligation to care. Originally, virtue ethics mainly concerned relations with humans and human institutions.

Accordingly, advocates of stronger protections for animals come in three varieties, corresponding to extensions of the three anthropocentric ethical frameworks:

- Animal rights: rights are based on a moral obligation that can override human demands by treating all subjects of a life as having rights that impose obligations on anyone working with or using them (Regan 1983)
- Animal welfare: following utilitarian moral theory, benefits and harms to individual animals should count in any decision affecting the well-being of humans and sentient animals (Singer 1975)
- Animal care ethics: virtuous humans should protect the welfare of sentient animals as an expression of a caring outlook on the world.

Each of these extensions of human ethics incorporates their individualistic assumptions. The individualism of traditional ethics is summarized in the "person-regarding principle" (PRP), which states that any moral harm must be a harm to an individual person (Parfit 1984). Extending ethics that embody the PRP to animals simply substitutes "sentient being" for "person," retaining the individualistic scope of moral obligations.[2] The important progress described by Maple and Segura (this volume) and embraced by most zoological institutions can comfortably be supported on one or another of the three extensions of human ethics. So far as zoos and aquariums can rest their activities on one of these principles, and so far as they and their supporters can articulate the extensions of human ethical ideas to animals, the institutions are to be lauded for bringing their practices under rationally justifiable principles and parameters.

There remains, however, a significant worry, since zoos have the goal of serving the conservation movement more broadly: Can stronger and stronger commitments to an ethic of animal welfare and care support a more holistic ethic—that is, an ethic emphasizing the care of ecosystems and habitats, not just individuals? Many publications have noted how different the ethical principles that would justify ecosystems and populations must be from principles guiding treatment of specimens (Sagoff 1984; Callicott 1989, chap. 1), and much of the literature of environmental ethics deals in one way or another with conflicts between individuals—human and nonhuman—and ecosystems. As I noted above, successful conservation activity in situ must protect habitats and ecosystems as well as individuals and populations, but adapting individual ethics to support activities that protect ecosystems—which seems to be demanded by the very challenge of in situ conservation—remains controversial. While conservationists wholeheartedly embrace a systems theory approach to understanding habitats, ethical thought has struggled to develop an ethic adequate to the challenge. Fortunately, obligations to protect ecosystems can rest either on a direct ethical commitment to protect ecosystems (Callicott 1989) or on the practical fact that to save animals and plants one must save their habitats (Norton 1987). So the need to protect species in situ can be justified either on an ethic for individual specimens (in which case the obligation to protect habitats is based on practical necessity) or on a holistic ethic that attributes moral standing to systems as well as individuals.

This question raises very complex and difficult moral issues: Will a purely individualistic ethic (whether utilitarian, deontological, or virtue theoretical) support the broader obligations of humans—and zoo professionals—to protect natural and functioning human/natural systems? I will return to this topic below.

A SUSTAINABILITY ETHIC?

At this point my discussion intersects with that of Cerezo and Kapsar (this volume), who argue that zoo contributions should be conceptualized and integrated into a broader program of sustainable development. Since the promulgation of the idea of sustainable development in *Our Common Future*, the report of the World Council on Environment and Development (United Nations 1987), known as the Brundtland report, greater and greater emphasis has been directed toward sustainability and policies that support it. I agree with the recommendations Cerezo and Kapsar provide (this volume) and will spend the rest of this chapter drawing out some consequences, for zoos and more generally, of framing the global conservation problem as one of sustainable development.

Sustainability can be thought of as one aspect of the broader idea of sustainable development; the Brundtland report was mainly directed at the goal of increasing resource availability and income necessary to address poverty, especially in third world nations. Sustainability adds a commitment to sustaining resources and ecological systems so as to support improved lives for all future humans. The Brundtland report emphasized addressing poverty; subsequent discussions have responded by emphasizing that constraints on current resource use may be necessary to support the ongoing availability of resources necessary for future development.

There is a huge literature on sustainable development and the concepts and definitions associated with it, and within this literature there is at least one glaring schism: the distinction between "weak" and "strong" sustainability. This key distinction turns on whether ethical obligations can only be to individuals (ethical atomism) or whether we accept obligations to protect ecological systems (some kind of communitarian ethical approach).

- Weak sustainability requires only that each generation maintain economic wealth (capital) so that subsequent generations of individuals have the chance to enjoy as much welfare as previous generations, since capital will support technological advance in the use and protection of resources.
- Strong sustainability advocates, on the other hand, argue that true sustainability must (in addition to protecting the welfare of future generations) require the protection of systems—more specifically, the *resilience* of functioning natural and natural/human systems.

Advocates of resilience thinking—who sometimes refer to themselves as members of the "Resilience Alliance"—define resilience as "the capacity of a system to absorb disturbance and still retain its basic function and structure" (Walker and Salt 2006, xiii). Resilience has become an important qualifier of sustainable development in that its advocates have convincingly argued that sustainability cannot be achieved unless the current generation can protect the resilience of ecological systems.[3] Since natural systems and mixed human and natural systems are constantly changing, resilience is required if humans are to adapt and live within natural systems that can equilibrate after disturbances (whether humanly or naturally caused).

Resilience is related to diversity in two senses. "Functional diversity" depends on the existence in a system of different functional groups of organisms with similar functions. "Response diversity," on the other hand, exists in a system when each element of a functional group (those that perform the same function) responds in different ways to disturbances and opportunities. Walker and Salt (2006, 69) explain the connection of diversity with resilience:

> When it comes to resilience, what's important is that the different organisms that form part of the same functional group each have different responses to

> disturbances. . . . In the case of the algae-eating species, resilience is enhanced if the different species, all of which are providing the same basic service of controlling algae, each respond in different ways to changes in temperature, pollution, and disease.

Understood in this way, the resilience of a system depends both on the existence of a diverse set of species and on their structures and their interrelations. Diversity and resilience both decline when species are lost and when disturbances interrupt the functioning of complex systems.

ZOOS, CONSERVATION, AND SUSTAINABILITY

Zoos face a significant challenge in contributing to the protection of biodiversity—a central goal of sustainable development—including the diverse habitats their specimens hail from, because long-term protection of fragile species almost always requires attention to the ecological system that supports those species in situ. Sustainability necessarily includes concern for systems as well as for populations and individuals, and the project of educating visitors and whole communities can hardly be considered an unqualified success unless that education includes a modicum of ecological literacy. Species will be saved for the long run only if public opinion and conservation management will support efforts to maintain functioning ecological systems.

Today the ecological systems involved are almost always hybrids of natural and human systems in the sense that they are partially determined by ecological forces and partially driven by human choices and activities. Historically, ecologists have tended to classify ecosystems as "pristine" or "natural" as opposed to "degraded," and they have focused on the former, taking human disturbance as decreasing systems' authenticity, and have directed their research toward learning about pristine systems as the only truly ecological knowledge. Of course this stance has meant that in many situations they have had to avoid or ignore at least some human effects.

More recently, restoration ecologists have moved away from this dichotomy and the hypocrisy it required by placing ecosystems on a continuum stretching from systems exhibiting little human influence to those heavily managed for agriculture and human settlement (Hobbs et al. 2014). Operationally, this shift has encouraged three strategies. First, where systems have been little altered and maintain their basic, historical structure and functions, the goal is often to maintain and restore the system as far as possible to its historical status. Second, at the other end of the spectrum are systems that are intentionally and aggressively managed for the fulfillment of human needs. These systems, such as agricultural and residential systems, are usually managed to perpetuate their benefits to humans (often referred to as "ecosystem services"), with the goal of protecting their healthy functioning. The third category, systems that have been heavily altered by human activities but are not intentionally managed for particular human goals at this time, have been called "novel ecosystems" (Hobbs et al. 2014). They have changed, as a result of historical or indirect human forces, into systems that once did not occur there (or perhaps anywhere else) but are developing on a trajectory governed neither by historically natural forces nor by human plans.

In these novel situations, managers have considerable freedom to consider a range of possible strategies. For example, if a species of frogs is threatened by a pathogen that endangers its entire population over its entire range, managers might propose relocating a breeding population to a novel ecosystem if the conditions for its continued survival are present there. This is not to say it would be wrong to use historical parameters to determine management of a novel system—the goal could be to move that system back as far as possible toward a historical norm. Rather, it means only that in some cases broader conservation goals as applied to a novel ecosystems may allow fruitful experiments and new opportunities for learning.

But the important point behind these considerations is that systems, regardless of how far they function independently of humans, *are systems*! And zoos generally exhibit and manage populations of individuals. Accepting that zoo conservation efforts should be integrated into the broader ethical framework of sustainable development, I argue that we need to embrace *strong sustainability* and to engage with other organizations and individuals to protect the *resilience* of functioning systems. Can zoos and aquariums, given their emphasis on individual specimens, bring their important strengths to bear on habitat loss and the ongoing degradation of ecological systems (and their resilience) in the face of human development?

We can summarize the argument and its holistic conclusion as requiring that sustainable development include obligations to protect resilient ecosystems as well as the well-being of present and future people. Perhaps we can now see why the future of zoo conservation presents such a difficult challenge. Most of the important successes in recent conservation efforts by zoos can be justified on an ethic of individual animal welfare. Scientific work on reproduction, the creation of assurance populations, and producing individual animals to augment flagging populations in situ all focus on individuals or populations of individual members of a species. In making management decisions and in developing a conservation message for visitors, zoos can be comfortable with an individualistic ethic based in obligations to avoid pain to animals, and this ethic can support and govern much that they do. Referring back to the three individualistic extensions of human ethics noted above, it seems that none of those extended ethics—so useful in supporting zoos' work with individual animals—will straightforwardly support a full-blown ethic of sustainable development.

Some environmental ethicists, such as Callicott (1989, 1999), have deployed similar holistic arguments to support the conclusion that we must attribute intrinsic values to ecosystems themselves, transcending both individualism and anthropocentrism by insisting that obligations to protect ecological systems can compete with and sometimes override obligations to either human or nonhuman individuals.

In my own work I have found these attempts to articulate and protect environmental values by embracing holistic nonanthropocentrism unnecessary, because a human-based ethic can fulfill the same function. Anthropocentrists can support the argument that sustainable development requires protection of resilient ecological systems by positing that humans currently living have an obligation to human communities: in addition to not impoverishing future people (weak sustainability), they must maintain ecologically resilient ecosystems. Following the British conservative philosopher Edmund Burke, who defines a society as including past, present, and future individuals, we can recognize that there is a public good beyond the aggregation of individual interests of those currently living. Our obligation to engage in sustainable development can thus be understood as a duty to protect the public good of an ongoing society or community (Norton 2015).

Thus, while Callicott offers an ethic that is apparently sufficient to support a holistic interpretation of sustainable development efforts by citing direct obligations to ecological systems, this nonanthropocentric and holistic move—which is anathema to many ecologists who object to the reification of ecological systems—is not necessary in order to do so. One can, as a less radical rejection of individualism, support an ethic protective of ecological systems that will be necessary for future generations to survive and prosper by attributing these obligations to an ongoing, multigenerational community of human individuals. On this anthropocentric, communitarian view, each generation will have obligations to protect resilient ecological systems (and strongly sustainable policies). These obligations can be understood as protecting ecosystems indirectly as a result of obligations to the multigenerational community in which one lives.

RECOMMENDATIONS

If I'm right that true sustainability must be strong sustainability, zoos must somehow transcend their laudable and comfortable ethical commitment to individual animals. The challenge is, How can zoos contribute to the broader goals of strong sustainability, given that their ethical touchstone extends only to animals of sentient species and yet those broader goals apparently require a more holistic ethic? More specifically,

- Can zoos expand their scientific work to provide ammunition for the protection of functioning, resilient systems?
- Can zoos develop education programs that will encourage learning about systems and the importance of resilience of natural and human/natural systems?

I have referred to these conceptual issues as a *challenge* to express my belief that zoos and aquariums can, in fact, rise to that challenge and be leaders in conservation. I believe, more specifically, that they can do so if they pursue the following understandings and recommendations—many of them explained and advocated in other chapters in this volume.

1. Emphasize education programs, especially programs that connect animals to their in situ habitats. Educational programs should complement the actual viewing of animals with material that encourages more holistic thinking in signage, brochures, and special programs.
2. While a case can be made for having both "global" zoos and "regional" zoos, I think this dichotomy should not be seen as a sharp one. Every global zoo could have regional features and emphasize local animals and the habitats they depend on, helping visitors understand the importance of habitat in their own backyards.
3. Join with conservation organizations that are more "holistic" in scope, developing ties to local, regional, and national organizations in order to integrate conservation science and conservation messages as they affect local management questions.

CONCLUSION

Zoos have made important progress in developing scientific endeavors, in improving conditions for zoo animals, and in improving their conservation message by displaying animals in more naturalistic settings. Most of these activities can be justified by a robust ethic of individual animal welfare. The challenge that remains is for zoos to expand their message about saving animals from extinction to a message that encompasses the broader goals of sustainable development. In doing so, zoos need to incorporate the ongoing science of system resilience into their messages and into their activities.

Insofar as zoos contribute to the broader conservation efforts to create sustainable systems and all layers of biological diversity (genes, populations, and ecosystems), it may turn out that this contribution will require them to shift emphasis somewhat from charismatic species to visitor education and perhaps stress local species and the natural habitats they depend on. An important determinant of the future contribution of zoos may well be the quality of their educational message; it remains to be seen whether zoos can maintain their broad educational and conservation message against community pressures to be drivers of local economic development.

PART V

The Science and Challenge of the Conservation Ark

CHAPTER TWENTY

Opportunities and Challenges for Conserving Small Populations: An Emerging Role for Zoos in Genetic Rescue

Oliver A. Ryder

The surest and most robust way to save species from extinction is maintaining functional ecosystems that provide the environment where their evolution has occurred and that intrinsically has the resources and interactions that typically afford population sustainability. Of course the number of individuals of a species is a critical factor in its survival. Systems of species risk assessment focus on direct or surrogate measures of effective population size or its rate of change that estimate the propensity for loss of genetic diversity. It is generally appreciated that low genetic diversity and low population size are associated with risk of extinction.

Zoos were not founded and developed envisioning the influence they may now exert in preventing species extinction. Similarly, many technological developments in biomedicine and animal science were not initiated with thoughts of preserving genetic diversity and contributing to sustainability of exotic species threatened with extinction. Genetic techniques such as whole genome sequencing and genetic engineering, and alliances of these technologies with advances in cell and developmental biology, especially stem cell biology, now portend a new form of husbandry that might enrich and sustain at least some of the extant species resulting from the evolutionary processes that produced their diversity.

Emerging possibilities for genetic rescue and recovery are the focus of this chapter. The main topic is extending and expanding ex situ programs for species survival and the potential use of advanced genetic and reproductive technologies, which are also discussed in chapters by Friese and Tubbs in this volume. I recognize that this topic may at the moment be beyond the comfortable grasp of many both within and outside the conservation biology community. Having noted this, I will discuss how well-considered scientific studies and applications of newly emerging technologies can be applied to aid species survival. "Science can help" is a hopeful and motivating message.

With the emergence of the era of genomics and the continuing developments in assisted reproductive technologies, we are poised to exploit biobanks and other new ex situ genetic resources to assist with managing and maintaining genetic diversity, contributing to population sustainability, and reducing extinction risk for ex situ and in situ populations of endangered species (Wisely et al. 2015; Saragusty et al. 2016).

At this moment these potentialities are largely unrealized, perhaps in part for lack of effort: no one has yet tried a particular procedure or intervention with a particular species. In many instances recognized and as yet unrecognized technical hurdles must be overcome. There is great opportunity for discussion and evaluation of both possibilities and hurdles. It seems, though, that there is intention to explore them through experimental or pilot efforts to better inform us of the possibilities for preventing extinctions when all the default options have been insufficient. The prospect of seeing many species move into a state of doom is a certainty based on current trends. What is to be our reaction?

CHANGING VIEWS OF EXTINCTION

I suggest that the nominative view of extinction—that extinction takes place when the last individual of a species dies, as when Martha, the last passenger pigeon, died (Barrow, this volume)—is impractical in somewhat the same way as defining the position of an electron around its nucleus. In both cases it is more meaningful to consider the probability distribution. We know that the probability of extinction increases as population numbers decline, as genetic drift overwhelms selection, as homozygosity decreases homeostasis, and as stochastic events contribute to what has been called the extinction vortex (Gilpin and Soulé 1986). Enormous work has been undertaken to develop, define, and use categories of risk; hence the IUCN Red List and the criteria that guide listing (Mace et al. 2008). Extinction in a functional if not a literal sense can take place when a population fails to have reproductive potential. But it is possible to reverse the status of a species from having critically limited reproduction potential (or none) with planning, cryobanking, and the application of advanced genetic and reproductive technologies.

Possibilities for intervention are broadening, creating both hope and concern. But just as captive breeding, population augmentation, and reintroduction have all faced scrutiny, sometimes harsh (Mendelson, this volume), it is appropriate to discuss and explore possibilities.

ZOOS HAVE HELPED

Perhaps an old story now, but still a true one, is the rescue of species from extinction by harboring a reproductive population in a managed (captive) environment. The Przewalski's horse, Arabian oryx, black-footed ferret, California condor, species of Galápagos tortoises, the 'alala (Hawaiian crow), and some poison arrow frogs, are notable examples (see Rothfels, this volume; Allard and Wells, this volume). The list is likely to grow as government agencies charged with protecting wildlife decide to bring more endangered species into captivity as a crucial part of efforts to prevent extinction. The Pacific pocket mouse, mountain yellow-legged frog, saiga and hirola antelopes, and Sumatran rhinoceros are all examples.

Zoos have been leaders in developing technologies for species management and recovery in the fields of behavioral biology, genetics, reproductive physiology, endocrinology, health and disease management, nutrition, and biomaterials banking. Some fundamental and case studies, particularly noting the work of my colleagues in San Diego, are presented in table 20.1.

TABLE 20.1. Disciplinary examples of in situ and ex situ conservation enhancements involving zoos (especially San Diego)

Conservation science discipline	*Topic*	*Literature example*
Behavior	Giant panda	Wei et al. 2015
Genetics	Population management	Romanov et al. 2009; Ralls et al. 2000; Hedrick et al. 2016
Reproductive physiology	Gamete banking Pregnancy monitoring	Liebo and Songassen 2002; Durrant 2008 Willis et al., 2011
Health and disease	Disease diagnosis, avian malaria, avian pox epidemiology	Braun et al. 2014; Witte et al. 2010; Schrenzel et al. 2011
Endocrinology	Diet optimization based on phytoestrogen content	Tubbs et al. 2014
Translocation	Optimizing success for heteromyid translocations	Shier and Swaisgood 2012; Linklater et al. 2012
Reintroduction	California condor lead toxicity	Rideout et al. 2012

Zoos have helped to develop and improve the technology of reintroduction (Berger-Tal et al. 2015), accrue accurate pedigree records (Field et al. 1998; Bowling et al. 2003), and develop methods for population management (Lacy et al. 1995), including reproductive management (Luskutoff 2003) and sustainability (Monfort and Christen, this volume; Mendelson, this volume). Many evolutionary, population, and transmission genetic studies have been contributed by zoo investigators or developed using samples readily obtainable only from zoos have laid the groundwork for the emerging field of conservation genomics (Ryder 2005; Allendorf et al. 2010; Steiner et al. 2013).

Zoos are places where visitors and zoo professionals alike can learn about animals. For the latter, increasing the understanding of diverse aspects of the biology of species in managed care is crucial to achieving conservation objectives as zoos are ever more likely to increase their role as refugia (Conde et al. 2015). Wildlife regulatory agencies in many countries have sought to establish ex situ populations to meet conservation objectives, including population augmentation, species survival, and reintroduction. San Diego Zoo Global has participated in forty-three such efforts to date (table 20.2).

SAN DIEGO ZOO'S FROZEN ZOO®

The San Diego Zoo, founded in 1916 as the Zoological Society of San Diego, a nonprofit corporation, and now doing business as San Diego Zoo Global, initiated the Frozen Zoo in 1975. We now believe it is the largest, most diverse, best characterized and most utilized collection of its kind. It consists of frozen cell cultures from nearly 10,000 individual vertebrates, including DNA extracts, frozen tissues, and sperm and oocytes: nearly 100,000 total samples representing more than 1,000 vertebrate taxa. Over 180 contributors of samples include zoos, nature centers, oceanariums, field biologists, and wildlife agencies. Samples have been provided to hundreds of investigators across the United States and in foreign countries under noncommercial transfer agreements.

Cryopreserved cells from the San Diego Zoo's Frozen Zoo have been used over many decades for a wide variety of basic and applied studies. Studies on aging, chromosomal evolution, and gene mapping constituted some of the earliest work, undertaken in the 1970s and 1980s. More recently, the biomaterials collections of the Frozen Zoo have been provided for more than fifty whole genome sequencing studies.

TABLE 20.2. San Diego Zoo Global's involvement* in reintroduction programs (FORTY-FOUR SPECIES)

Amphibians (1 species)	
Mountain yellow-legged frog	*Lithobates muscosa*
Reptiles (8 species)	
Anegada iguana	*Cyclura pinguis*
Turks and Caicos iguana	*Cyclura carinata*
Jamaican iguana	*Cyclura collei*
Grand Cayman blue iguana	*Cyclura lewisi*
Desert tortoise	*Gopherus agassizii*
Western pond turtle	*Emys marmorata*
Red-crowned roof turtle	*Batagur kachuga*
Palawan forest turtle	*Siebenrockiella leytensis*
Birds (19 species/2,313 individuals)	
Nene	*Branta sandvicensis*
Guam rail	*Gallirallus owstoni*
Light-footed clapper rail	*Rallus longirostris levipes*
White-bellied heron	*Ardea insignis*
Harpy eagle	*Harpia harpyja*
California condor	*Gymnogyps californianus*
Andean condor	*Vultur gryphus*
Western burrowing owl	*Athene cunicularia*
Southern ground hornbill	*Bucorvus leadbeater*
Rimitara lorikeet	*Vini kuhlii*
Ultramarine lorikeet	*Vini ultramarine*
San Clemente loggerhead shrike	*Lanius ludovicianus mearnsi*
'Alala	*Corvus hawaiiensis*
Puaiohi	*Myadestes palmeri*
'Akikiki	*Oreomystis bairdi*
Palila	*Loxioides bailleui*
Maui parrotbill	*Pseudonestor xanthophrys*
'Akeke'e	*Loxops caeruleirostris*
Mangrove finch	*Camarhynchus heliobates*
Mammals (16 species)	
Tasmanian devil	*Sarcophilus harrisii*
Giant panda	*Aleuropoda melanoleuca*
Stephen's kangaroo rat	*Dipodomys stephensi*

(countinued)

TABLE 20.2. (*continued*)

San Bernardino kangaroo rat	*Dipodomys merriami parvus*
Pacific pocket mouse	*Perognathus longimembris pacificus*
Los Angeles pocket mouse	*Perognathus longimembris brevinasus*
Przewalski's horse	*Equus przewalskii*
Black rhinoceros	*Diceros bicornis*
Greater one-horned rhinoceros	*Rhinoceros unicornis*
Peninsular pronghorn	*Antelocapra americana*
Saiga	*Saiga tatarica*
Giant Chacoan peccary	*Catagonus wagneri*
Tule elk	*Cervus canadensis nannodes*
Mountain bongo	*Tragulaphus eurycerus isaaci*
Scimitar-horned oryx	*Oryx dammah*
Addax	*Addax nasomaculatus*

* contributed animals and/or expertise

With advances in genetic technologies and in cell and reproductive biology, two endangered species have been cloned, both relatives of cattle: the Indian gaur, *Bos gaurus*, and the Javan banteng, *Bos javanicus*. One goal of the banteng cloning effort was testing if it would be possible to restore genetic variation lost from the genetically small North American zoo population by cloning the cells of a bull with particular genetic value. The male selected had the lowest mean kinship with the population of female bantengs of any living bull in the managed breeding program. Were this experiment to be repeated today and a reproductively capable banteng bull to be produced from the same cell culture that produced the previous clones, I believe such an individual would still be the most genetically valuable bull for breeding in the North American population. I suspect such an effort has not been reinitiated for a variety of reasons, one being the cost of population management for multiple species of bovids.

Cells from the Frozen Zoo provided the first examples of the feasibility of generating induced pluripotent stem cells (iPSC) from endangered species (Ben-Nun et al. 2011; an additional example is described in Ramaswamy et al. 2015). The potential use of the viable cell culture collections of the Frozen Zoo remains to be systematically explored, but it could have a profound impact on genetic rescue and population sustainability for targeted species such as the black-footed ferret and northern white rhinoceros, discussed below.

These same technological developments engender additional possibilities. They may enable preserving forms of life as banked cells for centuries, millennia, or even geological epochs. This is a novel prospect considering the history of life on Earth.

CONSERVATION GENOMICS

The continuing development of whole genome sequencing, alignment, and assembly concomitant with precipitous reductions in the cost of generating whole genome sequence data has launched a new era in the history of biology (Haussler et al. 2009). As genomic studies become incorporated into all arenas of biological inquiry, new opportunities are opening for conservation assessment, monitoring, and management. Genome sequence data from the San Diego Zoo's collections are available for over 120 species in GenBank entries.

Cryopreserved cell cultures also provided DNA extracts for the first comparison of genomewide genetic diversity, evolutionary comparisons at the whole genome sequence level, and genomewide assessment of genetic diversity of eastern gorillas, *Gorilla graueri*, from a sample of a wild-born male eastern lowland gorilla, *Gorilla graueri graueri*, M'Kubwa, who lived for many years at the Houston Zoo. The availability of M'Kubwa's genome for analysis has led to new insights into the genetic diversification of gorillas and has allowed evaluation of aspects of their genetic makeup that provide new insights into risks associated with the small population size of fragmented gorilla populations, including the mountain gorillas of the Virunga volcanoes and Bwindi Impenetrable Forest National Park. These new assessments utilize inferences of genetic load as presumptive deleterious genes that have implications for long-term population sustainability and managed gene flow (Prado-Martinez et al. 2013; McManus et al. 2015).

If conservation biology–based interventions for sustainable populations are to look beyond crisis-oriented recovery actions, developing a more informed view of long-term population sustainability will be essential. And it is likely that studies and assessments supporting such efforts will benefit from access to specimens that have been banked before populations reach critically low numbers (Ryder 2015). Museum specimens are crucial in this regard. However, since they typically yield low-quality DNA because collection, processing, and storage conditions contribute to DNA degradation, they are encumbered by significantly greater cost and effort and ultimately cannot be readily assembled by themselves without a reference genome.

The advantages of establishing, banking, maintaining, and utilizing viable cell cultures for a wide variety of purposes, including direct benefits to human and animal health, are already becoming apparent. It also seems likely that novel insights from the discoveries that will accompany the expanded view of organismal development, natural history, and evolution anticipated from comparative genomic studies will advance projects like Genome 10K (Haussler et al. 2009)

Accordingly, it becomes ever clearer that efforts to describe and conserve biodiversity will benefit from the immediate expansion of efforts to establish and bank viable cell cultures as a resource for current and future use. Efforts currently under way do not adequately emphasize the importance of banking viable somatic cell cultures. Because zoos have had special access, the progress in this important area is largely due to their efforts.

GLOBAL WILDLIFE BIOBANK: AN ACTIVE EFFORT FOR ESTABLISHING A COMPREHENSIVE NETWORK OF FACILITIES LIKE FROZEN ZOO

The wider conservation community's access to high-quality samples for a variety of population genetic, evolutionary genetic, and functional reproductive studies will benefit from expanded efforts to bank cells and follow the precedent of the Frozen Zoo. However, issues of national sovereignty, tenets of the Convention on Biological Diversity, and fears of bioprospecting have limited international exchanges of samples that would advance conservation-related studies and applications.

BUILDING CAPACITY FOR VIABLE CELL CULTURE BIOBANKING

In 2013, 2014, and 2016 the San Diego Zoo Institute for Conservation Research held Frozen Zoo Cell Culture seminars at the Institute's Beckman Center. Participants from Australia, Germany, Ecuador, Colombia, Mozambique, Argentina, South Africa, India, Italy, Puerto Rico, Denmark, and the United States spent a week engaging in hands-on laboratory activities and lectures from experts in the field of somatic cell culture, stem cell technology, and conservation genetics, gaining exposure, experience, and background that they could use to begin efforts like the Frozen Zoo in other countries.

Viable cell culture collections represent a regenerative source of biodiversity samples. Collecting samples and then establishing and freezing viable cell cultures is frequently feasible during conservation assessment and management, for example, when animals are captured for health assessments, tracking technology is deployed, or necropsy examinations and postmortem studies are made. Cell cultures can be thawed, propagated, and refrozen. These undertakings offer to significantly broaden the availability of resources for species conservation efforts if they can be expanded through a global network for global benefits. There is a clear need for the capacity building that may be accomplished by these continuing Frozen Zoo Cell Culture seminars. The strategic importance for conservation of biodiversity and sustainability of small populations underlines the importance of integrating viable cell culture banking into ex situ conservation activities and establishing an expanding network of practitioners.

GENETIC RESCUE

Genetic rescue is a recognized approach to species conservation (Whitely et al. 2015; Wisely et al. 2015). Genetic rescue is generally considered to be the result of undertakings that increase population fitness by infusing new alleles into a population, whether these alleles are restored (because we assume they were previously present) or possibly novel (because we cannot be sure whether they were previously present). The acuity of the distinction is a sign of the times: as we have entered the era of genomics, it is possible to identify and characterize genetic variation as never before.

Potential for Genetic Enrichment—Even Genetic Rescue—of the Black-Footed Ferret

The black-footed ferret, *Mustela nigripes*, is an iconic example of a species once on the brink of extinction, thought to have become extinct, then extant again as a remnant population was discovered. Immediately the quandary emerged regarding conservation management of the last known surviving population (Miller, Reading, and Forrest 1996; Jachowski 2014; see also Allard and Wells, this volume), and fortunately some wild black-footed ferret individuals were brought into captive management before the wild population was lost, apparently succumbing to canine distemper. The entire surviving population traces its ancestry to seven wild-caught individuals, and after more than thirty generations of careful genetic management, the levels of inbreeding have unavoidably risen to levels commensurate with half-sib mating and genetic variation has been lost through genetic drift

(Garelle, Marinari, and Lynch 2013). Fortuitously, when wild black-footed ferrets were being captured to initiate a captive population, State of Wyoming wildlife authorities sent skin biopsy specimens to the Frozen Zoo. Cells from two individuals were successfully cultured and frozen (Wisely et al. 2015). These cell cultures represent a source of genetic rescue if they harbor genetic variation that may be lost from the living population and if the frozen cells can be conduits for infusing genetic variation into the extant population. Wisely et al. (2015) suggested that "conservation cloning" of the cells in the Frozen Zoo might accomplish such a purpose and that whole genome sequencing studies can help evaluate the potential for the banked cells to enrich genetic variation in black-footed ferrets.

In an effort to develop a more empirical understanding of the genetic variation of individuals sampled over generations from a critically small population, and the potential impact of assisted reproductive technologies in genetic rescue, whole genome sequencing has been conducted on cell cultures from two black-footed ferrets. These came from the last wild population and two living individuals descended from the founder individuals that were also from the last wild population. One of the living individuals studied was produced by artificial insemination using frozen semen from a male of the first captive generation. Accordingly, the sire and dam of this black-footed ferret were from the earliest and most recent generations in the pedigree of the managed population. This study suggested that the potential for restoring genomewide heterozygosity—which could be considered a form of genetic rescue—might be accomplished by cryogenic preservation of sperm or cultured somatic cells.

The availability of banked viable cell cultures, such as those held in collections like the Frozen Zoo and other facilities, may ultimately provide resources for genetic rescue of species for which samples have been collected. This may be the only means of preventing the extinction of a species.

Genetic Rescue of the Northern White Rhinoceros

A dramatic example of this potential is represented by the plight of the northern white rhinoceros. Two forms of white rhinoceros survived into the twentieth century. Famously, the southern white rhinoceros recovered from a genetic bottleneck that left only one population of this widely distributed rhino in the Hluhluwe-Imflozi Reserve in South Africa, where numbers fell to an estimated thirty to one hundred individuals. In a remarkable story of species recovery involving the development of specialized capture and translocation technology (Player 1973), by 2013 there were more than twenty thousand living individuals encompassing numerous populations.

The northern white rhinoceros, once more numerous than the southern form (on the latter, see Tubbs, this volume), was also known as the Nile rhinoceros and occurred in Southern Sudan, Uganda, Democratic Republic of Congo and parts of Chad. The last known wild population was found in Garamba National Park, a protected area and World Heritage Site. The species is considered extinct in the wild; only three living individuals remain and the likelihood of their reproducing is remote. The two remaining females could contribute ova for in vitro fertilization, and semen is preserved from no more than five males. One of the males is the father of one of the females and the grandfather of another, raising seemingly impossible obstacles for preventing extinction of this form of rhinoceros. However, since 1978, twelve northern white rhinos have had viable cell cultures banked in the Frozen Zoo; eight of them are unrelated. Ben-Nun et al. (2011) published the first example of successfully obtaining induced pluripotent stem cells (iPSCs) for endangered species, including the northern white rhinoceros. In the laboratory mouse, the model most used for stem cell biology, iPSCs produced from skin cells have ben reported to produce functional sperm (Hayashi et al. 2011) and eggs (Hayashi and Saitou 2013), demonstrating pluripotency for iPSCs and documenting in one species the possibility of using iPSCs for genetic rescue.

The only currently imaginable means of preventing the extinction of the northern white rhinoceros will involve stem cell resources such as those banked in the Frozen Zoo. The challenges are daunting. But striving to seize the hope, address the numerous challenges, and try to save the northern white rhino from otherwise certain extinction has become a commitment of an international team of investigators (Saragusty et al. 2016).

The northern white rhino and the black-footed ferret are dramatic examples introducing the concept of using stem cell technology and other assisted reproductive technologies. They demonstrate that expanding the systematic collection and banking of viable cells from wildlife species can help conserve biological diversity and thus reduce extinction risk and contribute to long-term population sustainability. There can be little doubt that expanded collection and banking of viable cells will benefit future conservation efforts.

CONCLUSION

In an era of declining biodiversity and expanding capabilities in genome biology and genetic editing biotechnology, a new relationship with nature stands to emerge. Zoos now stand at the interface of attempts to educate people and thus encourage them to conserve wild species and their habitats and of efforts to mitigate the failures of attempts at conservation in situ. Their success in gaining public support has allowed zoos to engage in this dual role and simultaneously advance efforts for in situ and ex situ conservation. The unifying theme of small population vulnerability has focused attention on the potential of banking viable cells, assisted reproductive technologies, and stem cell biology for genetic rescue as a method of managing extinction risk.

Zoos were not founded and developed envisioning the role they may now play in preventing species extinction. Similarly, many technological developments in biomedicine and animal science were not initiated with the purpose of preserving genetic diversity and contributing to the sustainability of exotic species threatened with extinction. Genetic techniques such as whole genome sequencing and genetic engineering and alliances of these technologies with advances in cell and developmental biology, especially stem cell biology, now portend a new form of husbandry that may enrich and sustain at least some of the species that are products of evolutionary processes resulting in the diversity of extant species. The nascent efforts of zoos in recent time may emerge as fostering new opportunities and roles for humankind (Friese 2013, this volume).

ACKNOWLEDGMENTS

Supported by the John and Beverly Stauffer Foundation, Seaver Institute, and Caesar Kleberg Wildlife Conservation Foundation and San Diego Zoo Global. Revive & Restore organized and contributed financially to genetic rescue efforts for the black-footed ferret described here.

CHAPTER TWENTY-ONE

Cloning in the Zoo: When Zoos Become Parents

Carrie Friese

In *Cloning Wild Life: Zoos, Captivity, and the Future of Endangered Animals* (Friese 2013), I argued that there have been three kinds of cloning projects in zoological parks, all seeking to piece together nature and culture in different ways. One focused on developing assisted reproductive technologies so that zoos would have the skills increasingly required when reproducing animals of very small populations. The goal here is to make endangered animals, or "nature," through technique. The second project focused on genetically managing small populations of endangered animals using cloning to rescue genetic information lost through death but conserved in frozen zoos as previously nonreproductive cells (see, e.g., Ryder, this volume). Here the biology, or "nature," of endangered species has served as a guide as endangered animals and populations were remade through technique. The third kind of cloning project used interspecies nuclear transfer to better understand reproduction itself. The goal here is to know the biology of reproduction by changing it, and to evaluate techniques such as cloning based on that understanding of "nature." I argued that while cloning was a deeply contentious topic within and across zoos, "cloning" per se was not really at stake. Rather, the relation between nature and culture was being put to debate through the trope of cloning. Different ideas about how nature and culture should be patched together, or conserved, in remaking species on the brink of extinction are at the heart of debates over cloning. The imagined futures worked on through cloning include one in which humans can remake endangered animals so they can better survive a human-dominated planet, another in which humans shepherd endangered animals in the zoo by strategically privileging genetic definitions, and a third that reworks the element of surprise, which has long shaped environmental encounters with nature, through a specifically scientific and lab-based set of practices and discourses based on understanding the reproduction of endangered animals.

Since I researched and wrote that book a decade ago, the role of cloning in zoos has probably declined. The Audubon Center for Research of Endangered Species (ACRES) in New Orleans, which had devoted itself to developing cloning, closed in July 2015. The director, Dr. Betsy Dresser, had been a well-known proponent of developing assisted reproductive technologies with endangered zoo animals, and she successfully raised funds for this effort. She retired after Hurricane Katrina and the Gulf Coast oil spill, which is significant because both events sharply emphasized the importance of wetlands, fisheries, and coastal habitats. The new managing director was hired to raise funds for projects based in conserving wetlands and coastal habitats rather than in reproductive science. The institute decided it could no longer afford to be a benefactor for developing assisted reproductive technologies with endangered zoo animals, and the program closed (C. Earle Pope, pers. comm., October 7, 2015). The technical skill of this group of scientists and the laboratory system they created has thus ended.

Meanwhile the cloned banteng—a species of endangered wild cattle whose valuable genetic information was rescued through cloning—has not reproduced (Oliver Ryder, e-mail message, October 5, 2015). This was an example where zoo officials had placed great hope in the techniques as a path to rescue. Oliver Ryder, director of genetics at the San Diego Zoo, has said that the zoo does plan to continue using reproductive and genetic technologies, which Ryder discusses in this volume. So although the cloned banteng has not reproduced and his genetic information has not been distributed into the captive population, the experiment has nonetheless led to the possibility of merging cloning with the genetic management of small animal populations as part of conservation efforts.

Finally, the cloning that Bill Holt supervised at the Zoological Society of London did result in one cloned frog embryo (Bill Holt, e-mail message, October 7, 2015). It does not seem to have resulted in new knowledge regarding the nature of mitochondrial DNA in reproduction as the researchers had hoped. Rhiannon Lloyd, the postdoctoral researcher who worked with Holt on this project, has since shifted her mitochondrial research from reproduction to cancer.

Meanwhile, cloning has reentered the public imagination in recent years through interest in de-extinction. Here cloning is pursued alongside genome editing and back-breeding as a means to try to bring extinct species—from the passenger pigeon to the woolly mammoth—back to life (Sherkow and Greely 2013). This work is not embedded in a zoological park, but captive environments of some kind are nonetheless imagined as the final home for de-extinct animals. It is often assumed that the "new" development of de-extinction raises a whole set of new social and ethical dilemmas. However, as Claire Marris and I have argued (Friese and Marris 2014), many of the debates over de-extinction have occurred before, not only in the context of cloning endangered animals in zoos, but also in the context of developing and using assisted reproductive technologies more generally with zoo animals. In other words, zoological parks are a site of knowledge that can usefully guide discussions about de-extinction.

I thus ask what is new about cloning and zoos. In reviewing questions about cloning, I have come to realize that my own centering the debates on questions of nature and culture risks obscuring some important areas for discussion. This is particularly true in the context of de-extinction, but also in terms of cloning endangered animals. It is worth exploring how different ideas about reproduction undergird debates about cloning both endangered and extinct animals. To do this, I start by showing the problems with using nature and culture as an optic. I then seek to use reproduction as an alternative lens for seeing what is at stake when we talk about cloning in zoos. Specifically I ask, Exactly what does cloning reproduce when used in zoos?

THE PROBLEM WITH NATURE AND CULTURE

The claim that cloning is wrong because it is not natural has in many ways been the centerpiece of cloning discourses since at least the birth of Dolly the sheep, and more likely since the birth of John Gurdon's cloned frog (Callahan 1998; Franklin 1999; Kass 2002; Maienschein 2001; McGee 2002; President's Council on Bioethics 2002). It is a common criticism that is often linked with the idea that humans are "playing God." This criticism is also rooted in a long-prevailing Western notion that nature is the grounds for culture and society, a style of thought that has been troubled and critiqued over the past several decades. Because the position that cloning is unnatural is so common, it has also become rather standard to argue against this mode of criticism. My book *Cloning Wild Life*, for example, argues that we need to consider the implications of cloning in a different way. Rather than arguing against cloning because it changes nature, we should empirically explore how "nature," or the materiality of the world that includes animal bodies, is made through cloning projects. In turn we need to ask how certain social and cultural ideas shape those practices and thereby get into "nature" itself. I do this by building on the concept of "co-production" (Jasanoff 2004) that is an important idiom within science and technology studies.

Indeed, in reviewing the debates over de-extinction, I found that all discussions by ethicists first raised and then dismissed the objection that cloning is wrong because it is unnatural (Cohen 2014; Sandler 2013b). Here ethicists would argue that polarizing natural versus artificial is not sustainable (Cohen 2014). Human intervention is long-standing, as agriculture makes abundantly clear (Sandler 2013b). The use of cloning to bring extinct animals back to life is not in itself a problem so long as the animals produced increase the diversity and integrity of an ecosystem (Cohen 2014, 168). Playing God in conserving healthy ecosystems is considered an unproblematic social good.

I became worried about this line of reasoning, however, not because I want to preserve a line between nature and culture, but because of the kinds of arguments that were being dismissed as a result. Specifically, some have argued that extinct species and the de-extinct animal have two very different natural histories (Carlin, Wurman, and Zakin 2013). They cannot be considered the same because they have evolved in completely different milieus. De-extinction may create an animal that is like the passenger pigeon, but it won't bring the passenger pigeon back to life. Ben Minteer (2014), for example, has recently argued against de-extinction by stating, "The evolutionary toil and historical richness of the forerunner species, including their co-evolution with other species over time, has been lost. . . . A living species' natural history is not the only reason why we value them, but it's a profoundly important one to conservationists."

Other bioethicists have dismissed this line of argument by saying it presumes an ontologically prior nature. Cohen (2014, 176), for example, argued that this kind of argument exhibits a "historical aesthetic," valorizing a pure nature that is not philosophically or empirically sustainable. I disagree with Cohen's characterization. For example, environmental historian Paul Wapner (2010) has shown that environmentalism is finding a new way of doing politics, one that appreciates the "otherness" of nature without understanding it either as foundational or as existing outside social, historical, and cultural contexts. Most contemporary conservation arguments assume there is not a pure nature diametrically opposed to culture (Minteer 2006). But most environmentalists will nonetheless argue that there is a big difference between a de-extinct animal and an evolving species (Minteer 2014). What is of concern here is not so much humans playing God or the ontological primacy of nature as a worry that history, space, sociality, and behavioral learning are all being displaced, and that these processes are crucial for many species to develop biologically and socially over time (Oksanen and Siipi 2014, 12). These concerns cannot be dismissed as another version of the nature/culture divide.

Debates over nature and culture risk obscuring rather than illuminating the stakes of cloning in zoos. This is why I turn to reproduction as an alternative optic, one that might better capture what is being debated through the trope of cloning either endangered or extinct species. I will now explain what I mean by reproduction as an optic and why I think this is a better way of understanding the debates about cloning. I find reproduction helpful in part because it denotes both biological and social processes. This means we are less likely to get sidetracked by arguments regarding a pure nature, a theme that is particularly central to this edited volume. So rather than asking how nature and culture are patched together through cloning, as I did in my book, I instead ask what cloning reproduces.

REPRODUCTION

Feminist sociologists, anthropologists, historians, and lawyers have highlighted the significance of reproduction for understanding social life (Clarke 1998; Colen 1995; Franklin and Ragone 1998; Ginsburg and Rapp 1995; Jackson 2001; Laslett and Brenner 1989; Mamo 2007; Martin 1987; Petchesky 1987; Roberts 1997; Thompson [Cussins] 1998). In arguing for an anthropology that explores the politics of reproduction, Faye Ginsburg and Rayna Rapp (1995, 1–2), for example, argued in their now canonical book *Conceiving the New World Order* that a focus on reproduction allows us to see how cultures are produced and contested as people create the next generation of children (biological reproduction) who are necessarily born into and out of complex social relationships (social reproduction). The goal of this scholarship has been to disturb the bifurcations of nature/culture, woman/man, reproduction/production by putting reproduction at the center of social theory. Troubling Marxist conceptions in particular, this feminist scholarship has argued that (masculine and compensated) forms of production require and rely on (feminine and uncompensated) reproduction. Reproduction is thus an invisible form of labor.

It is important to point out that the notion of reproduction came into being only during the mid-eighteenth century and in conjunction with the concept of heredity; during the nineteenth century reproduction supplanted the notion of generation, wherein living beings were thought to be uniquely engendered in a manner that may be considered closer to production (Hopwood et al. 2015; Muller-Wille and Rheinberger 2007, 3–4). And so reproduction has become linked with notions of mechanization, as life multiplies from a kind of blueprint and follows a more or less standardized process (Hopwood et al. 2015, 384). Assisted reproductive technologies intervene in this presumably standardized procedure, creating both biological and cultural changes. These changes have in turn provided insight into the biological processes of reproduction itself, as in vitro fertilization has somewhat ironically become a model of sexual reproduction (Franklin 2013).

This section asks, What does cloning reproduce when it is used in zoos, with endangered animals? What cultures of nature are produced and contested as people create the next generation of endangered animals (biological reproduction) that are necessarily born into and out of complex social relationships (social reproduction)?

What is reproduced in the first kind of cloning projects, which focus on developing assisted reproductive technologies and making endangered animals or "nature" through technique? The nuclear DNA of endangered animals is reproduced through interspecies nuclear transfer, albeit in a slightly modified form through the introduction of mitochondrial DNA from the egg donor and presumably its gestational development within a different but closely related species. But more than DNA is reproduced in these cloning projects. Indeed, the focus seems to be on reproducing the technical skills of scientists. Each cloning experiment, as a more or less standardized procedure, allows scientists to develop the skills required to technologically reproduce genetic information. Through the reproduction of nuclear DNA, embodied by cloned animals, the act of zoo-based assisted reproduction is again carried forward in time. Cloned animals stand as witnesses to these skills, which are located within a zoo and can therefore be called on as needed by zoos.

What is reproduced in the second kind of cloning project, which focuses on genetically managing small populations of endangered animals by using cloning to rescue genetic information lost through death but conserved in frozen zoos as previously nonreproductive cells? The goal here is to reproduce lost genetic information—specifically lost nuclear DNA—through asexual reproduction in the form of an animal. That animal can then carry that nuclear DNA forward in time once again through sexual reproduction.

The cloned animal is a witness to the ways populations of varying kinds, including cryopreserved cells, ex situ populations, and in situ populations, can be managed in tandem. And so both frozen zoos and Species Survival Plans are also socially reproduced. In other words, the frozen zoos and the Species Survival Plans are the complex social relationships that cloned endangered animals are born out of and into.

What is reproduced in the third type of cloning experiment, where interspecies nuclear transfer is a means through which early phases of reproduction can be better understood? It is important to emphasize that cloning has been central to experimental physiology across the twentieth century. Jane Maienschein (2001, 2002, 2003) has historicized cloning in this context and notes that Hans Spemann is considered the initiator of the nuclear transfer technique, having first inscribed cloning as a potential means for answering questions about embryonic development. Spemann used experimental techniques, largely with frogs, to answer questions about cell differentiation. He was curious about what would happen if the nuclei of cells were transferred to enucleated eggs at different stages of development (Maienschein 2003, 115–16). By changing reproductive processes in this way, Spemann thought much could be inferred regarding normal reproduction. The social practice of changing biological processes in order to understand normal patterns of reproduction within scientific research is "reproduced" in contemporary cloning experiments involving endangered zoo animals.

So cloning not only reproduces DNA, it also reproduces skills and knowledge practices. Charis Thompson (2005) has argued that using assisted reproductive technologies for human reproduction does not simply make babies; these techniques also make parents. A similar line of argument can be made about assisted reproduction in zoos. Cloning does not simply make individual animals; it also makes technically skilled scientists, knowledge practices, and populations of animals. Further, cloning can promote the reproduction of the zoo as an institution committed to species preservation. By reproducing its animal populations rather than collecting animals from their native habitats, zoos can argue that they are creating and re-creating populations that stand as backups or reserves for more precarious populations in situ. In this way zoos can assert that they are involved in species preservation rather than contributing to species extinction through their collection practices. As a result, cloning is one among many techniques that is turning zoos into parents.

The socially reproductive practices that occur after birth are therefore crucial as well. Species are of course not only genomes but often also distinctive social behaviors that are carried forward in time through interaction and learning. What is typically delineated as "social reproduction" is also important for animals and is indeed at times species-defining. For example, in her classic monograph on monkeys, Thelma Rowell (1972) cites a study regarding the behavior of wild hamadryas baboons living in captivity. One of the distinguishing aspects of this species is that adult males herd females, and this has been considered a species-differentiating behavior (Rowell 1972, 80–81). Male animals born in situ and then moved into a captive setting will retain this behavior, but males born in captivity will not practice it. These differences are of course crucial for the zoo in the context of reintroduction, because potentially species-defining behaviors are lost through the socially reproductive space of the zoo. Indeed, a metareview has found that reintroducing carnivores tends to be more successful with wild-caught than captive-born animals (Jule, Leaver, and Lea 2008). Further, a study comparing reintroduction success among different generations of oldfield mice found that animals from a longer lineage of captive-born mice sought refuge less often than animals from a shorter lineage of captive-born mice. In other words, the more generations of oldfield mice born in captivity, the more likely the individual animal is to die from predation after being reintroduced (McPhee 2003). Based on these studies, it seems entirely plausible to argue that cloning endangered animals also reproduces captivity, and the behaviors associated with it, since these animals are neither domestic nor wild (see also Ritvo, this volume). In other words, the zoo is a space where social reproduction occurs, and these processes can make captivity a requirement for some species into the future.

I have so far focused on cloning endangered animals in zoos. But what is cloning reproducing in the context of de-extinction? To date, cloning has been used to bring only one extinct species "back to life," and only for minutes. Genetic information from the last surviving Spanish bucardo, or ibex, was preserved in the form of somatic cells, and these cells were used in a cloning experiment. However, the newborn died minutes after birth owing to problems with its lungs. Again, more than an animal was reproduced through this experiment. Specifically, the hope that extinct animals could be reproduced through cloning was carried forward into the future. In other words, this cloning project reproduced the project to revive extinct animals.

Reproducing hope is indeed a key part of the rhetorical work of de-extinction specifically and cloning more generally. Stuart Brand, the environmentalist-futurist who has become the key spokesperson for de-extinction, has stated that part of his intention with de-extinction is to provide an alternative narrative for species preservation, one focused on telling people good news. In his TED talk, Brand stated that "bad news bums people out . . . reviving extinct species is good news." The idea here is to get people excited about species preservation through technology. Indeed, I have myself argued that nature and technology are not oppositional (Friese 2013); technology has long been a way of generating interest among people who might not otherwise be concerned with nature and its preservation (see also Clark 1999; Franklin 2002; Yule 2002). Changing how we think about extinction, by emphasizing hope, is the focus here.

It is worth pointing out that reproduction is at times understood as a process denoted by continuity and sameness. For example, we see this in the sociological idea that the education system reproduces existing class structures (Bourdieu 1984). Biologically, we see this in the idea that cloning is the ultimate form of reproduction in that it carries the exact genome forward into the future. For example, an industry scientist I interviewed stated that cloning is useful because "if they [dairy farmers] had . . . for instance a Holstein cow that produces a lot of milk, the idea was to clone that animal so that the producer would potentially have a small, elite herd of identical animals" (interview July 1, 2005). Here we see the idea that cloning creates a copy not only of an animal's genome, but also of the phenotype of that valued animal. Continuity and sameness are thus, at times, understood as central to both social and biological reproduction.

However, others have emphasized that reproduction always emphasizes both continuity and change. For example, even cloning introduces difference at the genetic level. Somatic cell nuclear transfer requires an egg cell donor, almost always one different from the somatic cell donor. The cloned animal thus inherits mitochondrial DNA from the egg cell donor and so is genetically different from the "original." As such, both developmental biologists and sociologists have argued that we need to understand reproduction as a process involving both sameness and difference (Franklin 2007, 2013). There is no such thing as "purity" in the process of biological or social reproduction (see also Rothfels, this volume). Drawing on this idea of reproduction as recursive without being exact (Franklin 2013), how is de-extinction continuous with and different from the idea of "extinction" as it developed across the twentieth century? In other words, how does "de-extinction" reproduce the idea of "extinction"?

Extinction developed as a cautionary warning through which people were called on to change their practices in relation to other species. Stephanie Turner (2008, 60) has noted that wilderness preservation is based in the European romantic idealization of the natural order as divine and human experience of that order as transcendent.[1] She states that while conservation to some extent requires domesticating the natural, most conservationists have done so to reestablish natural orders that existed before human intervention. "Conservation and preservation efforts are, in short, efforts to make right, through new technologies, the unintended wrongs resulting from older technologies" (Turner 2008, 61).[2] De-extinction draws on this long-standing moral understanding of species preservation, albeit with a stronger sense of technological optimism than often seen among those concerned with the environment. De-extinction is thereby rooted in environmental philosophies and ethics that emphasize the inherent worth of nature and of living things beyond humans. But it adds a restorative justice element to this morality (Cohen 2014; Sandler 2013b) and also attempts to change the discourse of extinction. In the context of de-extinction, the spectacle replaces the warning in order to call on people to support science as opposed to changing their practices in relation to the land and to other animal species. Indeed, remaking animals and species through science is the means of carrying the act of righting past wrongs forward into the future. It could be said that de-extinction thereby reverts to ideals of the traditional zoological park in that, like the traditional zoo, de-extinction is primarily rooted in an entertainment function. Both the zoo and de-extinction proponents can therefore be criticized for focusing primarily on entertainment, and the utility of both projects within species conservation can be questioned.

As many have noted, the ways de-extinction proponents seek to change "extinction" as a discourse are risky because people may come to believe that extinction is reversible, and this could undermine rather than promote conservation efforts (Cohen 2014, 168). In this context, Thom van Dooren and Deborah Rose (2013) have used another statement from Stuart Brand's TED talk as a starting point for critiquing the rhetoric of de-extinction. In contrast to Brand's command, "Don't mourn: Organize," Van Dooren and Rose argue that what we need now, more than anything, is time to mourn the massive scale of death that has occurred across so many species. Turning to the work of psychologists, Van Dooren and Rose emphasize that mourning is a transformative learning process wherein people relearn their selves and their place in the world in the context of significant loss. They state, "Mourning is about dwelling with a loss and so coming to appreciate what it means, how the world has changed, and how we must *ourselves* change and renew our relationships if we are to move forward from here. In this context, genuine mourning should open us into an awareness of our dependence on and relationships with those countless others being driven over the edge of extinction. In short, dwelling with extinction in this way—taking it seriously, not rushing to overcome it—might be the more important political and ethical work for our time." Through mourning, Van Dooren and Rose seek to sustain the rhetorical valence of "extinction" as it developed over the course of the twentieth century while changing the human exceptionalism that has long been central to this discourse (Van Dooren 2014). They want to reproduce extinction as a warning. Assisted reproductive technologies cannot right the wrong of extinction in this instance; only a change in human behavior can do that, insofar as this would allow other species to continue to reproduce alongside humans into the future.

What we see here are different versions of the problem of extinction. Is it a problem of human behavior that must be addressed by changing the socially reproductive practices of humans? Or is it a problem of lost species that can—in part—be addressed by changing the biological reproductive practices of animals?

CONCLUSION

I have sought to explain what is at stake in using cloning in zoos by asking what cloning reproduces. I have argued that cloning reproduces far more than animals. It carries forward technical skill, technical practices, knowledge practices, populations, and captivity. It also carries forward hope in technology for righting past human errors. Reproducing these skills, practices, populations, and rhetorics means that other things are not reproduced.

What should be carried forward in time, and what should not, is therefore central to the debates. This is, after all, what conservation is about. My previous understanding of cloning in zoos, rooted in questioning how assisted reproductive technologies are used to patch together nature and culture, asked what kinds of worlds we want to make through science. Shifting the optic to reproduction has instead emphasized cloning as a form of invisible labor, a kind of work, and like any other form of labor it and its products need to be *nurtured* if they are to be carried forward.

ACKNOWLEDGMENTS

I thank Bill Holt, C. Earle Pope, and Oliver Ryder for taking the time to update me on their cloning research. An earlier version of this chapter was presented at the conference "De-extinction: Rescue or Boondoggle?" at

Oregon State University, where I received helpful comments from other panelists and workshop attendees. Thank you especially to Luis Campos, George Estreich, Susan Haig, and Anita Guerrini for enriching my thinking on the topic. Sarah Franklin, Hannah Kietinger, Barbara Prainsack, and Manuella Perrota read an earlier draft of this chapter and helped me substantially refine the argument. Careful comments and suggestions from Ben Minteer, Jane Maienschein, and James Collins on a later version of this chapter were crucial in solidifying the arguments around reproduction.

CHAPTER TWENTY-TWO

Advancing Laboratory-Based Zoo Research to Enhance Captive Breeding of Southern White Rhinoceros

Christopher W. Tubbs

INTRODUCTION

As the chapters in this volume make clear, the role of zoos and aquariums in plant and animal conservation is evolving. No longer are these institutions focused primarily on exhibiting animals, and many have taken leadership roles in conservation-centered programs. The scope of conservation efforts specific to zoos is broad, ranging from developing web-based applications designed to inform the public about sustainable harvest of food resources (e.g., Cheyenne Mountain Zoo 2016; Spring, this volume), to in situ field projects aimed at the conservation of critical habitat or species, to using laboratory-based approaches to study the biology of species of concern.

While field studies typically receive the most media attention, the lab has long been valuable in advancing conservation science and helping zoos and aquariums reach their conservation goals. Yet the potential for laboratory studies to advance "zoo science" is unreached: the techniques employed by lab-based researchers at zoos can be decades behind those of researchers performing analogous research at universities, government agencies, or in the private sector (see Knapp, this volume). However, as zoos and aquariums dedicate more resources to conservation science, access to new technologies is allowing for more sophisticated approaches. Here I provide one such example of how we are doing this at the San Diego Zoo Institute for Conservation Research by conducting novel, noninvasive in vitro studies to investigate the role of diet in the long-standing problem of low fertility in the southern white rhinoceros (SWR; *Ceratotherium simum simum*) in zoos.

Our goal is to apply the findings from our studies to the management of SWRs, to enhance the captive breeding of a threatened species whose wild counterparts are currently facing a poaching crisis.

THE SOUTHERN WHITE RHINOCEROS

SWRs are a conservation success story, but their future is uncertain. By the end of the nineteenth century, hunting rendered this species scarce in habitats where it once was abundant. In 1882 a bull that was shot and killed was believed to be the last individual in what is now northern Zimbabwe (Renshaw 1904). Additional SWRs were later discovered, but by the beginning of the twentieth century the population had become extremely small. The precise number that remained is unknown, but estimates range from just a few animals to about two hundred (reviewed in Rookmaaker 2000). Regardless, the likely disappearance of SWRs was recognized, prompting the protection of this species and tracts of native habitat, which today continue as some of South Africa's oldest wildlife reserves. These highly successful in situ efforts were the driving force in the rebounding of the SWR population, and SWRs living in the wild now number more than twenty thousand individuals (International Rhino Foundation 2014).

Ex situ efforts played a smaller role in the conservation of SWRs. Compared with other species, SWRs were relatively late arrivals to zoo collections. The first animal brought into captivity was an orphaned calf found in 1946, and by the 1970s exports of larger groups of wild rhinos to institutions in Europe and the United States were occurring (Rookmaaker 1998). Under proper management conditions (e.g., large enclosures, large group size) early exported SWRs reproduced well and served as a valuable assurance population for a wild population that at the time numbered fewer than five thousand individuals (Emslie and Brooks 1999). Over time, however, it became clear that captive-born females were not reproducing as well as founding female SWRs. With most of those founders now dead or past reproductive age, the captive population is no longer self-sustaining (Emslie and Brooks 1999; Schwarzenberger et al. 1999; Swaisgood, Dickman, and White 2006). At this time SWRs have seen a hundredfold increase in poaching in less than a decade, to the point that the population may soon once again be in decline (Ferreira, Botha, and Emmett 2012). With current in situ efforts to curb poaching failing, having a healthy ex situ assurance population of SWRs is more important than ever to the conservation of this species. To achieve this we must broaden our understanding of SWR reproductive health, particularly focusing on why captive-born females fail to reproduce, in order to restore this population's fertility so zoos may maximize their contributions to the global conservation of this species.

POTENTIAL CAUSES

The specific factors that cause or contribute to the poor fertility of captive-born female SWRs have yet to be identified. At the San Diego Zoo Safari Park, which has experienced a marked decrease in fertility of its captive-born females compared with females imported from the wild, behavioral observations eliminated aberrant reproductive behaviors as a possible cause (Swaisgood, Dickman, and White 2006). Wild- and captive-born females copulate with equal regularity, and male SWRs do not prefer one group of females to the other. In an investigation into a possible role of stress in reproductive failure, glucocorticoid (stress hormone) levels in female SWRs at sixteen North American institutions showed no relation between stress and fertility (Metrione and Harder 2011). Perhaps the most telling observations came from reproductive health examinations of fifty-four captive white rhinos from institutions around the world. In that survey, both wild- and captive-born females exhibited a high incidence of early onset fertility-compromising reproductive tract pathologies consistent with exposure to high levels of estrogen (Hermes et al. 2006). The development of pathologies was attributed to long nonreproductive periods associated with high estrogen production, which is dampened when an animal is pregnant (Hermes et al. 2006). Although this paradigm is not unprecedented, it does not account for why SWRs fail to get pregnant in the first place, or for the possibility that environmental sources of estrogenic chemicals interfere with SWR endocrine (hormone) systems to impair fertility.

ENDOCRINE-DISRUPTING CHEMICALS

It is well known that environmental chemicals can interfere with animals' endocrine function and compromise reproduction (Colborn, vom Saal, and Soto 1993). Broadly classified as endocrine disrupting chemicals (EDCs), they are most widely recognized as anthropogenic, but there are also naturally produced EDCs. One group of natural EDCs are the phytoestrogens: chemicals produced by plants that mimic actions of the hormone estrogen when eaten in large quantities. In some laboratory and livestock species, eating legumes such as soy, clover, and alfalfa, which are high in phytoestrogens, reduces fertility and leads to reproductive pathologies similar to those observed in SWRs (Jefferson, Patisaul, and Williams 2012; Adams 1995). Interestingly, both soy and alfalfa are common components of zoo diets, including those fed to SWRs, establishing phytoestrogens as a likely exogenous estrogenic EDC they are exposed to.

An important aspect of phytoestrogen and other EDC effects is that the timing of exposure can profoundly influence the effects observed. When adults are exposed, the effects are typically less severe than when exposure happens during development. Moreover, developmental exposures are more likely to be permanent, affecting endocrine function and fertility for a lifetime, whereas adult exposures can often be reversed (Adams 1995; Guillette et al. 1995). In the captive SWR population, both wild- and captive-born females at some institutions are fed soy- and alfalfa-based diets, but while wild-born females reproduce relatively well, captive-born females do not. This observation provides additional support for the hypothesis that exposure to phytoestrogens, particularly during development, is a component of the SWR fertility problem, providing a possible explanation of why captive-born females are more strongly affected than females born in the wild.

STUDYING EDC EFFECTS ON A THREATENED SPECIES: PHYTOESTROGENS AND SWRS

Despite a long-standing awareness that EDCs can compromise reproduction in humans, laboratory animals, livestock, and wildlife, zoos have not been much involved in this area of research. Conducting EDC research is a zoo setting is not practical for many reasons. For example, populations of zoo animals are typically small, and potential experiments needed larger sample sizes to obtain significantly powerful data. But more important, the studies needed to provide the strongest links between EDC exposure and adverse reproductive outcomes involve controlled exposure of animals to harmful chemicals and invasive tissue sampling (not welcome practices in zoos). Where the developmental consequences of EDC exposure are suspected, as with SWRs, long-term, multigenerational studies are warranted, but these are simply not feasible, particularly for slow-reproducing species. Nevertheless, as I describe below, increased accessibility to techniques appropriate for zoo-based laboratory research has excellent potential to broaden the scope of reproductive and EDC-focused zoo research and enhance breeding of challenged species.

To study the possible effects of phytoestrogens on SWRs, we focused on the molecular mechanisms by which phytoestrogens and other EDCs interfere with endocrine function. For a hormone to elicit an effect, it must bind to a receptor—a protein that becomes activated to catalyze the biochemical changes that ultimately result in a physiological response. EDCs function similarly by interacting with hormone receptors to activate them at inappropriate times or to block their activation. Since EDCs function through interaction with a receptor, in vitro methods that measure receptor-EDC interactions can be used to predict organismal response to EDC exposure without ever exposing an animal to a potentially harmful chemical (Tubbs et al. 2014).

We began our studies by isolating estrogen receptors (ESRs) from both SWRs and greater one-horned rhinoceros (GOHRs; Tubbs et al. 2012). We accomplished this by cloning the genes encoding each species' estrogen receptors from small pieces of reproductive tissue sampled opportunistically during necropsy of collection animals that died of natural causes. We then inserted rhino ESR genes into a common laboratory cell line that produced the functional receptor protein. We chose to study the two rhino species because both receive similar diets in captivity but GOHRs reproduce much more successfully than SWRs. After treating the cells expressing the rhino ESRs with the various phytoestrogens commonly found in soy- and alfalfa-based feeds, we measured ESR activation, looking for differences in receptor responses that may explain the differences in fertility between the two species. Somewhat to our surprise, we did find such a relation by discovering that SWR ESRs were significantly more sensitive to phytoestrogens (activated to a higher degree) than GOHR ESRs.

Our initial study provided a promising association between species differences in receptor sensitivity to phytoestrogens and reduced fertility, but observations from a test tube do not prove phytoestrogens compromise a rhino's fertility. We conducted a follow-up study aimed at providing additional layers of information to assess whether dietary phytoestrogens contribute to poor SWR reproduction. Specifically, we collected samples of diets from nine of the most successful SWR breeding institutions in North America and measured their ability to activate SWR ESRs as described in Tubbs et al. (2016). We then used the International Studbook for White Rhinoceros (AZA 2014d) to calculate fertility for all females at each of the institutions, and in doing so we produced our most convincing piece of evidence to date. For females born in the wild, we found no significant relation between phytoestrogen content of the captive diets and fertility.

In contrast, for females born in captivity there was a significant negative relation between diet estrogenicity and fertility, suggesting that exposure to phytoestrogens during development is contributing to the poor reproductive success in this species.

Given that our data suggest that developmental exposure to phytoestrogens is involved in low SWR fertility and that developmental EDC exposures tend to be more severe and permanent, an obvious question is, Can fertility be restored in any captive-born females if diets are changed? While it will likely take years to know for sure, some recent developments at the San Diego Safari Park are promising. By the beginning of 2014, we had reduced the amount of soy- and alfalfa-based pellets fed to our SWRs. After approximately one year on the new diet, there were confirmed pregnancies in two of our five nonbreeding females of reproductive age, one in a female that had not reproduced in nearly a decade and one in a fourteen-year-old female that had never reproduced (Tubbs, Durant, and Milnes 2017). With more institutions following suit and making their own diet changes, we will be closely monitoring the captive SWR population for more encouraging news.

CONCLUSION

As the roles of zoos and aquariums change and their efforts become more conservation-focused, so must the tools and approaches they use change to promote their mission effectively. Although zoo-based laboratory research has greatly advanced conservation science in areas like genetic management and the development of techniques to enhance captive breeding (see Barrow, this volume, for a historical account of captive breeding efforts of zoos), it has not kept pace with other research arenas. Zoos therefore have tremendous opportunity to expand their research to additional avenues to apply to the conservation of threatened and endangered species. Our example here, where contemporary science is applied to improve reproduction of a species in need, illustrates some of the potential benefits of embracing novel and evolving laboratory approaches to conservation. Moreover, the techniques at the cutting edge of science are nicely suited for zoo researchers, who are often limited by sample availability and the inability to use standard research approaches. With recent technological advances, only small biological samples are needed to sequence entire genomes, conduct in-depth gene expression studies, or develop stem cells (see Ryder, this volume). In addition, the costs of these approaches are shrinking, making them accessible to more and more researchers with the result certainly being a very bright future for laboratory science in conservation research.

ACKNOWLEDGMENTS

I am especially grateful to Dr. Barbara Durrant and Dr. Thomas Jensen for their thoughtful review of this manuscript.

CHAPTER TWENTY-THREE

Beyond the Walls: Applied Field Research for the Twenty-First-Century Public Aquarium and Zoo

Charles R. Knapp

INTRODUCTION

The ability of zoological organizations to make meaningful contributions to conservation is increasingly recognized within and outside the zoo community (West and Dickie 2007; Penning et al. 2009). Recognition of how zoological organizations can fulfill their conservation mandate, however, is limited primarily to social and veterinary science (Wharton 2007), along with a major emphasis on conservation breeding (Zimmermann et al. 2007; Conde et al. 2011a; Fa et al. 2011; Conde et al. 2013) and, more recently, engaging diverse and large audiences in pro-environmental actions (Grajal, Luebke, and Kelly, this volume). Fa et al. (2011) have argued that zoos should use an interdisciplinary approach to conservation by drawing on captive management and breeding, small-population biology, reintroduction biology, and science education. This approach is effective for advancing conservation in certain circumstances, particularly in zoos, but there are important differences between aquariums and zoos (Penning et al. 2009). The differences in their character, operational requirements, stakeholder communities, and challenges associated with conserving aquatic environments compel aquariums to broaden their approach to conservation relative to that of zoos. For example, the difficulty of breeding and rearing biologically complex marine species (see Rising Tide Conservation, http://risingtideconservation.org/) makes it challenging to apply the conservation-breeding approach at scales needed to make immediate contributions to declining populations.

One solution for addressing the need for conservation at public aquariums includes developing capacity for conducting in situ applied research. Public aquariums have generally lagged behind zoos in developing staff-led in situ conservation research programs, perhaps because many early public aquariums were founded on fisheries principles (see Muka, this volume) or established more recently to aid economic development (Hill, Iliff, and Brailsford 2001) or support urban revitalization (Devenney 2011). Recently, however, both zoos and aquariums are facing an influential landscape that includes the recent biodiversity extinction crisis, shifting perceptions of nature experiences, and increased questioning of the value of animals under managed care (Grajal, Luebke, and Kelly, this volume). As such, zoological organizations are increasingly pressed to demonstrate their relevancy in the twenty-first century. In response to this changing landscape, as well as from a genuine desire to positively affect wild species and habitats, zoological organizations are diversifying and expanding their conservation portfolios (Ivanyi and Colodner, this volume). Consequently more aquariums–and zoos–are developing capacity for conducting meaningful and high-quality in situ research that focuses on detecting, diagnosing, and halting population declines in the wild (see Penning et al. 2009).

Increasing capacity for in situ applied conservation research is compelling for aquariums because along with problems pertaining to conservation breeding of marine fish species, successful reintroduction programs for fishes are generally associated with larger propagation facilities and methods (George et al. 2009) that may be difficult to replicate under daily husbandry schedules. Conducting applied in situ conservation research driven by staff scientists at aquariums–and zoos–is also compelling because, unlike academic institutions, here the effect of the work can be amplified by capitalizing on the direct link between the conservation issue, the research conducted by staff scientists, and the millions of visitors to aquariums and zoos.

Modern zoological organizations are obligated, through requirements for accreditation by the Association of Zoos and Aquariums (AZA) and the World Association of Zoos and Aquariums (WAZA), to fulfill a growing number of responsibilities including scientific research, wildlife conservation, public recreation, and education. Each zoo and aquarium is unique, and thus priorities must be set in relation to goals, objectives, and missions of an individual organization (Roe, McConney, and Mansfield 2014). Although conservation activities will always be multifaceted among zoological organizations and tailored individually to organizational priorities, in situ research programs will continue to expand as zoological organizations become more prominent as field research practitioners (Minteer and Collins

2013). The theoretical motives of integrating in situ research within the conservation portfolio of an organization are intuitive, but the practical approaches to developing a new conservation research program must not be underestimated. Indeed, aquariums and zoos are unique venues that differ from academic institutions and traditional environmental nonprofits. As such, there are opportunities and challenges associated with developing conservation research programs for zoological organizations generally, as well as specifically to each organization.

The John G. Shedd Aquarium (Chicago) has engaged in conservation research for two decades, but it significantly increased its in situ research capacity in 2012 by establishing the Daniel P. Haerther Center for Conservation and Research. With this institutional commitment, Shedd now supports a team of PhD scientists who conduct field research around the globe. For over thirty years Shedd has owned a live-aboard research vessel based in Miami, used predominantly for sustainable collecting and as a teaching space. The expansion of Shedd's commitment to in situ research, however, has changed its primary use to conservation research. The commitment to maintaining a research vessel enables in situ investigations and training opportunities beyond the capacity of many academic institutions.

Since more aquariums–and zoos–are expected to develop capacity for in situ conservation research, in this chapter I will share lessons learned from the development of Shedd Aquarium's relatively recent conservation research program from concept to design and finally to implementation. This overview is meant primarily for new research programs and is not a prescriptive framework, since supposed best practice in one situation, or at one organization, might be less appropriate for another. Instead, I hope that elements of this overview can serve as a scalable model for other public aquariums and zoos interested in expanding their in situ conservation research. In that context, I will discuss the evolution of strategic planning for conservation research including selection criteria, evaluation methods, staffing considerations, and organizational adaption.

CONSIDERATIONS WHEN SELECTING RESEARCH PROGRAMS

Research at zoological organizations aimed at conserving biodiversity can be applied to the species or landscape level as well as being influenced by thematic considerations (e.g., climate change or sustainable fisheries). The merits of such approaches are discussed within the zoological community

(Penning et al. 2009) and in the conservation literature (e.g., Boughton and Pike 2013; Veríssimo et al. 2013; Breckheimer et al. 2014). The processes for establishing research priorities should be informed by an organization's unique attributes, including geographical location, collection composition, institutional culture, and financial restrictions. Though Shedd Aquarium's ex situ research primarily concerns animal biology, behavior, and husbandry, the applied in situ research program addresses projects that have clear implications for saving species, populations, and habitats in the wild.

Shedd Aquarium has criteria for selecting field research programs. Not all criteria must be met for program approval, but the considerations serve as a useful rubric for evaluating options. Two related criteria include ensuring that a field program aligns with the aquarium's collection and featured ecosystems and that it offers a story that is accessible, inspiring, and meaningful for the general public. Stories that connect animals, and presumably the in situ research associated with them, to visitors can elicit evocative responses of support for conservation initiatives (Clayton, Fraser, and Saunders 2009). Concentrating on a single charismatic species within the collection is easier for any zoological organization because people typically associate animals with aquariums and zoos. In reality, a robust institutional interpretive and communication plan will enable nuanced discussion around most field research programs, but having at least one charismatic "hook" for the program can help to generate interest (Moss and Esson 2010).

A mixed portfolio of research programs will broaden opportunities to discuss organizational research. For example, Shedd Aquarium works with a variety of charismatic species including the Nassau grouper (*Epinephelus striatus*) in the Bahamas and the arapaima (*Arapaima gigas*) in Guyana. Both species serve as conservation flagships while also acting as gateways to discuss the need for sustainable fisheries and habitat preservation. Shedd also works with less traditionally charismatic species such as the queen conch (*Lobatus gigas*), for which overharvest is a critical conservation issue in the Caribbean. Working with this species presents more problems from an interpretive aspect but does allow deeper explanation of sustainable fisheries and the importance of studying multiple life history phases and integrating data into conservation management (e.g., Marine Protected Area design).

The degree of collaboration is also considered when making program decisions. Methods used to plan and implement conservation initiatives cannot implicitly assume that actions should be undertaken by just one organization, because in reality multiple organizations usually act in a particular region (Bode et al. 2010). It is therefore critical to identify regional partners to maximize conservation efficiency (see also Ivanyi and Colodner, this volume). In addition, organizations with newer research programs must be sensitive to other regional players, particularly because it may take time for other conservation and research stakeholders to become aware that quality field research is part of an organizational mission. All in situ research projects at Shedd Aquarium nurture partnerships to varying degrees with academic institutions, nongovernmental organizations (NGOs), government organizations, and local stakeholders. In some cases, acting in a supporting role is the best strategy for advancing conservation. Shedd's nearly two-decade partnership with Project Seahorse (http://seahorse.fisheries.ubc.ca/conservation) exemplifies this approach. To broaden and maximize scientific capacity for executing projects, academic institutions and their faculty members are significant collaborators in other projects, particularly conservation research projects conducted in the Great Lakes region. For international projects, working with in-country partners such as the Bahamas National Trust–the NGO mandated to oversee the national park system for the country–has enabled applying research results to direct conservation management. For example, data from Shedd's long-term iguana research program (e.g., Knapp and Owens 2005; Knapp, Alvarez-Clare, and Perez-Heydrich 2010), along with recommendations from Shedd scientists, were used to help demarcate boundaries for a national park expansion on Andros Island in the Bahamas.

Other conservation partners to consider when implementing programs should include fellow zoological organizations. The recent Association of Zoos and Aquariums' Saving Animals from Extinction (SAFE) initiative allows accredited members to collectively convene with scientists and stakeholders to identify threats, develop action plans, raise new resources, and engage the public in the conservation of selected species (Grow, Luke, and Ogden, this volume). In addition, the International Union for Conservation of Nature (IUCN) Species Survival Commission (SSC) has staff members dedicated to encouraging partnerships and furthering collaboration between the IUCN SSC Specialist Groups and the zoo, aquarium, and botanical garden communities. This approach involves supporting and enhancing existing relationships as well as identifying new opportunities for more effective collaboration between the IUCN SSC Specialist Groups and the ex situ community for stronger conservation outcomes.

Shedd considers the learning and training openings that a program can provide to local and international stakeholders when selecting new projects, and it prioritizes scientific capacity building within government entities and NGOs in range countries. This approach is important given that an acute lack of infrastructure, educational resources, and chances for professional development are viewed as critical obstacles to developing the capacity to manage biodiversity in less developed countries (Ceballos et al. 2009). Indeed, developing the knowledge and skills of individuals and organizations has been widely recognized as an important element in implementing conservation globally (Salafsky et al. 2002; Rodriguez et al. 2005). Shedd Aquarium has a unique opportunity to provide capacity training to in-country stakeholders, facilitated by our scientific team and through the use of our research vessel. The vessel makes frequent scientific explorations to the Bahamas and the Caribbean, and we invite and encourage participation in research cruises among our partners, including local resource managers and university students (e.g., Yates 2015).

Zoological organizations are uniquely positioned to advance conservation awareness and learning by maximizing the collaborative potential of their science, education, and communication departments. Research programs can expand these learning occasions for stakeholders through engagement and citizen science projects (Evely et al. 2011; Spring, this volume). Shedd's local fish monitoring and amphibian studies, as well as the Bahamian iguana research program (Knapp 2004), allow researchers to work with the Education Department and involve high school and university students in field research. Shedd's research team has been working with communications and marketing staff to promote two citizen-science smart phone applications that were enhanced (iSeahorse), or developed (Great Lakes Fish Finder) by Shedd. Involving laypeople in science projects can enhance civic engagement and activity and thus has the potential to raise public awareness about environmental concerns (Silvertown 2009). Citizen science opportunities also provide zoological organizations an engagement strategy for reaching new audiences and building an affinity for aquariums and zoos.

There is growing interest among conservation planners in considering the probability of success and conservation costs when setting conservation priorities and allocating effort (Naidoo et al. 2006; Joseph, Maloney, and Possingham 2009). These trade-offs are inevitable and must be considered to minimize the inefficient allocation of conservation resources and unsuccessful outcomes (Murdoch et al. 2007). Shedd uses triage (Bottrill et al. 2008) to consider biodiversity benefits, probability of success, and cost when making program decisions. Actions that provide the greatest benefit to biodiversity persistence are higher priorities. Net biodiversity benefits are measured as the difference in outcomes with and without conservation action. If a species/population/habitat is likely to persist without a particular action, then the intervention will have a lower net biodiversity benefit than if a species/population/habitat would not likely persist without conservation action. In other words, Shedd evaluates the conservation urgency of engaging in a program relative to the threat facing the targeted asset (species/population/habitat).

Uncertainty concerning whether an action will achieve its stated goal is arguably the most overlooked parameter in conservation investments, so resources can be wasted on improbable endeavors with little chance of success (Bottrill et al. 2008). As such, when selecting programs Shedd considers the probability that conservation action will succeed. Factors used to determine the probability of success include the biological potential of a species/population/habitat to recover or persist, existing social or legislative conditions, and the willingness or capacity of relevant management groups to facilitate the action. Finally, conservation projects are constrained by limited budgets, and thus the cost of conservation action is a crucial component of decision making, though it is rarely considered explicitly (Bode et al. 2008). Shedd does consider the cost of engaging in a program and generally (but not always) favors a cheaper conservation action if all program parameters are equal. This approach allows Shedd to maximize research support for a variety of conservation needs.

PROGRAM EVALUATION

Investments in conservation can be costly, and as a consequence of decreasing conservation funding and the relative lack of progress in halting the decline of global biodiversity (Butchart et al. 2010), conservation organizations including aquariums and zoos now face increasing expectations that investments should result in effective outcomes (Nicholls 2004). Evaluation of conservation programs is demanding because realizing conservation success requires time frames beyond an annual review cycle. Thus conservation performance evaluations have historically focused on inventories of inputs (e.g., investment dollars) and outputs (e.g., scientific publications, workshops, stakeholders trained) and less on outcomes (e.g., species and habitat improvements) (Ferraro and Pattanayak 2006). Various methods for evaluating conservation programs have been developed, including business excellence models (Black, Meredith, and Groombridge 2011) and program evaluation processes (Ferraro and Pattanayak 2006) that make inferences about an unobserved counterfactual event: What would have happened had there been no intervention?

Assessments of conserving biodiversity using counterfactual cases on a global scale can reveal a pattern of conservation success while underscoring the inefficiency of current conservation efforts in offsetting the main drivers of biodiversity loss (Hoffmann et al. 2010). An analogous model for zoos and aquariums is the index used by the Durrell Conservation Trust (http://www.durrell.org/wildlife/durrell-index/explore/), which is a mix of performance indicators that includes inputs, outputs, and outcomes (using counterfactual assessments of the IUCN Red List index without Durrell intervention).

New in situ research programs are difficult to assess using indexes such as the IUCN Red List Index, because realized outcomes require many years of consistent conservation intervention. Based on discussions with staff members and stakeholders, Shedd currently sets annual scalable objectives that must be met to achieve a larger conservation goal for each research program (see Carroll et al. 2007). Programs are evaluated based on meeting targeted annual objectives, which include inputs, outputs, and outcomes.

STAFF CONSIDERATIONS

The need to conserve biodiversity has stimulated the creation of diverse conservation entities including local and national government departments and a rapidly increasing number of NGOs and zoological organizations (WAZA 2005; Bode et al. 2010). As in most work environments, organizational cultures differ among conservation groups. Difference in work culture is a critical consideration when hiring conservation research scientists at zoological organizations, since most candidates will be formally trained, coming from academia. Candidates for research positions at zoological organizations must understand and accept the unique challenges and opportunities of working with nonacademic organizations. In academia, productivity is judged based on quantitative indexes such as publications and citations and grant funding (Grant et al. 2007), while multidisciplinary or interdisciplinary efforts have traditionally been undervalued and underrewarded. Working in conservation requires considering both ecological and social factors to achieve desired outcomes. However, most highly disciplinary academic programs, though improving, have struggled to develop approaches that integrate these perspectives and allow a deeper understanding of the socioecological systems in which they occur (Andrade et al. 2014). Thus the conservation goals and objectives of a position must be made clear, and candidates should demonstrate experience or a sincere desire to conduct interdisciplinary research that prioritizes advancing conservation.

In addition, working at an aquarium or zoo will most likely require the ability to speak effectively to a general audience (e.g., visitors, donors, trustees), to disseminate results using various approaches (e.g., peer-reviewed journals, scientific conferences, blogs, organizational newsletters, social media platforms, broadcast and print media), and to work interdepartmentally on various projects (e.g., planning and design, development, marketing, education, husbandry). A research scientist at an aquarium or zoo must be responsive to the culture and have a genuine desire to work in an interdisciplinary manner to achieve conservation outcomes in a nontraditional venue. Ignoring, or not disclosing, differences in culture and daily responsibilities risks having a conservation scientist who is ill-equipped to fulfill organizational expectations or feels unsatisfied in the role. Worse still, a conservation scientist with an unyielding academic mentality can prove disruptive to an interdisciplinary team.

To be sure, the advantages of having a conservation field biologist work in an aquarium or zoo are significant. Rarely do researchers have the opportunity to work collectively, under one roof, with diverse skill sets to advance conservation. Working with development, communications, animal husbandry, veterinary health, education, and exhibit design staff members maximizes the effectiveness of advancing conservation through field research while also increasing organizational support for conservation in general. Working in this environment can thus be particularly compelling to academically trained scientists, but it takes administrative effort to guide them.

ORGANIZATIONAL CONSIDERATIONS

Aquariums and zoos must also consider that effective conservation requires time. Conducting good science, nurturing stakeholders, implementing and evaluating strategies, and realizing success can take decades. Institutional support for conservation research must be absolute to realize conservation goals. As research departments within aquariums and zoos continue to expand, organizations with new programs must adapt to meet the demands of research staff. For example, the accounting department at Shedd modified institutional expense reports in response to researchers' spending several months in other countries. Procedures for transferring funds were adapted to meet researchers' needs for cash and for working extended periods in remote locations. Since advancing conservation initiatives based on science requires in-person visits and participation at meetings, budgets were expanded to accommodate travel. The development team learned to respond to collaborative grant opportunities, and to erratic deadlines, with academic organizations. The communications team developed strategies to promote research and reach new audiences. The planning and design department incorporated research stories into exhibit spaces as well as modifying workflow procedures to accommodate the unpredictable need for outputs such as videos (http://iseahorse.org/trends-how-to-videos), downloadable assets (Loh, Knapp, and Foster 2014), posters, and shirts that advance field conservation. The exhibit fabrications team helped to design, and also crafted, customized research equipment. Shedd's husbandry and animal health professionals modified work schedules to participate in field research and train staff members about veterinary procedures.

The structure of the research department must be thoughtfully considered and will vary depending on individual organizational priorities and projects. When developing a research team, Shedd initially experimented with a postdoctoral model consisting of three-year term positions only, managed by a full-time supervisor. This strategy provided a scaled approach to developing a permanent team and allowed time to evaluate projects and personnel that were a good fit for the organization. However, high staff turnover associated with term positions can be a problem. The time required to develop partnerships and build trust, to acquire consistent funding streams and develop the capacity to submit large, multiyear proposals, and to incorporate research stories into other organizations' experiences (e.g., exhibit spaces, education programs) is often longer than a postdoctoral term. Another benefit of recruiting full-time research staff members is that it fosters support for the organization and its mission. Limited term positions can still be effective in certain circumstances, however, such as training future conservation biologists and answering targeted conservation questions. For example, Shedd currently supports a graduate student studying an amphibian species associated with a temporary special exhibit. The work is important from a conservation perspective and also maximizes the ability to communicate conservation issues concerning amphibians within and beyond the special exhibit space.

The satisfaction and retention of a conservation science team also depends on the culture of conservation at an organization. Conservation research scientists are incredibly passionate, dedicated, and intelligent. Their sense of fulfillment and job satisfaction is predicated on understanding that other staff members, boards, and governing authorities of zoological facilities are fully committed to the conservation of the natural world. Indeed, an institutional ethos and ethic of care for the natural world (as in any conservation-oriented NGO) will help inspire the work of a research team and provide further meaning to their personal efforts.

CONCLUSION

Zoological organizations in the twenty-first century are afforded manifold opportunities to advance biodiversity conservation. Indeed, these organizations are unique spaces serving as informal learning centers that can inspire visitors to connect with the natural world and encourage behaviors that support biodiversity conservation. As zoological organizations evolve in their role of addressing the biodiversity extinction crisis as well as proactively demonstrating relevance in a changing public opinion landscape, their conservation portfolios are expanding. An increasingly popular approach among zoological organizations for addressing these issues more directly is the support of in situ conservation research teams. The elements a zoological organization can bring to bear on conservation issues (e.g., outreach, learning opportunities, communications, veterinary expertise, and exhibit interpretation), as well as the corresponding research to mount a response, are unique and go beyond the capabilities of many academic institutions or NGOs. However, the distinct nature of zoological organizations relative to academic institutions and traditional environmental nonprofits presents opportunities and challenges that must be considered when developing conservation research platforms.

Zoological organizations that consider supporting in situ conservation research teams must consider the organizational changes required to be successful as well as identify research candidates who can adapt to a nontraditional research organization. Organizational support must to be unwavering and reflected in the culture of an organization from governing authorities to all staff members. Budgets and financial priorities need to be reassessed. Using established criteria for selecting research programs should be an objective and transparent process that focuses on the principal conservation goals and objectives of an organization. Organizations must also prioritize evaluating programs to ensure that objectives are met and funds are used responsibly. Because productivity is evaluated differently in zoological organizations than in academic institutions, the conservation goals and objectives of a position must be made clear, and candidates should demonstrate experience or a sincere desire to conduct interdisciplinary research that advances conservation.

Aquariums and zoos have adapted through time to meet the welfare needs of animals and the expectations of people (Zimmermann et al. 2007). Though aquariums and zoos face increasing scrutiny, their role in advancing conservation is critical today. If managed correctly, incorporating active in situ research into the many conservation opportunities afforded to aquariums and zoos is another mechanism for ensuring that the counterparts of the species in their collections remain safe in the wild while demonstrating the relevance of such organizations for contributing to scientific research and protecting biodiversity.

ACKNOWLEDGMENTS

I am deeply grateful to the many people who provided generous counsel on developing and maintaining a research team at Shedd Aquarium. I am particularly grateful for constructive discussions with Allison Alberts, Lisa Faust, Steve Thompson, Anna George, and Michael Tlusty. Andy Kough and Rebecca Gericke provided thoughtful comments on earlier versions of the chapter.

CHAPTER TWENTY-FOUR

Frogs in Glass Boxes: Responses of Zoos to Global Amphibian Extinctions

Joseph R. Mendelson III

THE BACKGROUND OF AMPHIBIANS IN DECLINE

In retrospect, the history of the amphibian crisis will be a fascinating case study of science, society, and conservation—a timeline of paradigm shifts in regard to global conservation (Mace 2014). Collins and Crump (2009) wrote the first chapter of the story, and their book is highly recommended. After the first anecdotal discussions of problems with amphibians (Rabb 1990), we spent the next fifteen years slowly realizing the fact and extent of the onset of a contemporary mass extinction event (Vredenburg and Wake 2008). Lacking any contemporary precedent for biodiversity losses at such a scale, the academic tradition of reliance on hard data—which largely did not exist—had to be reconciled with the anecdotal realities of one-time observations of dead amphibians or, more typically, amphibian populations or species simply gone missing. Retrospective studies indicated that some declines had actually occurred decades ago but had gone unnoticed (Lips et al. 2004; Cheng et al. 2011). An academic debate raged through the 1990s regarding the root cause(s) of such unimaginable situations. Eventually a previously unknown pathogenic chytrid fungus (*Batrachochytrium dendrobatidis, Bd*) was identified as a primary cause of declines (Berger et al. 1998; Longcore, Pessier, and Nichols 1999).

A conservative estimate is that one-third of amphibian species are threatened, with hundreds more classified as data deficient, which often means they simply cannot be found (Stuart et al. 2004). We now know that *Bd* was a direct cause of many of these declines, though habitat loss also has devastating effects. Since the biology of *Bd* was still poorly known, a breakthrough study in Panama (Lips et al. 2006) finally demonstrated the direct effects *Bd* can have on otherwise nonthreatened populations. This was the dataset to rally the stakeholders to face the daunting realities of this devastating emerging infectious disease in wildlife.

Based on Lips's as yet unpublished data, a small group of international stakeholders met in 2004 to consider the implications of her findings. Her data indicated a clear spatiotemporal spread of epidemic chytridiomycosis in Central America and thus allowed predictions of new epidemics at currently healthy sites. The meeting emerged with a suggestion that, in the absence of any ability to control these epidemics in the wild, emergency captive collections should be established to protect at least some of the diversity at sites before the invasion of *Bd*. Such safe-house colonies seemed a logical match for the mission of zoos, and thus the concept of an ark for amphibians was formed. A pilot project was launched in El Valle, Panama—just ahead of Lips's predicted pathogen trajectory—to assess the logistics of quickly establishing multiple species of amphibians in captivity. The lessons learned from this project (Gagliardo et al. 2008) were sobering and emphasized that many amphibian species are challenging to maintain and reproduce in captivity. The handful of amphibian species that predominate in the pet trade and in zoo exhibits do not serve as informative proxies for communities as taxonomically and ecologically diverse as lower Central America (see also Smith and Sunderland 2014). The project also indicated that real challenges of diplomacy, logistics, and animal health dictate that captive survival assurance colonies should be developed in the range country of the species in need.

To begin the preparations for a global zoo response to safeguard amphibians, the IUCN Conservation Breeding Specialist Group (CBSG) convened an international meeting to produce initial guidelines and protocols (Zippel, Lacy, and Byers 2006). These guidelines emphasized pathogen biosecurity and development of in-country programs and encouraged the innovative use of commercial shipping containers as affordable, somewhat portable, pod units for housing amphibians in the absence of existing zoo infrastructure. Maintaining extensive proactive survival assurance colonies in captivity as a conservation strategy was included in the IUCN Amphibian Conservation Action Plan (Gascon et al. 2007) and was communicated widely (Mendelson and Rabb 2005; Mendelson et al. 2006; Zippel and Mendelson 2008).

RESPONSE OF ZOOS TO AMPHIBIAN DECLINES

To help establish the global network necessary to accommodate the hundreds of critically endangered species anticipated to be brought into captivity, CBSG and the World Association of Zoos and Aquariums launched the Amphibian Ark (AArk 2016a) program in 2007. The priority amphibian species for which ex situ conservation programs are recommended have been identified by regional or national working groups in many parts of the globe using the AArk Conservation Needs Assessment framework (AArk 2012). Materials to inform global preparations for amphibian captive programs emerged in various forms: CBSG published guidance on infrastructure and biosecurity (Zippel, Lacy, and Byers 2006) and a manual for management of infectious diseases in captivity (Pessier and Mendelson 2010); AArk produced guidelines for taxon management in captivity (Schad 2008); AArk and the Zoological Society of London offered amphibian husbandry training workshops worldwide, and the Association of Zoos and Aquariums (AZA, the North American zoo-accrediting organization) did the same in the United States; and in 2012 AZA published a manual on basic husbandry (Poole and Grow 2012).

These new attentions to the plight of amphibians culminated in a worldwide "2008 Year of the Frog" campaign sponsored by the world's major zoo associations. The campaign aimed to raise funds and awareness and to build the capacity across zoos and local partner organizations to establish the approximately five hundred rescue colonies in the range countries identified by regional prioritization workshops. A clear message to zoos was that amphibian programs could be maintained for decades in relatively small spaces, and for a fraction of the cost of even a single new exhibit for a large mammal. AArk encouraged each institution to commit to just a single species of amphibian in need; this would ensure that global capacity would quickly approach the scale of the necessary response. The public relations offices of zoos worldwide waved the banner of the 2008 Year of the Frog, highlighting zoos' important role in conservation.

Zoos broadcast the message that they were going to step up to secure hundreds of threatened species. Yet the reality is that they have missed their own stated goal by large margins (Dawson et al. 2015; Harding, Griffiths, and Pavajeau 2015). For example, the panels using the AArk prioritization tools (AArk 2016b) have so far identified 196 species globally for rescue programs (species that are in imminent danger of extinction and need ex situ management); at the same point, AArk listed 55 of those species actually in rescue programs (AArk 2016c). Dawson et al. (2015) reported that by

2014 zoos maintained 6.2 percent of the world's threatened amphibian species, and they commented that this figure was far lower than those for threatened birds and other reptiles or mammals. Importantly, the values from Dawson et al. (2015) are underestimates because NGO-style programs were not included in the surveys, or are overestimates because the surveys merely indicate that an endangered species is present in the collection, not whether it is part of a successful breeding program. More optimistically, Harding, Griffiths, and Pavajeau (2015) reported an increase in amphibian programs following the publication of the IUCN Amphibian Action Plan (Gascon et al. 2007), but they also commented on the imbalance with respect to birds and mammals. Efforts by the AZA Amphibian Taxonomy Advisory Group (ATAG; Barber and Poole 2014) allow a detailed evaluation response by AZA institutions:

- Between 2009 and 2011:
  - 24 percent of institutions increased isolated space appropriate for amphibian conservation programs
  - 57 percent had not developed such isolated space
  - 75 percent did not anticipate adding any such space over the next five years
- In 2014:
  - 38 percent had isolated space
  - But only 50 percent of that space was being used for species prioritized for an ex situ conservation program
  - 70 percent still did not anticipate adding any such space

THE CULTURE OF AMPHIBIAN PROGRAMS IN ZOOS

Highlighting zoos' failure to respond to a global crisis is not very constructive. Rather, I am interested in the evident disconnect between zoos' rhetoric and their response to the conservation needs of amphibians, and I approach the issue in the framework set by Barrow (this volume). Emulating his questions, we can ask, Why aren't modern zoos more engaged in amphibian conservation? Or why have a few become so notably engaged when others have not? How should we interpret the discrepancy between the consensus-based needs assessments and the low response of zoos? The "one species per zoo" message was practical and well received, yet very few zoos followed up. Are there realities specific to amphibians that make them less attractive or less tractable for zoos than other threatened taxa?

Let's acknowledge a few realities. Most amphibians make poor exhibit species because of their small size, drab coloration, or secretive behaviors. Most zoos have settled on a short list of species that exhibit well (Barber and Poole 2014, table 8). While many zoo-based conservation programs maintain animals in off-exhibit space, they typically also display a few "ambassador animals" to educate visitors about the species and the zoo's often considerable investment in its conservation. Few amphibians can serve as effective ambassadors for their own species, but Palmer, Kasperbauer, and Sandøe (this volume) discuss the issue more generally and provide counterarguments. Conway (2011) criticized zoos for their poor record of investing resources into such uncharismatic species in need of conservation in favor of charismatic exhibits for nonthreatened species. With regard to amphibians, Conway's criticism becomes more complicated because many zoos display endangered, but charismatic, species (e.g., poison frogs, family Dendrobatidae) that are bred and traded among zoos and private breeders. The founders often were derived from the black market (e.g., any species from Brazil) or the commercial pet trade, without reliable locality information. They are maintained without genetic management and without connection to reputable conservation programs. These animals conflate surveys by Dawson et al. (2015) because they represent endangered species being bred in zoos, but in the absence of any conservation efforts. In a position statement letter opposing any government regulation of transport of *Bd*-infected amphibians, AZA stated, "Our source of founder animals, and fresh breeding stock, will largely disappear if the commercial trade dwindles" (Olson 2010). Barongi (this volume) reinforced the importance and effectiveness of connecting zoo exhibits with actual conservation programs, yet this is not being done with poison frogs or with most other amphibians. The same letter claims that "AZA fully supports the control of invasive species, however we also recognize that our accredited zoos and aquariums are not the source of the problem and thus must not be unduly penalized by any regulatory remedy . . . for amphibians." This position ignores the facts that *Bd* was first discovered in a dead poison frog in an AZA institution and that Schloegel et al. (2012) found multiple genetic strains of *Bd* among AZA collections. Yet AZA denied any role in the spread of *Bd* and defended the commercial trade as essential for cooperative breeding in the absence of investment in conservation for those species. Meanwhile, institutional investment in secretive, dull brown endangered frogs (e.g., family Strabomantidae) has been essentially nil (fig. 24.1).

Another disparity exists in the tradition of herpetology departments at most zoos, in accordance with the academic tradition of studying amphibi-

FIGURE 24.1. A dendrobatid poison frog and a strabomantid streamside frog for comparison of charismatic charms. Images by Brian Gratwicke, used by permission.

ans and a subset of reptiles that excludes birds, creating mixed departments combining reptiles and amphibians. The issue here is that zoo herpetology programs typically have an extraordinary bias, by any measure, for nonavian reptiles, in terms of both their collections and managed breeding programs (AZA SSPs: seventy-seven reptiles versus seven amphibians; aza.org, 9 December 2015). Based on personal experience, I gauge that the level of interest in amphibians among zoo herpetology staff sometimes borders on antipathy. I think Detroit Zoo got it right in developing a dedicated amphibian program.

A BROADER PERSPECTIVE

The emphasis on developing amphibian programs within the range country or region (Gagliardo et al. 2008; Carrillo, Johnson, and Mendelson 2015) highlights the perennial problem of insufficient infrastructure and training in many countries. Capacity-building workshops by AArk, AZA, and others aim to reduce these deficits. Zoos do contribute considerable funding to conservation programs worldwide; for example, AZA reported giving over $160 million annually toward conservation across about one hundred countries (AZA 2016b). In amphibian programs, these resources typically are in the form of funding or in-kind labor/expertise donated to *existing* foreign programs. Few US zoos, evidently, can or will commit to developing and maintaining a facility in another country, at least for amphibians. But other successful models include biosecure use of AZA zoos to fuel reintroduction programs in other countries, such as for the Puerto Rican crested toad (Bufonidae: *Peltophryne lemur*) or the Kihansi spray toad (Bufonidae: *Nectophrynoides asperginis*) in Tanzania. These exceptional programs appear to be working, with at least one population of toads in Puerto Rico being reestablished (Beauclerc, Johnson, and White 2010) and estimates of toads in Tanzania rising from zero to approximately six hundred in the wild (J. Pramuk, pers. comm.).

Lacking the will or the ability to create foreign in-country amphibian ex situ programs, AZA should still be on track to meet the conservation needs of amphibians in its jurisdiction of North America and the Caribbean, fitting the in-range/in-country model. However, Barber and Poole (2014, 10) suggested that AZA institutions may be misappropriating their amphibian conservation resources, highlighting a trend where zoos are working with regional populations of species identified by state wildlife agencies rather than with species identified and prioritized globally. For example, in response to state agencies, several AZA institutions invest considerable resources into ex situ programs for hellbender salamanders (*Cryptobranchus* spp.), although the salamanders are listed only as near threatened and were prioritized by AZA for in situ and ex situ research (only); coincidentally, these are popular display species. This embodies well the "think globally, act locally" ethos, and it is understandable for zoos to cooperate with requests from state agencies. The question to consider here may be, Is the amphibian community sending mixed messages among its stakeholder institutions? The AZA appears to be criticizing its members for following guidelines to develop in-country or in-range programs instead of investing resources for foreign species. In contrast, Australian zoos are heavily invested in native amphibian programs, to the near-total exclusion of nonnative species even for display. Is the Australian model flawed for ignoring nonnative species? Amid the confusion, considerations should be, How many species are being missed entirely? and At various scales, are resources being used most effectively? Is it more effective to import founders to developed countries and then export the progeny for reintroduction programs, or to develop and maintain capacity in the range country? The former model is expensive and is challenged by logistics and diplomacy; few zoos likely can accomplish it. The latter model also is challenging but follows general recommendations. By my estimate, the native species programs at AZA institutions are well conceived and implemented and are appropriately aligned with the goals of US state and federal agencies. Yet the report by Barber and Poole (2014) seems to diminish some of the programs for native US species and urge zoos toward the out-of-range model for foreign species. This mixed message from the influential AZA leaders does not help member zoos raise support for new local programs, nor does it encourage development of foreign programs. A case study on the golden frogs (Bufonidae: *Atelopus*) from Central America highlights these as well as other pervasive issues.

A CASE STUDY ON THE REALITIES OF AMPHIBIAN CHYTRIDIOMYCOSIS

The realization of emerging infectious diseases as direct threats to biodiversity was a paradigm shift in conservation programs because it revealed no way to safeguard species in protected areas (Mendelson et al. 2006; Berger et al. 2016). Realities of disease also changed the realities of ex situ programs. Project Golden Frog (Project Golden Frog 2016) exemplifies the confounding role that chytridiomycosis played during the evolution of a conservation program. The project launched in 1998 as a consortium of AZA institutions to confront the threats of habitat loss and overharvesting for two species (*Atelopus varius* and *A. zeteki*; fig. 24.2) in Panama, as well as in anticipation of the future effects that *Bd* may have on populations. At the beginning of the program *Bd* was not affecting either species in central Panama, but its threat was understood (Zippel 2002). By 2002, populations of *A. varius* in Costa Rica and western Panama were virtually eradicated by

FIGURE 24.2. A captive Panamanian golden frog (Atelopus zeteki). Image by Brian Gratwicke, used by permission.

Bd. Just before this, founder animals were collected from the wild and exported to AZA institutions, with the intention of breeding them and returning the progeny to Panama for reintroduction and to establish a breeding and education center there. In the United States, breeding was successful in carefully managed genetic bloodlines. This is an attractive, active display species, with a backstory as a cultural icon in Panamanian culture as well as the dramatic conservation success the captive program appeared to represent.

The project was on track to be a conservation success until *Bd* invaded central Panama in 2004 (Lips et al. 2006), when all wild populations were virtually eliminated (along with other species). The ex situ populations of *A. zeteki* and *A. varius* in the United States were secure, but all wild populations of *A. zeteki* disappeared, while very small numbers of wild *A. varius* can be found (Perez et al. 2014). This scenario seems to be a perfect justification for an ambitious reintroduction program to repopulate the areas devastated by *Bd*. This is where a number of issues arise:

1. *Atelopus* are extraordinarily susceptible to chytridiomycosis, showing very high mortality. *Bd* is now endemic across the original range of these frogs (Perez et al. 2014), and all evidence suggests that reintroduced individuals likely will simply die.
2. Experimental evidence (DiRenzo et al. 2014) shows that *A. zeteki* is a *Bd* "supershedder" and that reintroducing animals into the wild may create new local epidemics that would affect both the reintroduced frogs and other local amphibians.

These issues pose significant ethical challenges, and the most recent conservation plan (Estrada et al. 2014) reveals no progress toward their resolution, but not for lack of discussion. How should conservationists proceed in a scenario where reintroduced animals will certainly suffer high mortality and where reintroduction may make the original problem worse? The result has been the inevitable standstill. As of 2013 there were approximately fifteen hundred golden frogs (mostly *A. zeteki*) in AZA institutions, with annual breeding. Space constraints and genetic management are such that surplus eggs or larvae are culled, with some juveniles being provided to *Bd* research labs. Both of these options have their harsh ethical critics (see reviews by Hutchins, Dresser, and Wemmer 1995; Lindburg and Lindburg 1995). Progress toward reintroductions, however, is stymied by the reality of chytridiomycosis. The case study reveals a well-intentioned and gener-

FIGURE 24.3. El Valle Amphibian Conservation Center in 2013. Generations of frogs in glass boxes, just one brick wall and ten meters of grass away from the forest to which they may not be returnable. Image by Julia Paull, used by permission.

ally well-managed conservation breeding program that has no exit strategy in sight because of the intractable nature of fungal diseases. Discarding optimism for the moment, there is a chance that *Bd* may simply prevent successful reestablishment of golden frogs in the wild. Is it possible that these frogs will persist for eternity only in glass boxes? (See fig. 24.3.)

These frogs in glass boxes are ecological ghosts, with some perverse implications. This scenario becomes the model of "protecting individuals of endangered species" rather than the "improving management of larger ecological systems" that Norton (this volume) encourages of zoos. It also presages some of the criticisms of the so-called de-extinction enterprises (Ehrlich 2014; see also Friese, this volume), in that it leads to the same "OK, now what?" conundrum, but from the preextinction rather than postextinction side of the timeline. Geist (1995) approached the issue conceptually, arguing for maintaining species in captivity even in the most pessimistic environmental situations—holding on to hope for some positive development at some unknowable time in the future. It is difficult to argue against such unqualified optimism, but the conversation needs to be grounded in objective realism.

MOVING FORWARD

This problem is larger than the golden frogs. The rise in fungal diseases in wildlife globally has dramatic implications for the future of biodiversity and for human and global health in general (Fisher et al. 2012). The precedent of emerging infectious diseases in wildlife, learned from amphibians, is now playing out in taxa as diverse as bats (Turner, Reeder, and Coleman 2011) and sea stars (Hewson et al. 2014). Some guarded optimism is evident in recent efforts suggesting that some amphibian populations originally decimated by *Bd* may be recovering (Abarca et al. 2010; González-Maya et al. 2013; Newell, Goldingay, and Brooks 2013; Perez et al. 2014); thus the creative programs under way in Australia (Scheele et al. 2014) may hold promise. But this also implies that we may be off base in setting our species priorities. Perhaps the amphibians that are showing evidence of tolerating *Bd* should be prioritized for assisted recovery and we should just accept that some species will perish. This approach seems more realistic and efficient, even if it is not the most satisfying. This is different from prioritizing a species because it has no hope against disease in the wild. But accepting that "we cannot save them all" is difficult for some, and suggesting that nothing can be done in tragic cases will be controversial.

Epidemics of chytridiomycosis were caused by globalization involving movements of amphibians and potentially other materials containing *Bd* (Rosenblum et al. 2013; James et al. 2015); other wildlife epidemics appear to be similar. What is lacking are robust national and international policies to reduce the likelihood of transporting wildlife pathogens, although many such policies exist with regard to agricultural plants and animals (Voyles et al. 2014). In 2009 Defenders of Wildlife petitioned the US Department of the Interior to invoke the Lacey Act to declare amphibians infected with *Bd* injurious wildlife and therefore prohibited in interstate trade and importation. This would require that health certificates accompany transported amphibians, as is common for many animal species. The petition was never acted on, in part owing to opposition from AZA (Olson 2010) that dismissed the risks of *Bd* in the United States and claimed that a permit process would eradicate zoo-based conservation and education efforts. Farmers and state governments in the 1900s similarly opposed federal programs to eliminate Texas fever in cattle (Olmstead and Rhode 2015); that legislation passed, and the disease is no longer a major concern. Emerging wildlife diseases—involving both known and novel pathogens—are increasingly common and serious, yet putative conservation organizations such as AZA do not appear consistent or current in their practices. The recent discovery of an additional pathogenic amphibian chytrid, *B. salamandrivorans* (*Bsal*; Martel et al. 2013), in salamanders underscores this problem, as the pathogen appears not yet to be present in North America (Grant et al. 2015), where it is certain to affect a number of species. In spite of opposition from AZA (Vehrs 2015), the US Fish and Wildlife Service implemented an interim rule in 2015, using the Lacey Act, to regulate transport of salamanders.

Conway (2011) made blunt recommendations that zoos, if they are to call themselves conservation organizations, need to reevaluate their missions and operations. Applied to amphibians, Conway's thesis indicates that zoos need to aggressively resolve the following issues regarding resources:

- Preference for charismatic species that make for nice displays over the less attractive species of conservation concern
- Maintaining populations of threatened species that have no real connection to conservation programs
- Disinclination to support non-US conservation programs

I would add to this that some regional zoo associations need to consider their mandate and reconcile mixed policies with regard to the state, federal, and international conservation priorities. Palmer, Kasperbauer, and Sandøe (this volume) review the advantages to locally based zoo conservation programs.

Conservation practitioners have entered a new realm of complex ethical challenges for which we are wholly unprepared (Minteer and Collins 2005, 2013). The difficult discussions must be initiated immediately, because they are even more difficult when the situation is at hand: the last golden frogs are sitting in glass boxes, and the last individual of Rabbs' fringe-limbed tree frog passed away in a glass box (Mendelson 2016). While the issues will be focal at zoos, diverse-stakeholder discussions are necessary to fully grasp the implications of the conservation realities on a rapidly changing planet. At stake for zoos will be their willingness to step up as leaders in the research (Minteer and Collins 2013), some of which may be invasive or terminal, in order to address considerable conservation difficulties. Also at stake will be their willingness to address and resolve Conway's (2011) challenges and to seriously reconsider how a local zoo can best contribute to a global problem involving thousands of species. Optimism and realism inevitably will clash when some species are knowingly left off the ark or unsuccessful programs are phased out. As the first taxon to signal the current sixth mass extinction (Wake and Vredenburg 2008), amphibians are challenging zoos and others to find solutions to conservation issues that are fundamentally new and overwhelming. The concept of an ark is no longer simple, whether in the form of a zoo or as a protected area.

ACKNOWLEDGMENTS

I am grateful to Karen Lips, Kevin Johnson, and Anne Baker, who provided insightful conversations or comments on drafts of this chapter. Thanks also to the editors of this volume for their comments and suggestions.

PART VI

Alternative Models and Futures

CHAPTER TWENTY-FIVE

Sustaining Wildlife Populations in Human Care: An Existential Value Proposition for Zoos

Steven L. Monfort and Catherine A. Christen

INTRODUCTION

In the face of widespread concerns about biodiversity and habitat losses, accredited zoos of the late twentieth century were motivated by appeals to become modern centers focused on holistic conservation alongside their traditional exhibition and entertainment functions. Conservation holism in zoos was conceived as encompassing excellent care of individual animals, including reinforcing their natural behaviors, and scientific management to improve the long-term sustainability of genetically and demographically viable populations of endangered species. Also expected was zoos' support of external initiatives to protect and restore the habitats wildlife need for survival in nature (International Union of Directors of Zoological Gardens 1993; Rabb 1994; Conway 1995a). In response, zoo-based programs such as Species Survival Plans, Taxon Advisory Groups, Fauna Interest Groups, and Scientific Advisory Groups were inaugurated in the 1980s and 1990s, amid rising confidence that the zoo community was on an appropriate new path emphasizing sustaining species in human care to help avert an apparently impending biodiversity apocalypse.

Fast-forward to the present decade. Like every other conservation-minded organization, zoos have been overwhelmed by catastrophic declines in biodiversity fueled by unrelenting human impacts on the natural world. Zoos have responded by renewing their commitment to holistic missions centered on animal care, public engagement, research, and conservation (Barongi et al. 2015; see also Barongi, this volume). Yet in meeting these mission goals the zoo industry is dogged by a wicked problem: compromised reproductive management among too few founder lineages and, ultimately, unsustainable zoo populations of endangered and even common species (Lees and Wilcken 2009). In 2011 the Conservation Breeding Specialist Group (CBSG) concluded that "most zoo populations are not being managed at adequate population sizes, reproductive rates, genetic diversity levels, and projected long-term viability that would allow them to contribute positively to species conservation" (CBSG 2011). It is a paradox that modern, accredited zoos provide exemplary veterinary and keeper care and management for the individual animals in their collections, including increasingly sophisticated reinforcement of natural behaviors yet, lacking adequate intellectual and physical resources, struggle to sustain the genetic and demographic viability of their managed populations of species (see Ryder, this volume; Tubbs, this volume; Traylor-Holzer, Leus, and Byers, this volume).

ZOOS AND SPECIES SUSTAINABILITY

In zoo biology terms, "sustainability" means achieving zoo-managed populations of species that retain at least 90 percent of the genetic diversity originally derived from their wild source population(s) for a minimum of one hundred years (Lacy 2013; also Cerezo and Kapsar, this volume). The acceptability of this standard, which simply slows the inevitable loss of genetic diversity in "closed populations," continues to be debated among present-day zoo and population ecologists (Lacy 2013). That discussion might draw our attention more closely here if zoos today generally approached this established sustainability goal, but most fall far short.

We contend that zoos' ongoing failure to sustain the genetic and demographic viability of their critical populations of endangered species effectively devalues animals in ways that are bad both for zoo business and for conservation. We propose that to attain a viable future the zoo industry must recommit to its own existential value proposition of being the foremost authority in understanding and sustaining the living biodiversity in its care. Zoos now espouse conservation but struggle to sustainably manage their own critical populations. Abrogating this management responsibility—not because meeting it is impossible, but because it is difficult and expensive—also prevents zoos from retaining their own moral and ethical high ground. It forces them to oversell many other zoo functions, including education, inspiration, entertainment, even the use of animal exhibits to generate field conservation funding, as ostensible—but highly debatable—justifications for holding animals in captivity (see also Mendelson, this volume).

Without question, achieving genetic and demographic sustainability for zoo-managed populations is expensive and fraught with challenges. Plus, regardless of their level of investment, zoos alone will rarely be credited with "saving" a species. In fact, rarely will any one category of institution "save" a given species. The successful recovery of thousands of "conservation-reliant species" will necessarily entail ongoing, active management, most often expressed in collaborations among various institutions and conservation specialties (Scott et al. 2005). Lasting survival for these species will require strong, effective leadership, major new financial investment, and a commitment to working collaboratively along a continuum of ex situ to in situ conservation, with differing levels of intervention, from intensive biotechnology applications, to controlled captive breeding, to metapopulation strategies, managed extractive reserves, and protected areas (CBSG 2011; Conway 2011; Lacy 2013). A functional continuum will depend on zoos' stepping up to fulfill an irreplaceable anchoring role at the level of single-species biology and management, in captive breeding, in reintroduction, and at other levels of intervention.

Can zoos step up? The zoo industry already recognizes that it has a species sustainability problem. Yet, worryingly, this community has not aggressively invested the conservation capital essential for sustaining the genetic diversity and stable age distributions of zoo-managed populations. Such investments include greater production and deployment of new science knowledge and applied management expertise as well as the construction of improved animal management facilities and flexible "sustainability space" in zoos and related venues. Below we briefly address each of these topics.

Knowledge and Expertise

As a community, zoos are not alone in having historically underestimated the biological complexity and diversity of the animal kingdom, its constituent species, and its ecosystem dependencies. In general, surprisingly little is known about the biology of the thousands of species of wildlife whose survival now extensively depends on human care or management. Additionally, surprisingly few researchers working on related topics—basic or applied—are based at zoos. Today only about 20 percent of accredited European and American zoos have research departments, and a much smaller percentage employ full-time PhD-level scientists and conservationists (Reid et al. 2008; see also Knapp, this volume). Zoo science research to date has focused on too few species (Wildt et al. 2003; Monfort 2014) and has been taxonomically biased, reflecting similar skewing in zoo collections toward birds and large mammals (Martin et al. 2014). In short, this level of investment is insufficient for creating the fundamental new knowledge needed to sustain the plethora of species in human care and to support the work of ensuring their survival in nature.

During the past several decades, as many of the chapters in this volume discuss, scientists and animal husbandry experts have seemingly performed "miracles" by rescuing black-footed ferrets, Przewalski's horses, California condors, and a few other animal species from extinction. Yet waiting until founder populations have already fallen to perilously low levels before taking action is extremely risky. During these same past decades, the standards of zoo practice have been successfully elevated to include across-the-board excellence in animal health care, and the best zoos employ qualified full-time veterinary specialists as a matter of course. We suggest that now the industry must similarly greatly increase investment in zoo and conservation biology research and practice, either by regularly employing staff scientists or by forging partnerships or cost-sharing agreements with universities or conservation organizations. The time to do the needed science and integrate it into standard zoo population management is when a species is still relatively abundant.

Facilities and Infrastructure

The greatest barriers for zoos in achieving sustainability are the deficiency of space, dearth of appropriate management tools needed to routinely handle and manipulate animals, lack of flexibility to mimic or manipulate social groupings, and inadequate capacity to deal with multiple-male aggression or surplus animal production. In general, zoo design has emphasized animals' well-being, visitors' safety, and education and entertainment while giving short shrift to investment in acquiring and building adequate functional space, either front or back of house, to support sustainability goals.

Zoos can pursue many promising avenues, either separately or collectively, to increase functional "sustainability space" that can be used to manage highest-priority conservation-reliant species to improve their genetic and demographic sustainability. While not an exhaustive list, suggested approaches include: (1) expanding on-site space, acquiring off-site space, or both; (2) improving the quality of existing space to optimize efficiency of animal management; (3) specializing, that is, reducing the number of species managed in existing space; (4) forming regional consortiums to reduce geographic separation and facilitate resource sharing; (5) partnering with others to share space, facilities, and expertise (e.g., Conservation Centers for Species Survival, or C2S2—a consortium of six conservation centers around the United States that cooperatively manage more than twenty-five thousand acres devoted to endangered species study, management, and recovery) or simply to share expertise alone (e.g., Amphibian Ark; see Mendelson this volume); (6) investing in sustainability spaces owned or operated by others (e.g., Turtle Survival Alliance's Turtle Survival Center); (7) developing and operating new population management facilities for the direct benefit of collaborative zoo owner-investors (e.g., the National Elephant Center); (8) establishing or assisting with the management of range country zoos and conservation centers; and (9) establishing, supporting, or managing metapopulations and extractive reserves.

CAN ZOOS AFFORD TO BE SUSTAINABLE?

Zoos have been referred to as "chimeras . . . composed of a business, a conservation body, and a school" (Fa et al. 2014). Without animals, they cease to function as any of these. The success of the zoo industry depends on a steady supply of animals to drive attendance and thus generate the revenues needed to sustain capital, operations, and investments in public engagement, education, and conservation. Any business at risk of losing control of its supply chain and product pipeline has every reason to be concerned. Zoos should be very concerned. The inconvenient truth zoos face today is that failure to innovate and develop effective species sustainability programs will have catastrophic consequences for the zoo industry.

Impressively, zoos worldwide invest more than $350 million a year in field conservation (Gusset and Dick 2011) despite no expectation of return on investment except for demonstrating tangible support for conservation consistent with their stated missions. In North America alone, AZA-accredited zoos collectively contribute more than $160 million annually for field conservation (3 percent of operating budgets) from an aggregate operating budget of $3.1 billion (pers. comm., Grow, AZA, 2015). Looked at another way, AZA zoos devote upward of $2.95 billion (more than 95 percent) of their budgets annually to unrivaled efforts to ensure the health and well-being of wild animals in their care. Yet, despite tens of billions of dollars invested in infrastructure and billions more each year in operational costs that directly or indirectly support animal care, zoos are unable to draw a straight line between this outlay and a manifest and uniquely zoo-based conservation outcome: the consistent and successful management of genetically and demographically sustainable populations of species, to help hedge against extinction as part of a continuum of conservation efforts.

If zoos globally were owned and operated by a single management entity, the choice would be clear: regroup, retool, and build the effective, resilient supply chain needed to manage species sustainably and provide the animals needed to support both traditional "zoo business" and conservation. In many ways the financial costs and technical barriers to achieving sustainability may be easier to overcome than the formidable problems associated with diverse business models and governance structures, distributed decision making, and obstacles such as geographic separation and an overreliance on volunteers to achieve species sustainability. Regardless of the difficulties, failure to act means surrendering to the inevitable laws of supply and demand, with associated whipsaw effects on animal availability, which will surely drive up zoos' business costs. Few industries gladly welcome increased costs, but zoos have too long operated under the false premise that "if you build it, animals will come" at little or no cost. Those days are waning, and future business models are sure to move toward recovering the very real costs of animal production. Finally, the burden of animal importation and production has for too long been disproportionately borne by a relatively small number of large zoological institutions. As market forces adjust, smaller zoos will have every reason to be concerned that such largesse will not be everlasting.

CONCLUSION

Some in the zoo community apparently have concluded that species sustainability is an intractable problem and that zoos should just double down on efforts to use whatever animals are available to inspire and educate their publics and to generate increased funds to support field conservation. We contend that the core zoo functions of animal care, public engagement, and supporting field conservation are inseparable from the equally important core function of ensuring genetic and demographic sustainability of critical populations. Sustainability of zoo-managed populations should be considered a cornerstone of the industry, one of the foundational elements that helps support the legitimacy of zoos, fuels public engagement, and inspires caring and empathy for species and the ecosystems that sustain them. Likewise, the implicit social contract between zoos and their publics requires that zoos connect the species in their own care to the work they undertake to champion those species' survival in nature. Sustainability is one of the essential elements zoos need to make that connection a reality.

Failure to act on ensuring this sustainability may have limited consequences for zoos (and conservation) in the near term, but over time adverse effects are likely to be cumulative, and public questions regarding the role and value of zoos in society and for conservation are sure to intensify. Ensuring sustainability of zoo populations is the zoo industry's vital niche; without this, zoos are missing a huge opportunity to be an irreplaceable part of the conservation continuum that, in turn, can ensure sustainability of conservation-reliant species in nature. There remains an extraordinary interest in and passion for zoos' becoming holistic conservation organizations. Together, excellence in the care *and* the sustainable management of animals in situ and ex situ exemplifies a holistic perspective of "health and well-being"—both for the individual animals requiring human care and collectively for the species that require sustainable management for survival. We suggest that zoos—still—have an amazing opportunity to expand the definition of "conservation" by developing proven competency in sustaining the genetics and demography of critical populations of species in human care.

CHAPTER TWENTY-SIX

Reflections on Zoos and Aquariums and the Role of the Regional Biopark

Craig Ivanyi and Debra Colodner

> Learn One Big Thing, teach it masterfully, and teach it so well that it rallies the world behind solving the single most intractable dilemma of our time: how Homo sapiens will ever learn to share a sinking ark with any other species but himself.
>
> MIKE WEILBACHER

ZOOS AND THE CONSERVATION CHALLENGE

Zoos may have originated as entertaining displays of personal influence and wealth, but gone are the days of postage stamp menageries assembled from the far corners of the world merely for our amusement (Braverman 2012). Today zoos and aquariums face continuing scrutiny from a public increasingly skeptical of the need for keeping animals in captivity, and they are challenged to demonstrate that their influence is for the greater good (Grajal, Luebke, and Kelly, this volume; Maple and Segura, this volume). In response to this demand, and with a sincere desire to help wild populations, progressive zoological organizations are increasingly focused on conservation.

As described elsewhere in this volume, zoos and aquariums (abbreviated as zoos for most of this chapter) contribute directly to wildlife conservation through both zoo-based (ex situ) work with captive populations and field-based (in situ) work to protect wild populations. Much of the field-based work is accomplished by supporting other conservation organizations working in the region of interest. In 2014, US zoos spent over $150 million on more than three thousand field conservation projects in 130 countries (AZA 2014b). While having significant impact, the amount spent on in situ (in the field) conservation remains relatively small (Gusset and Dick 2011) compared with the overall budgets of zoos and aquariums. During the same year, based on information from 194 AZA-accredited and -certified related institutions, the average percentage of budget spent on direct conservation was 3 percent (Shelly Grow, director of conservation programs at AZA, pers. comm., October 14, 2015). As argued by Grajal, Luebke, and Kelly (this volume), zoos also have an effect on conservation via their influence on the knowledge, attitudes, and behaviors of their visitors. However, the nature and magnitude of this effect is more indirect and is hotly debated.

In spite of this relatively new and commendable orientation toward conservation, zoos still have many detractors. Some critics don't believe there is a legitimate reason for zoos to exist in modern times (see Palmer, Kasperbauer, and Sandøe, this volume; PETA 2015). Others are not completely opposed to animals in captivity but question the industry's effectiveness at delivering conservation outcomes (Fravel 2003). To maintain public support in the future, zoos will need to achieve excellence in animal care, in situ and ex situ conservation, and conservation education including behavior change. This will require more research on animal behavior to inform the design of captive environments that meet not only animals' needs, but also their desires (Maple and Segura, this volume). Likewise, we need more research on how zoo visits affect visitors' knowledge, attitudes, and behaviors, especially over the long term and in the context of other social factors (see, e.g., Clayton and Le Nquyen, this volume). Not surprisingly, we also need more research on the effectiveness of different direct (in situ and ex situ) conservation strategies.

All these research areas are complex, multidimensional, and not well funded, but as evinced by many of the chapters in this volume, research is currently being conducted in them. Accredited zoos continually study animal behavior and incorporate new knowledge into new expectations and standards for the industry (Knapp, this volume). Regarding conservation education, zoos commonly use audience size as a measure of their success. Nearly 180 million people visit zoos each year in the United States (Fuller 2011), and over 700 million do so worldwide, representing a huge *potential* for change (Gusset and Dick 2011). Zoos also point to surveys that demonstrate positive effects on biodiversity literacy and on visitors' intentions to act in ways that will contribute to conservation (Moss, Jensen, and Gusset 2014; Falk et al. 2007; Ballantyne et al. 2007). However, despite significant investment in conservation education, the evidence for positive effects on subsequent conservation behavior is generally weak (Ogden and Heimlich 2009), and we lack a shared definition for what constitutes positive effects. Nonetheless, slow progress is apparent, and new methods are being applied to measuring the effects of zoo visits (Grajal, Luebke, and Kelly, this volume).

Additionally, the industry has not explored the influence that organizational focus (regional versus global) has on mission success. It is this dimension—the regional zoo or aquarium and its role in conservation—that we explore in this chapter by highlighting one of the best-known regional organizations in the United States.

THE ARIZONA-SONORA DESERT MUSEUM: A MODEL REGIONAL BIOPARK

Most zoos and aquariums have worldwide collections, allowing visitors to experience exotic animals and environments that few of them will ever visit. Only a handful of accredited zoos and aquariums are regionally focused or holistically interdisciplinary, exploring the interconnectedness of biotic and abiotic components. The Arizona-Sonora Desert Museum (ASDM) is one such organization, founded on the principle that good stewardship may be best served by interpreting the region where the institution is situated. ASDM interprets its region's natural and cultural history, including plants, geology, oceans, and people, so that visitors can gain a deeper appreciation for the interrelatedness of these components. Regional visitors to ASDM often are learning about their own backyards, making it easier for them to apply the lessons to help wildlife and wild places in their everyday lives.

Since it opened in 1952, the ASDM has become a model for the regional zoo and aquarium. Although it's nestled in the Sonoran Desert, the concept was incubated twenty-five hundred miles away, along the Hudson River. In the mid-1940s William Carr moved to Tucson, Arizona, from southeastern New York, where he had worked for almost two decades to develop and grow Bear Mountain Trailside Museum and Zoo, an organization very similar to the Desert Museum. It too is regionally focused, with a holistic approach to exhibits and interpretation, although its animal facilities are smaller and more rudimentary, since it was developed in the 1920s. After arriving in Arizona, Carr marveled at the beauty of the Sonoran Desert but found few people who shared his reverence. As an environmental educator, he worried that if something wasn't done to preserve this unique corner of the world, it might be lost forever to the rapid development he saw happening all around him. He set out to change this by developing a place where people could learn about the area where they live. An early quotation from him clearly shows this concern, and it set the tone for the vision of the

ASDM: "We must somehow instill in the minds and hearts of individuals a true awareness of this life around them, and their potential for destroying or preserving it. Upon this awareness, in the final analysis, depends the survival of man himself" (William Carr, *Summary of Concepts*, Arizona-Sonora Desert Museum, early 1950s, exact date unknown).

Carr had a groundbreaking concept but not the cash to break ground. He teamed up with like-minded philanthropist Arthur Pack, and they approached the local County Board of Supervisors with their idea. Though not universally embraced (some are said to have thought it was crazy, perhaps underscoring the thin line between grand vision and lunacy), Carr and Pack were granted permission to create the museum on leased property within a county park fourteen miles from Tucson, accessible only by dirt roads, with the primary road winding precariously through a mountain pass (fig. 26.1).

With Pack's backing, Carr and a handful of others set about creating the Arizona-Sonora Desert Trailside Museum, which subsequently became the Arizona-Sonora Desert Museum. Considering the size and scale of the

FIGURE 26.1. Excavating large-animal enclosures at the Arizona-Sonora Desert Museum in the early 1950s. Photo courtesy of the ASDM, used by permission.

museum in the beginning, it's doubtful that the founders realized they had forged an entirely new path in natural history interpretation that would be copied around the world.

The museum's approach (then as now) is to focus on a single region—the Sonoran Desert—and to do so holistically. In a unique blend of zoo, botanical garden, aquarium, natural and cultural history museum, art gallery, and nature center, this organization revealingly chose not even to call itself a zoo. We believe this was an outgrowth of the founders' mind-set, which was firmly rooted in education and conservation, along with avoiding limitations that the word zoo typically conjures up in the visitors' minds. Though it is an AZA-accredited zoo (something we are *very* proud of), not using zoo in its title may have allowed it to more easily step outside the typical image. Locals embrace the Desert Museum as a place of pride, which has helped them develop pride of place. It has developed into the top attraction in southern Arizona, drawing visitors from around the world (see fig. 26.2). We know that many credit it with having considerable influence on local development. We've repeatedly heard that people have come to the museum for a variety of reasons. Some want to learn about the desert and how they could survive in it, while others want to know which plants they should grow in their yards. Still others tell us they've gained appreciation for desert environments and are more conservation-minded after their visit. This information has come to us through testimonials as well as from recent

FIGURE 26.2. The entrance of the Arizona-Sonora Desert Museum today. Photo courtesy of the ASDM, used by permission.

survey data, where 50 percent of local conservation leaders who responded to our study ($N = 22$) considered the Desert Museum to have had moderate influence on their engagement in conservation (Ivanyi and Colodner 2015a).

In the arena of conservation, the ASDM occupies a special niche in regional efforts. Over its history, it has played a vital role in Mexican wolf recovery; in protecting islands in the Gulf of California and tropical deciduous forest in southern Sonora, Mexico; in direct salvage, augmentation, and reintroduction of at-risk reptiles and amphibians; and in establishing regional coordination centers for invasive species management as well as recognizing the importance of pollinators and documenting the effects of climate change on mountain "sky island" biota.

Despite the Desert Museum's success in these areas and its considerable influence on several organizations that have drawn inspiration from it, including but not limited to the Living Desert Zoo, Virginia Living Museum, High Desert Museum, and Monterey Bay Aquarium, very few accredited zoos and aquariums today have adopted a regional approach. Perhaps this is due to the popularity of worldwide charismatic megafauna and concern that local species would be less of a draw, or to a desire to interpret global biodiversity and global issues. It may also be attributed to the inertia of historical legacy (Morse-Jones et al. 2012). Alternatively, maybe it occurs because some zoos are in locations where the natural environment is perceived as less compelling (although as Minteer [this volume] notes, zoos developed in Yosemite and Yellowstone National Parks in the 1920s failed miserably). Zoos and aquariums may also feel that the most important conservation issues involve organisms and environments outside their state or national borders.

Developing and maintaining a regional zoo certainly presents some unique challenges. Collection planning and sustainability are more difficult when you work with plants and animals that few other institutions have in their collections. This can lead to undesirable diffusion of effort (Monfort and Christen, this volume). It also limits participation in large multi-institutional projects that often focus on exotic species, and it affects the institution's status among colleagues. In a nutshell, if you replace elephants, great apes, and killer whales with bison, small mammals, and minnows, can you have viable collections, can you still collaborate with mainstream zoos and aquariums, will you have the respect of your colleagues, and perhaps most important, will people still come?

Certainly there are many questions that must be considered before creating a regional zoological park or aquarium to ensure economic success, as well as success in mission delivery and alignment with other organizations. But based on the Desert Museum's experience, we offer these aspects of the successful regional organization, elements that may better align with evolving public opinion and still offer a viable business model:

1. It offers a unique experience compared with mainstream zoos and aquariums.
2. The setting is an extension of the native, natural environment, allowing greater immersion and a more authentic experience. Essentially, you can borrow the view to create a better context for exhibits, portray a seamless integration of wild and designed, and reveal the unseen through deeper immersion in fauna and flora.
3. Regional collections are better adapted to the local climate, making them easier to maintain year-round, especially outdoors.
4. It can be a trusted local advocate for local conservation, well positioned to engage the community in promoting wise stewardship within a framework of sustainable development (Cerezo and Kapsar, this volume).
5. It can offer direct protection of a wild space in which the organization is placed.
6. It becomes a site for developing a sense and pride of place via hands-on, active, place-based education.
7. There is lower cost for field conservation, whether via research or education.
8. There are more opportunities to involve local visitors and communities in citizen science and conservation, including surveys, identification and removal of invasive species, and salvage of native species at risk.
9. Regional organizations can make deep, broad, long-term commitments to projects. (ASDM has several projects it has been engaged in for decades.)
10. Local and regional partnerships can be used to advance a common conservation mission.

To be clear, we are not implying that this is guaranteed to work for others, since we can't say with certainty why the museum flourished. To some degree it may have been timing and good luck, but we believe its success was due to much more than this; likely it benefited from everything that makes it and Tucson unique, including the aesthetics and overall appeal of its location and the unique flora and fauna. It also "grew up" during a time of rapid immigration and growth of tourism in the region, with an accompanying high level of interest in getting to know this unfamiliar place. Coupled with the museum's holistic approach and its early and steadfast commitment to both conservation and educational outreach, the regional approach proved to be a successful business model.

Which type of organization best serves the conservation role accredited zoos and aquariums self-identify with—regional or mainstream? We encourage the research community to address the many facets of this question through comparative study, and we suggest that Tucson, with both regionally and globally focused accredited organizations, is a candidate as a study site. For example, it may be possible to study which type of collection more effectively motivates visitors to take various types of conservation actions, or which type of collection raises more or less discomfort in people concerned about animal welfare. Without peer-reviewed research to guide us, we think there is likely a place and a need for both regional and global zoos and aquariums that are dedicated to conservation. Regarding their effects on institutional leadership, a recent survey of directors of AZA-accredited institutions supports this assertion. Almost 41 percent of twenty-two respondents felt that mainstream zoos had a greater influence on their conservation perspective, while 32 percent believed regional organizations had more influence (Ivanyi and Colodner 2015b).

CONSERVATION AND THE FUTURE OF ZOOS

Zoos and aquariums are still popular, but generalized criticism of them is growing, as is targeted criticism related to specific, highly sentient animals (e.g., elephants, primates, whales), especially among younger audiences. This is an ideal time for zoos and aquariums to experiment with new approaches, as many are doing. Rather than digging in our heels, it might be time to divest ourselves of species that do not do well in captivity, or at least to recognize that some institutions are ill suited for some species. This has been the approach of several zoos and aquariums, including most recently the decision by SeaWorld to stop breeding orcas to use in theatrical shows. At the same time, the public may be more open to regional organizations and approaches, owing to many of the factors outlined above. This is one direction that could allow the industry to capitalize on public opinion and begin to reinvent itself.

Moreover, in this day of changing paradigms, we might reexamine the continuum of conservation organizations that zoos and aquariums generally collaborate with. Perhaps expanding our notion of our colleagues beyond other zoos and aquariums and partnering more with the full spectrum of nature-based conservation education entities, including small nature centers, environmental education centers, regional organizations, and bioparks as well as colleges and universities, would be an appropriate path to consider. If we view all of these organizations as a conservation collective we might get more done, stretch our dollars further, gain public support, and underscore one of the main tenets of conservation of nature—being interconnected and interdependent.

Local organizations and nature centers may be well positioned to extend our mission locally or regionally. In a recent survey, although the sample size was small, 91 percent of zoo directors reported having a nature center within twenty-five miles of their institutions, yet only 41 percent stated they have a formal collaborative relationship with the nature center (Ivanyi and Colodner 2015b). A small amount of funding and attention from a large zoo or aquarium could provide critical support for these smaller organizations, which are often underfunded and understaffed, along with providing validation of their work and increasing their ability to protect local habitat. In this way mainstream zoos and aquariums might have the best of both worlds—conservation of both international and regional biodiversity.

Since the day the Desert Museum was conceived, its message has clearly been about "saving place" in order to protect biodiversity, environmental quality, cultural diversity, quality of life, among other ends. This is in contrast to the main messages of other zoos, which generally lead with "saving species." Although research tells us that people empathize more with animals than with place-based concepts like ecosystems and habitat, in the case of the Desert Museum animals and landscape are so seamlessly integrated that visitors appear to make the connection easily. Does this change the way they perceive the museum's conservation mission compared with their feelings about other zoos? We'd love to know! This integrated, place-based approach was innovative at its founding, and the museum has maintained its pioneering ways ever since, leading to admiration by colleagues and community members alike. Who knows, perhaps one way to accomplish more of the goals zoos are charged with is for some of them to narrow their focus in a similar fashion, or to call themselves something other than zoos.

CHAPTER TWENTY-SEVEN

Today's Awe-Inspiring Design, Tomorrow's Plexiglas Dinosaur: How Public Aquariums Contradict Their Conservation Mandate in Pursuit of Immersive Underwater Displays

Stefan Linquist

I yelled at the Lorax, Now listen here, Dad!
All you do is yap-yap and say, 'Bad! Bad! Bad! Bad!'
Well, I have my rights, sir, and I'm telling *you*
I intend to go on doing just what I do!
And, for your information, you Lorax, I'm figgering
on biggering
and BIGGERING
and BIGGERING
and BIGGERING.

DR. SEUSS, 1971

THE MODERN MEGA-AQUARIUM

The past few decades have been described as a revolutionary period in aquarium design. Gone are the recessed glass panels arranged like paintings along a museum wall. As one aquarium engineer put it, "the watchword today is total immersion" (Smith 1994). Developments in the fabrication of Plexiglas and other acrylics allow for two-foot-thick panels that tower above the awestruck visitor. For example, the Okinawa Churaumi Aquarium in Japan boasts an underwater viewing panel 8.2 meters high and 23 meters wide (fig. 27.1). It was outdone in 2008 by the panel 8.3 meters by 32.8 meters at the Dubai Mall Aquarium. Similar exhibits are popping up throughout Asia, Europe, and North America (table 27.1). Older aquariums,

FIGURE 27.1. At the Okinawa Churaumi Aquarium (2002) technological advances take the theme of immersion to a large scale. Image by Leungchopan/Shuttersstock.com, used by permission.

whose displays seem outdated by comparison, are undertaking major renovations to keep up with the industry standard (see table 27.1).

Another popular feature is the walk-through viewing tunnel. An industry brochure produced by Reynolds Polymer Technologies (2014) explains its benefits: "The sight of the underbelly of a shark swimming overhead is certain to bring out delight in any crowd. However, another advantage of aquarium tunnels is their ability to direct traffic. . . . Nothing can clog an exhibit or walkway like a handful of guests milling around without progressing to the next station."

This comment also reveals how the visitor experience in modern aquariums has become choreographed. The "cinematic" approach to aquarium design was pioneered by Peter Chermayeff, legendary designer of the New England Aquarium (1969), who went on to construct massive aquariums in Baltimore, Osaka, and Chattanooga. "In Baltimore," Chermayeff explains, "we orchestrated the visitor experience in a linear path, like a musical piece, a book or a film. We carried people on travelators and escalators up to the top, gradually, through a central space, working upward, gallery by gallery" (Smith 1994, 57). In a further effort to orchestrate the visitor experience, many modern aquariums broadcast thematic music in their display areas, creating what Chermayeff describes as "a low-key sensory collage" (Smith 1994, 57).

TABLE 27.1. Sample of large aquariums constructed since 1990, noting size of largest fish tank (excluding mammal enclosures)

Year/ location	*Largest fish tank (liters)*	*Year/ location*	*Largest fish tank (liters)*
1990 Osaka	1,892,000	2008 Stralsund	2,600,000
1992 Chattanooga	1,700,000	2009 Istanbul	4,900,000
1998 Long Beach	1,324,000	2010 Seoul	6,000,000
1998 Lisbon	4,000,000	2010 Dubai	10,000,000
1999 Cincinnati	1,430,000	2012 Singapore	12,000,000
1999 Fukushima	2,500,000	2012 Sujeong-dong	6,000,000
2000 Charleston	1,400,000	2013 Copenhagen	4,000,000
2002 Shanghai	2,200,000	2013 Toronto	2,800,000
2002 Okinawa	7,500,000	2014 Hengquin	22,700,000
2003 Valencia	7,000,000	2015 Chendu	9,464,000
2005 Atlanta	23,850,000		

Note: Data collected from institution websites and press releases.

Biomass levels within display tanks are also on the increase. For instance, the six-million-gallon tank at the Georgia Aquarium contains more than one hundred thousand animals. This feat is achieved, in part, by 218 pumps working continuously to drive 261,000 gallons of water per minute through approximately sixty-one miles of pipe. Such advances allow today's simulated coral reefs to stock a higher density of fishes than their natural counterparts. They also make it possible to display larger, more charismatic specimens. At both the Georgia Aquarium and the Okinawa Churaumi, visitors encounter circling whale sharks, the world's largest fish. Recently constructed aquariums in Seoul and Singapore display schooling manta rays, apparently the latest "must have" specimen. The Monterey Bay Aquarium has pushed the envelope on several occasions by attempting to display a great white shark. This spectacle generated a significant boost in attendance, but the first animal died after just sixteen days in captivity, and subsequent attempts were also unsuccessful (Rogers 2011).

Unlike zoos, the rate of new aquarium construction is on the upswing. According to the World Association of Zoos and Aquariums (Penning et al.

2009), of the approximately 315 aquariums that exist globally, over a hundred have been established since the year 2000. In China alone, eleven new aquariums were built in the past decade. It is estimated that 450 million people visit an aquarium annually–more than will attend a professional soccer match.

The rest of this chapter argues that this trend toward larger, more immersive displays contradicts the environmental values most aquariums profess. I begin by tracing the historical origin of the "theatrical" approach to the visitor experience, contrasting it with a more authentic approach. The theatrical mode of engagement is so entrenched that people today rarely question it. I highlight it here to gain some critical distance on our experience as aquarium visitors. We expect an aquarium to simulate the world's oceans and are accustomed to playing along with the pretense of encountering wild nature. That we are actually engaging with a highly artificial, technologically mediated environment is something most visitors effortlessly overlook. It is helpful to consider how we got here.

I then turn to the educational mission endorsed by public aquariums. Immersive underwater exhibits are often justified by the hypothesis that they influence visitors' behavior positively. I argue that this claim lacks adequate empirical support and that, historically, it served to deflect attention from animal welfare concerns.

Moreover, there is a direct conflict between the growing public concern over CO_2 emissions and the trajectory of modern aquariums. At a time when conservationists are calling for reduced emissions, these institutions are increasing their rate of pollution while imposing greater stress on marine habitats. Although aquarium organizations acknowledge these problems and promise reform, I believe these efforts stand little chance of success. Mega-aquariums are locked into a boom-bust-expand economic cycle that makes growth necessary for their survival.

The final section of the chapter points to an alternative theme in aquarium design that aims to avoid these pitfalls. The emergence of the regional mini-aquarium provides a sustainable example of how people can enjoy more authentic experiences of marine life.

STRIVING FOR IMMERSION

Visiting a modern public aquarium has become like attending a theatrical performance. Both require a willing suspension of disbelief. As theatergoers, we allow two-dimensional scenery to stand in for buildings. We regard cast members as characters, not as ordinary people with bank accounts and commuting schedules. Such pretense is necessary if we are to enjoy the performance. Similarly, aquarium visitors are encouraged to pretend they are on a tour of the world's oceans. Artificial coral sculptures and naturalistic backdrops evoke wild habitats, and extensive pumps and filtration systems are hidden. Visitors are encouraged not to inquire about husbandry or collection techniques. Instead, an aquarium visit is typically billed as a quasi-wilderness experience, where visitors connect emotionally with animals as if they were encountering them in nature.

For some visitors, however, these efforts are largely wasted. Members of the aquarium hobbyist community enjoy a different kind of aquarium experience. Hobbyists are attuned to the difficulties of maintaining fishes artificially. They approach an aquarium visit the way an amateur stagecraft designer might view a Broadway production: keenly interested in what happens behind the scenes. The hobbyist might wonder, for instance, how delicate specimens were obtained from the wild. She might be interested in the feeding schedule that enables predators to cohabit with prey. And so on. This mode of engagement is undeniably more authentic than that of the typical aquarium visitor, because it recognizes that many species have ecological requirements that are costly and difficult to simulate.

The hobbyist's perspective predates the theatrical mode of engagement. The first public aquarium was an outgrowth of an already thriving home aquarium industry. Historian Bern Brunner (2005) documents the period in Victorian England when home aquariums became a craze. Their chief popularizer was naturalist Philip Henry Gosse, whose enthusiastic descriptions of fish and invertebrate behavior drove a generation of hobbyists to the seashore to gather specimens for private collections. Gosse's *The Aquarium* (1854) is partly a work of popular nature writing, partly a how-to guide for the hobbyist. It offered detailed instructions on aquarium construction, specimen compatibility, maintaining water quality, and a host of other technical concerns. In keeping with this dual focus on both technology and natural history, many commercial home aquarium systems accentuated the mechanics of life support. We can imagine that at Victorian social gatherings, where such items were admired, conversation focused as much on the problems of animal husbandry as on natural history.

It is therefore no surprise that when Gosse was commissioned to build the first public aquarium in London, about 1853, there was little attempt to construct an immersive underwater exhibit (fig. 27.2). Visitors wandered freely around each tank, inspecting life support systems as well as the contents of each display (Holdsworth 1860). This was a public aquarium designed for an audience of hobbyists, not theatergoers.

FIGURE 27.2. Philip Henry Gosse's "Fish House" (London, ca. 1875) made little effort to disguise the way display specimens were being artificially maintained. This theme was consistent with his natural history writings, in which ecological principles were interwoven with animal husbandry suggestions for the home aquarist. Image from London Zoological Society, used by permission.

However, by the 1860s, when public aquariums began appearing in continental Europe, immersion had become a popular design goal. An immersive exhibit (whether in a zoo, a natural history museum, or an aquarium) aims to transport visitors to the location represented in the display. In an aquarium this involves disguising plumbing and filtration systems with naturalistic scenery and props. Adjustments to lighting also create a sense that one is undertaking an ocean descent. This technique was pioneered by French naturalist Arthur Mangin, who explicitly adopted design principles from the world of stagecraft to produce what he described as a "new kind of theatre" (Brunner 2005, 103). His aquarium at the Jardin d'acclimation in Paris (ca. 1860) had visitors enter a darkened corridor lined with tanks illuminated from above (fig. 27.3). His efforts were celebrated in an article in a German industrial magazine: "One forgets about how many artificial resources had to be employed to create such an exposition" (Brunner 2005, 104)—a comment that would apply equally to the mega-aquariums of today.

The 1860s saw several attempts to take immersion to its technological limits. The 1867 World Exposition in Paris boasted a twenty-thousand-gallon aquarium that visitors viewed from below through a glass ceiling. As with today's acrylic tunnels, the aim was to create the impression of walking along the ocean bottom. However, Brunner (2005) explains that the exhibit was poorly conceived. Visitors complained of neck strain, and wind and rain clouded the water, so the vantage from below offered suboptimal views of marine life. The idea of having visitors look upward at passing fishes would have to wait for technological advances of the late twentieth century.

Perhaps the pinnacle of nineteenth-century immersion was achieved at the Berlin Aquarium, completed in 1869. This two-story grotto simulated a Jules Verne adventure. Visitors descended into a series of stone cham-

FIGURE 27.3. Arthur Mangin's Aquarium du Jardin d'acclimation was one of the first attempts to create an immersive underwater display by incorporating theatrical design elements: visitors entered a darkened corridor in which tanks were lit from above and life-support systems were hidden from public view. Bertrand 1863 (PD-1923). From Le Monde, January 10, 1863; public domain image (Wikimedia Commons).

bers evoking a system of underwater caverns. Brunner explains that "the glass panels of the water basins on the first floor looked like random breakthroughs in the rock" (2005, 110). Through these portals visitors observed diverse animals, including sea anemones and octopuses as well as crustaceans and fishes. Initially the Berlin Aquarium was a great success, with one hundred thousand visitors in the first three months. However, this immersive masterpiece was difficult to maintain both economically and structurally. Tanks burst. Pipes corroded. Rats colonized grotto cavities. Visitors eventually lost interest. The Berlin Aquarium soon resorted to more pedestrian streams of income, using tanks to stock fishes and lobsters for local restaurants. In a desperate effort to draw crowds, in 1876 a gorilla was put on display. The aquarium eventually closed in 1910.

As public aquariums started appearing in North America, efforts at immersion took a more modest turn. The South Boston Aquarium (ca. 1912) is typical of this period. The building consisted of a darkened corridor surrounded by viewing panels five feet high. Display tanks were illuminated from above "to give the visitor the impression of being underwater" (Ryan 2011, 62). What this aquarium lacked in architectural flair, it made up in size. The eight-thousand-square-foot building was twice as large as originally planned. The South Boston Aquarium also contained an impressive collection of specimens for its day. A 1927 inventory lists sixty-two tanks containing 2,835 fishes along with a few marine mammals, reptiles, and amphibians. The aquarium was also popular. On opening, it attracted fifteen thousand visitors in a single day, and even fifteen years later its average daily attendance was 811 visitors. However, after the wars the building fell into disrepair and was eventually closed in 1952 (Ryan 2011).

American aquariums became more ambitiously immersive starting in 1969 with the opening of the New England Aquarium. The main exhibit involved a four-story "Giant Ocean Tank" that was the largest of its kind at the time. The New England Aquarium website recalls that this exhibit "was notable for its size but also for how it created an intimate space for visitors to really see, experience, and connect with an ocean world that was largely unknown to them" (New England Aquarium 2016). The 1980s saw another step in this direction with the creation of the Monterey Bay Kelp Exhibit. Julie Packard (1989) recounts the technical challenges associated with its construction. The tank was at the forefront of acrylic construction, with panels seven inches thick rising to twenty-eight feet, enclosing 335,000 gallons of water. To sustain living kelp, natural seawater was brought in each evening, and surge pumps circulated it continuously throughout the day.

High-pressure sand filters prevented an accumulation of debris in the water column. It was a considerable undertaking, but the efforts were rewarded in attendance and fanfare.

This series of historical snapshots suggests that today's mega-aquariums are the expression of a design ideal that traces back to 1860s Europe. Going back just a little further, however, to the very first example of a public aquarium, we see an alternative design model. It is fascinating to contemplate how the history of aquarium design might have unfolded had it followed Gosse's original creation. Presumably these institutions would place as much emphasis on communicating the conditions for sustaining life as on displaying the wonders of the ocean.

The rest of this chapter will explore whether the theatrical approach to aquarium design is in conflict with the conservation ideals that aquariums have (more recently) embraced. It is helpful to keep in mind what I have called the hobbyist perspective (perhaps better described as an *authentic* mode of visitor engagement), though I am not suggesting that the aquarium hobby industry is environmentally benign. Gosse's influence, in fact, led to the denuding of the Devonshire seashore. On a much larger scale, today's aquarium hobby industry is responsible for significant destruction of habitat. Nor do I claim one must keep fishes to gain a more enlightened perspective on public aquariums. I hope to emphasize that there is a way of appreciating aquariums that does not depend on generating a sense of spectacle through immersion. This more authentic approach opens up a range of design possibilities that do not require building increasingly larger, ecologically harmful mega-aquariums.

AQUARIUMS' CONSERVATION MESSAGE

The 1980s and 1990s saw growing alarm about the loss of global biodiversity. Avoiding the impending "biodiversity crisis" required that consumers, industries, and governments become mindful of their impact on species extinction. Aquariums, with their expertise in displaying exotic specimens, were well positioned to capitalize on this narrative. As Susan Davis (1997) observes, visitors to SeaWorld encounter–amid the dolphin and whale shows–a fairly urgent message: the world's oceans are in peril, and conservation action is necessary. Similar values are expressed in the mission statements of most public aquariums (see, e.g., Spring, this volume). Such claims give the impression that aquariums are leaders in conservation science. Visitors are comforted by the (highly questionable) idea that simply visiting an aquarium qualifies as a form of environmentally responsible action (see Davis 1997).

Embracing this role as conservation educators allowed aquariums to deflect a different criticism. The publication of Peter Singer's influential *Animal Liberation* (1975) saw growing concern for the welfare of captive animals. For a burgeoning aquarium industry, the biodiversity crisis came at a convenient time. The tradition of exhibiting sentient animals could now be defended on the grounds that these "animal ambassadors" (Taylor 1995) helped promote awareness about the importance of marine conservation. The issue of animal welfare is particularly relevant for aquariums that display sentient organisms in cramped quarters–as it is, of course, for zoos (see, e.g., Palmer, Kasperbauer, and Sandøe, this volume). These ethical issues have been addressed in a number of recent publications.[1] Without minimizing the questionable ethics of displaying captive marine mammals, I now turn to more general inconsistences between aquarium operations and their conservation mandate.

Aquariums claim to fulfill their conservation mandate through three activities: research, conservation outreach, and public education. As exemplified by some of the larger, more established institutions, research has become an important component of aquarium operations (see Muka, this volume). However, critics note that most of this research focuses on animal husbandry issues that pertain to captive populations (Lawson, Ogden, and Snyder, 2008). In fact aquariums are even less productive than zoos in research output. An analysis of 395 articles in *Zoo Biology Journal* over fifteen years found that aquariums accounted for just 5.5 percent of total publications (Wemmer, Rodden, and Pickett 1997). It is also important that aquariums derive a number of institutional benefits from their research programs. For example, funding can be used to maintain animal husbandry experts on staff. It can help to build multipurpose holding facilities. Research excursions double as collecting trips. Most significant, perhaps, the appearance of a viable research program is a useful public relations tool. For example, the Vancouver Aquarium continues to point to its research on captive beluga whales to deflect mounting criticism from welfare advocates (Lupick 2014; Kane 2015). One should perhaps not be overly cynical about the conservation benefits of aquarium research, but these programs tend to serve institutional interests as much as they do nature.

A more direct form of conservation involves various outreach programs (see Knapp, this volume). The New England Aquarium, Monterey Bay Aquarium, and Shedd Aquarium are world leaders on this front. Funding for such projects derives in part from ticket sales at the door. Hence one might argue that the conservation efforts of modern aquariums must be assessed in terms of their outreach, not just their displays. In reality, however, few public aquariums engage in meaningful outreach programs. Most aquarium websites I reviewed for this chapter reported minimal efforts. Fraser and Sickler (2009) concur that "although some [zoos and aquariums] are able to contribute to field conservation as part of their mission-driven activities, the maintenance of zoological parts is still the predominant route by which [they] work towards their mission" (103).

Hence the key question is whether displaying exotic marine life, particularly in an immersive setting, has a meaningful effect on visitors' conservation behavior. Public aquariums often take such effects for granted. For example, Peter Chermayeff justifies the construction of immersive exhibits in terms of their educational value:

> The designs basically rely . . . on biophilia-that human bond with other species-and on the power of the experience when a child or adult, eyes open with wonder, peers at the gills, jaws, and fins of a passing shark, or the soft pulsing forms of jellyfish. The stimulation of curiosity, we hope, will encourage discussion, reading, environmental consciousness, and direct involvement in the protection of natural resources. (1992, 56)

This makes for a fairly compelling story, but it enjoys little empirical support. A study conducted by the American Zoo and Aquarium Association (Falk et al. 2007) is widely cited as the first direct evidence of a positive effect of attendance on visitors' conservation behavior. However, this study has been soundly refuted for its numerous methodological flaws (Marino et al. 2010). To date there is no reliable evidence supporting the claim that aquariums contribute to lasting conservation behavior, let alone that increasing the size of exhibits or their degree of immersion enhances this effect. At the same time, an alternative hypothesis proposes that aquariums instill precisely the opposite set of values. Visitors realize they are viewing captive animals under human control. It is possible that such encounters reinforce a sense of entitlement in their domination over nature (see Minteer, this volume). The trend toward more immersive displays, containing ever more spectacular inhabitants, might only promote this attitude.

AN EVEN GREATER CRISIS

The threat of human-induced climate change has become the dominant environmental concern of our time. With this new crisis upon us, it is very difficult to see how public aquariums can maintain credibility as environmental educators (see the chapters by Norton and by Cerezo and Kapsar, this volume). Aquariums have a much larger ecological footprint than zoos, partly because fishes-unlike mammals—are unable to internally regulate body temperature. This means that specimens from different parts of the world require a constant regimen of heating and cooling, drawing considerable energy. Massive pumps must also continuously provide adequate filtration and circulation. The lighting and humidity control required to keep visitors comfortable also is a significant electricity draw. One must also consider the transport of food and display specimens from around the world. It is difficult to obtain reliable data about the energy demands of modern mega-aquariums. For obvious reasons, aquarium representatives tend not to be forthcoming on this issue. In a rare moment of candor, however, the director of the Shedd Aquarium in Chicago compared that institution's CO_2 emissions to "an endless 2,200-car traffic jam" (Wernau 2013).

On top of these energy demands are two more direct ecological impacts that aquariums impose on natural habitats. Most aquarium specimens are carnivorous, relying on a diet of herring, smelt, or other wild-caught fish. These animals occupy a fairly high rung on the trophic ladder. Hence there is a cascade of ecological effects from nourishing the more than one hundred thousand fishes contained in a typical mega-aquarium. The second direct impact involves the procurement of display specimens. Unlike zoos, aquariums rely primarily on wild populations to stock tanks. The turnover rate for these animals can be significant. For example, the Okinawa Churaumi Aquarium reportedly lost sixteen whale sharks over a ten-year period (Mihelich 2005).

A proper environmental accounting is sure to cast a sobering light on the theatrical approach to aquarium design. At a time when conservationists are calling for a reduction in global CO_2 emissions, aquariums keep growing larger. As the rate of habitat loss due to coral bleaching increases, aquariums are ramping up the number of wild animals on display. While environmental leaders call on everyone to become more mindful of our global impact, aquariums continue to disguise their destructive practices. How could such hypocrisy possibly be consistent with an effective role in environmental education?

PLEXIGLAS DINOSAURS

In 2009 the World Association of Zoos and Aquariums published a policy document outlining "a global aquarium strategy for conservation and sustainability." The report describes a number of environmental issues aquariums should focus on. These include overfishing, pollution, the aquarium pet trade, and ocean acidification caused by CO_2 emissions. The report goes on to propose that "all zoos and aquariums will serve as leaders by example, [by] using green practices in all aspects of their operations and by demonstrating methods by which visitors can adopt sustainable lifestyles" (Penning et al. 2009, 40). This statement suggests that aquariums are planning to improve on traditional practices.

However, economic considerations suggest that these goals cannot be achieved by mega-aquariums. These institutions are trapped in a growth cycle dictated by the economics of their industry. Aquarium professionals are well aware that "most aquariums experience lucrative attendance numbers in their first few years, but unless they begin to insert additional exhibits, even the most remarkable facilities will see a dramatic decline" (Macdonald 2006, 17). The Newport Aquarium in Cincinnati is an example of this boom-bust-expand cycle. During its opening year in 1999, it had 1.25 million visitors. By 2003 attendance had dropped by 50 percent. This inspired plans for expanding by twenty-one thousand square feet and including a sea otter exhibit, among other attractions. Director Timothy Mullican justified this expansion by saying: "You need to stay fresh. You need to give people a reason to come back, and you do that by adding new exhibits" (Crowley 2003). In 2015 the aquarium announced plans for a new "shark bridge," which is being touted as "the only suspension bridge in North America on which visitors can walk just inches above nearly two dozen lurking sharks" (Shafer 2015).

With each turn of the wheel, aquariums become saddled with ever larger, costlier exhibits. Those displays often require maintenance at about the same time that their appeal begins to wane. This means aquariums inevitably land in an ever more desperate financial predicament with each downturn in the cycle, leaving little option but to expand yet again. One is reminded of anthropologist Franz Boas's (1937) criticism of the excessive costs invested in the immersive museum dioramas of his day. He complained that those "dinosaur-like" displays committed museums to a particular educational message that was bound to become outdated with the marching progress of science (Rader and Cain 2008). Arguably, modern aquariums are engaged in an even more capricious pursuit. It is a truism that society's threshold for spectacle is constantly being raised (see, e.g., Minteer, this volume). Special effects in movies that enthralled us a decade ago now seem quaint compared with the latest and greatest. Why should it be any different for underwater exhibits? Apparently some aquariums are now offering snorkeling trips into "shark infested" tanks (Cater 2010). Talk about immersion! How enthralling will today's large viewing panels or Plexiglas tunnels appear by comparison?

AN ALTERNATIVE MODEL

Where does this all leave the many people who enjoy, and perhaps derive inspiration from, viewing captive marine animals? Perhaps in this day and age no one is entitled to enjoy a simulated stroll through the world's oceans. However, an alternative model might also be considered.

I have suggested that the earliest example of a public aquarium provides an alternative to the theatrical approach to aquarium design. Smaller-scale institutions, providing authentic experiences to smaller groups of visitors, are the contemporary embodiment of Gosse's model. In fact there appears to be a growing counterculture of "regional" zoos and aquariums that are rejecting traditional design ideals. One such example is the Arizona-Sonora Desert Museum. Another is an innovative "mini-aquarium" established in 2005 on the west coast of Vancouver Island. As one of the founding directors of the Ucluelet Aquarium Society, I was involved in formulating the design ideals this institution is based on. Most of its displays are touch pools (fig. 27.4). No attempt is made to construct an immersive underwater experience. Instead, visitors are informed about the open-flow system that brings water directly from the inlet into tanks and back again. Displays are therefore a physical extension of the surrounding ocean. This approach aims to provide visitors with a grounded sense of the importance of maintaining water quality in the local harbor. It also allows the aquarium to display a broader range of species (e.g., filter-feeding invertebrates) that do not survive in large filtration systems.

The Ucluelet Aquarium displays only locally collected fishes and invertebrates—no mammals. Tank sizes are modest. Visitors' thirst for novelty is satisfied by adapting displays to seasonal changes in the local environment. For example, when squid show up to spawn in the inlet, a suitable display can be created on the fly. On hatching, larvae flow back into the ocean with minimal interruption. Little attempt is made to conceal plumbing and filtration—in this respect the aquarium feels more like a lab than a museum or amusement park-and many visitors express interest in learning about the operation of life support systems. Most distinctive, perhaps, is that all the animals on display at the Ucluelet Aquarium are released back

FIGURE 27.4. The Ucluelet Aquarium (2013) abandons the theme of immersion in favor of a more transparent approach where tanks serve as interaction centers, life-support systems are on display, and the animals are eventually returned to the ocean. Photo by Kumiko Bruecker, used by permission.

into the wild at the end of the season. This feature encourages people to view display organisms as fellow visitors, not as curios for our possession and amusement.

Admittedly, we currently do not know whether this alternative model of aquarium design is more successful than its larger counterparts in instilling love of nature. This is an area for future research. But it is perhaps encouraging to think that there are alternative, more ecologically responsible ways to bring people into contact with marine life.

CHAPTER TWENTY-EIGHT

Zoo Conservation Disembarks: Stepping off the Ark and into Global Sustainable Development

Adrián Cerezo and Kelly E. Kapsar

INTRODUCTION

Noah's ark has provided a powerful grand narrative for zoological institutions. Over time, the general notion of the "zoo as ark" metaphor—a living collection that displays the wonder of God's creation—has aligned more closely with the biblical script and now depicts zoological institutions as a refuge for fauna in a corrupt, doomed world (fig. 28.1). This simple, familiar, and compelling narrative has served as a ready-made statement of purpose in response to mounting ethical concerns and shifts in public spending priorities. But the ark metaphor does a poor job of reflecting the complexity of natural and social systems and thus provides a weak foundation for the very mission it implies: the conservation of wildlife.

In their chapter, Monfort and Christen (this volume) consider the science, technology, and management challenges posed by zoo-based wildlife conservation programs. Norton (this volume) expands the context to describe a multispecies, ecosystem-scale approach to conservation. This chapter expands this context to consider the future of zoo-based conservation as an element in the global enterprise of sustainable development. As zoological institutions reimagine themselves as wildlife conservation organizations, we believe it is time to disembark from the ark metaphor and consider new narratives that embody our passion for wildlife and our role as engaged, active participants in the global sustainable development movement.

Our chapter opens with a brief history of the concept of sustainable development, followed by an explanation of the recently adopted "Agenda

FIGURE 28.1. Zoo-based conservation has to juggle the dual mission of protecting species and engaging humans in biodiversity conservation. Design by A. Cerezo, 2015.

2030," also known as the Sustainable Development Goals. We then examine zoos' wildlife conservation goals within the context of the sustainable development agenda and describe how the success of zoos' wildlife conservation is inextricably linked to the successful implementation of the sustainable development agenda. As other contributors to this volume show, the scientific and technical difficulties posed by species management are immense and ever-increasing. But even these immense challenges are minuscule compared with the ones posed by the full agenda of global sustainable development.

In an environment where the available resources are inadequate even to support the core mission of care and management of captive populations (Monfort and Christen this volume), why would zoological institutions want to take on the full sustainable development agenda? We will explain how engaging the global network of sustainable development creates the conditions for increased institutional resources as well as long-term wildlife conservation success.

BIODIVERSITY CONSERVATION AND SUSTAINABLE DEVELOPMENT

It is a story well known by field conservationists: a conservation biologist disembarks into a place of wonder, brimming with unique organisms, ecosystems, and people, all under siege. After years of training, research, fundraising, and gearing up, it is time to come face-to-face with an incongruent landscape where natural and cultural richness coexist with widespread deprivation and crisis. Quickly enough, experience teaches our conservation biologist that ecosystem degradation and systemic biodiversity loss are inextricably connected to generations of poverty, malnutrition, low-quality education, inequality, injustice, and (in many cases) institutionalized corruption and violence (Dudley et al. 2002; Fisher and Christopher 2007; Hoole and Berkes 2010). It becomes clear that the fate of individual species is intertwined with the quality of life of humans within and beyond these ecosystems (Adams et al. 2004; Berkes 2007; Carter et al. 2014; Waylen et al. 2010; Wilshusen et al. 2011). More important, when wildlife conservation initiatives do not consider human development, they are doomed to fail or collapse over time. This insight, experienced by conservation biologists across the globe, evolved into a new paradigm for conservation, environmental protection, and human development. This approach is now known as sustainable development.

Defined as "development that meets the needs of the present without compromising the ability of future generations to meet their own needs," sustainable development seeks to address "three pillars": economic development, social equity, and environmental protection (United Nations [UN] 1987). The sustainable development paradigm provides a rich conceptual framework that truly represents the biosphere. But by virtue of this richness, it also opens up a new set of challenges in plotting a way forward. Foremost is the concern that to understand, predict, and develop sustainable approaches, it is important to engage with ecosystems and social systems as they are: complex (Folke et al. 2002; Liu et al. 2007) and dynamic (Rohde 2005). It is also necessary for policies and programs to understand that entropy and uncertainty are facts to be embraced rather than failures to be avoided (Armitage et al. 2009; Berkes 2007; Folke et al. 2002; Holling and Meffe 1996; Holling, Berkes, and Folke 2000; Ostrom 2007).

In the three decades since the Brundtland Report proposed sustainable development as the platform for conservation (UN 1987), multiple attempts have been made to describe the complex network of interactions that constitute global sustainable development and to provide practical policy and program recommendations. Early approaches to sustainable development grappled with the balance between technical detail and accessibility but lacked a nuanced definition of the concept itself (UN 1992, 2000). Later approximations struggled to expand beyond the role of material wealth as the sole measure of quality of life and development (UN 2002, 2012).

Developed by the United Nations, the most recent version of sustainable development is designed to be ambitious, comprehensive, and nuanced while being accessible and practical (UN 2015). This version, known as "Agenda 2030," consists of seventeen sustainable development goals (SDGs)

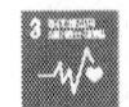

FIGURE 28.2. The seventeen Sustainable Development Goals (SDGs) of the United Nations Agenda 2030. Note that biodiversity conservation is considered in both goal 14 (Life below water) and goal 15 (Life on land). Design by A. Cerezo, 2015.

that describe the fundamental aspects of sustainable development and serve as "a plan of action for people, planet, and prosperity" (fig. 28.2; UN 2015). These fundamental aspects include nutrition, health, and the elimination of poverty, but they also consider equality, freedom, and justice as keys to improved quality of life and an improved interaction between humans and the ecosystems they are part of. Beyond being comprehensive in content, the SDGs also refer to and synthesize other relevant UN agreements, such as the Convention on Biological Diversity and its Aichi Biodiversity Targets (CBD 2010a), as well as the UN Framework Convention on Climate Change (UN 1998). Agenda 2030, which was adopted during the 2015 UN General Assembly, brings the sustainable development agenda to a new era by providing a thorough synthesis, along with concrete strategies for implementation. Not only do the SDGs create a comprehensive vision for the possible future, but on a more practical level they provide a map for institutions to visualize the diversity and extent of institutions that contribute to sustainable development and to expand their network of collaborators. It is now up to UN member states and institutions to make the most of Agenda 2030, and up to all who are interested in the future of the planet to help move the agenda forward.

Sustainable development—as framed by Agenda 2030—provides a blueprint and a rallying cry for improved quality of human life along with improvements in environmental quality and biodiversity conservation. In some developed countries, particularly the United States, the institutional and popular perception has placed particular emphasis on the economic and environmental elements of sustainable development (examples of this perspective in the zoo world include Dickie 2009, Hanson 2015, and Landman and Visscher 2009), creating a variant of the concept known as sustainability. Sustainability is focused on providing technical solutions to issues of energy, water, transportation, manufacturing, construction, waste management, and pollution reduction (Daly and Cobb 1989; Pearce 2012), which represents only six of the seventeen SDGs. This variant of sustainable development has become institutionalized, and it is now normal for corporations, government agencies, and educational and cultural institutions (including zoological institutions) to have "sustainability coordinators" on their staffs. The term and the idea of sustainability needs to be highlighted, because it has led to the misconception that "sustainability" and "sustainable development" are the same. The focus of this chapter is sustainable development in its most comprehensive definition as described by the SDGs and Agenda 2030.

ZOO-BASED CONSERVATION AND SUSTAINABLE DEVELOPMENT

For over three decades, sustainable development has been considered the overarching frame for biodiversity conservation (UN 1987). Yet zoological institutions are still struggling to expand their narratives and practice beyond "avoiding the end of wildlife" and to embed the broader agenda of sustainable development into their missions and institutional philosophies. To date, a significant number of zoological institutions (including the Association of Zoos and Aquariums) have adopted elements of the sustainable development agenda into their mission statements. As institutions move from aspirational statements to practical reality, many have chosen the path of "sustainability," installing all manner of energy efficiency and water conservation measures but largely steering away from the social, political, and cultural elements of sustainable development. In most cases zoos and aquariums are still places of contradiction where the content of exhibitions consistently alludes to wildlife conservation and sustainable behavior, while the visitor experience is designed to encourage consumerism, waste, and unhealthful eating. Beyond the ethical considerations of making revenue generation a priority over the social and conservation mission, these contradictions (and the sidelining of the larger frame of sustainable development) negate the positive effects of zoo-based wildlife conservation programs. As zoological institutions grapple with the future of zoo-based wildlife conservation, they have to broaden the scope to consider the future of zoological institutions as a whole.

This institutional introspection should pay special attention to the roles and responsibilities of zoological institutions in the communities that serve as their homes. Even though field conservation biologists understand and try to address local perspectives and engage in a broad range of social issues when doing projects in situ, these practices have failed to transfer to programs in the communities where zoos are physically located. A divergence of approaches at home and away from home is not a unique feature of zoological institutions: it reflects a generalized view in the United States that sustainable development is something to be done in developing countries with support, financing, expertise, and experts provided by developed nations. A recent example of this approach are the UN Millennium Development Goals (Sachs 2012; UN 2000). In this program, developed nations agreed to contribute financing and expertise to less developed countries in order to address eight key areas of sustainable development. The process delivered some significant improvements in quality of life among citizens of the participant countries (e.g., increased access to primary education and reductions in poverty and child mortality), but it did little to address the policies and practices that made developed nations the largest contributors to global biodiversity loss and climate change. To address this incongruence, Agenda 2030 has introduced the concept of "universality." Universality emphasizes that problems such as poverty, malnutrition, low-quality education, inequality, and injustice are damaging to all human beings, whether or not they live in a less developed nation. Universality also incorporates the understanding that global issues such as climate change, biodiversity loss, affordable and clean energy, and clean water can be addressed only with the collaboration of all countries.

Universality is particularly important for zoo-based wildlife conservation. Many of the threats to species conservation in the field reflect unsustainable policies and practices in developed nations (including the home communities of many zoological institutions; see E. L. Pearson et al. 2014).

An extreme example of the need for universality in wildlife conservation is the case of polar bears. The most significant threat to this species (as well as to Arctic ecosystems and human communities) is the reduction in area and duration of Arctic sea ice, which in turn is a consequence of global warming (IPCC 2014a; US Department of Interior 2008). Unsustainable practices in the United States make this country one of the world's largest contributors of greenhouse gases (IPCC 2014a). While it is certainly important to promote sustainable development and conservation practices in the Arctic, the viability of wild polar bear populations is dependent on whether the United States (and other developed nations) adopts sustainable development as a model and reduces its carbon footprint in time to avoid a collapse of Arctic ecosystems.

The principle of universality underscores how, in order to achieve their conservation goals, zoological institutions must involve all relevant actors, not just in far-off and exotic locales, but also within the communities immediately surrounding the zoo itself. More important, by becoming advocates and participants in the sustainable development of their surrounding communities in the developed world (particularly in the United States), zoological institutions can have a more significant effect on the conservation of some species than by working only in the locales these species come from.

PRACTICAL CHALLENGES AND GLIMPSES OF A POSSIBLE FUTURE

Zoo-based wildlife conservation programs lack the necessary resources to fulfill their mission (Gusset and Dick 2010, 2011; Miller et al. 2004). As part of their chapter, Monfort and Christen (this volume) discuss resource deficits in the context of maintaining genetically viable zoo collections. This section will expand the frame to compare current resources devoted to zoo-based wildlife conservation with the cost of protecting global biodiversity. This exercise will highlight the immense discrepancy between resources and needs. In doing so, it underscores how the frame of sustainable development is not only conceptually valuable, but also the most realistic path to achieving the long-term mission—and fulfilling the aspirations—of zoo-based wildlife conservation.

There is broad consensus that existing resources for zoo-based wildlife conservation are not adequate to address global biodiversity loss. What is not so widely known is the magnitude of this gap. A 2012 study by McCarthy et al. calculated an approximate cost of $76 billion for stabilizing and saving all endangered species globally, which is far more than the $154 million zoological institutions in the United States currently invest in field conservation (about 2 percent of the total budget; Association of Zoos and Aquariums 2014b). Even if zoos and aquariums devoted 100 percent of their budgets to field conservation, the resulting $8 billion or so would still miss the estimated need by 90 percent. Even in the extremely unlikely scenario where zoological institutions were able to generate the $76 billion needed to fully resolve species protection, this static figure does not include any species that would be added to the endangered species list in the future. In reality this cost would be perpetually increasing, a permanent expense (figs. 28.3-28.5).

FIGURE 28.3. Budgets, policies, and activities for collections management, public programs, management and financing, and conservation programs are disconnected from each other as well as from the core mission of animal care and conservation. Design by A. Cerezo, 2015.

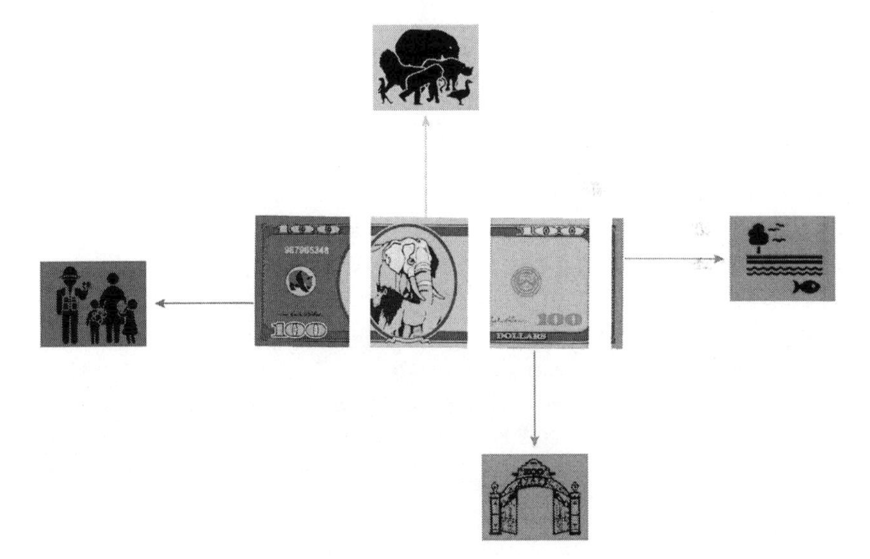

FIGURE 28.4. Zoological institutions (members of AZA) invest $155 million annually in conservation programs. This represents about 2 percent of their total budgets. Design by A. Cerezo, 2015.

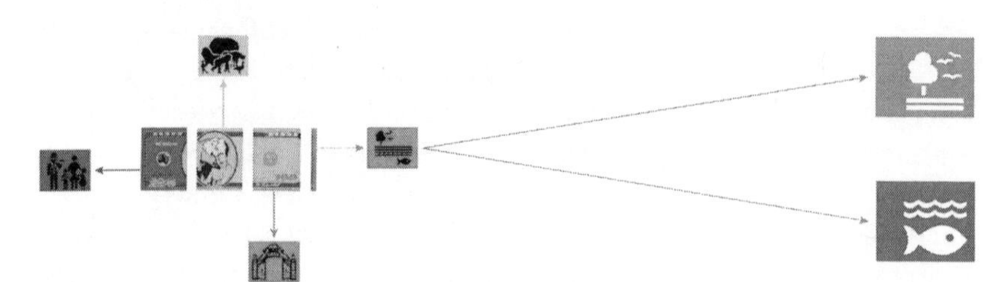

FIGURE 28.5. It is estimated that the cost of protecting all endangered species in the world is $76.1 billion. The $155 million invested by zoological institutions contributes .002 percent to that total. Design by A. Cerezo, 2015.

It is when we add the considerably higher cost of addressing the remaining fifteen SDGs that the limits of zoo-based resources come fully into focus. Just as an example, the estimated cost of addressing poverty (goal 1) annually is $66 billion, while addressing the infrastructure aspects of goal 9 (industry, innovation, and infrastructure) is estimated at $7 trillion annually (UN 2014; fig. 28.6).

Beyond the enormous deficit in funding, zoological institutions lack the diversity of expertise needed to take on the full agenda of sustainable development and therefore tackle wildlife conservation effectively and appropriately. One ingenious way zoo-based wildlife conservation practitioners have made the most of their limited resources is by reducing the geographic scope of community-based interventions and expanding their personal knowledge base to take on as many aspects of sustainable development as possible. Undeniably, many of these efforts have succeeded in protecting wildlife while improving the standard of living in the targeted community (e.g., Low et al. 2009). But as the conservation professionals who have undertaken these efforts know well, they are time consuming, difficult to implement, and very fragile. Given the limited financial resources of zoo-based wildlife conservation programs, this is not a model that lends itself to scaling so as to have long-term global impact.

But there is another concerning aspect to this zoo-driven, community-based conservation approach. The time devoted to taking on the coordination and delivery of multiple aspects of sustainable development is time that is not devoted to biological research and species management. As Monfort and Christen (this volume) have shown, if key zoo-based collections are to be viable in the long term, it is urgent to redirect existing resources to improving the management of captive species. In an environment of limited resources, zoo-based conservation professionals must figure out how to integrate their projects into the larger agenda of sustainable development without risking the very captive collections that make zoological institutions relevant in biodiversity conservation.

The purpose of this extremely sobering exercise is to show that no existing zoological institution (or group of zoological institutions), no matter how well funded or forward thinking, has the resources or expertise needed to address biodiversity conservation. This challenge is compounded in that biodiversity conservation can be successful only when it is taken as part of the exponentially larger agenda of sustainable development. The fact is that no existing conservation, social, economic, or political institution has the capacity to undertake the full agenda of sustainable development. Not on its own.

To undertake such a comprehensive and complex agenda it is necessary to forge a global network of diverse institutions reflecting the full scope of sustainable development (UN 2012, 2015). The long-term success of zoo-based wildlife conservation is tied to the success of sustainable development, just as the success of the sustainable development agenda is tied to

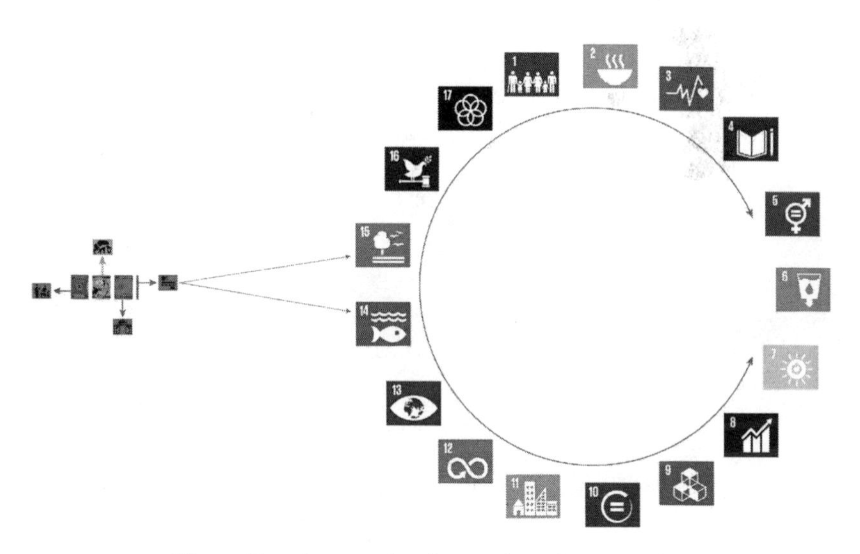

FIGURE 28.6. If we add to the cost of endangered species protection just the annual cost of addressing goal 1 (no poverty), $66 billion, and of goal 9 (specifically the targets for improved infrastructure), $7 trillion, the investment in zoo-based conservation contributes .000021 percent to that total. Design by A. Cerezo, 2015.

the individual success of all the institutions that are part of the network. It is imperative for zoological institutions to become better acquainted with the sustainable development agenda and fully active in the network of institutions supporting this agenda (fig. 28.7).

Understanding and engaging the network of sustainable development provides zoological institutions with a global, interdisciplinary platform for collaboration. Zoo-based conservation can benefit from the perspectives of other institutions when developing programs and conservation strategies. Institutions can also build partnerships where all participants can coordinate and contribute their specialized expertise and resources in truly multi-

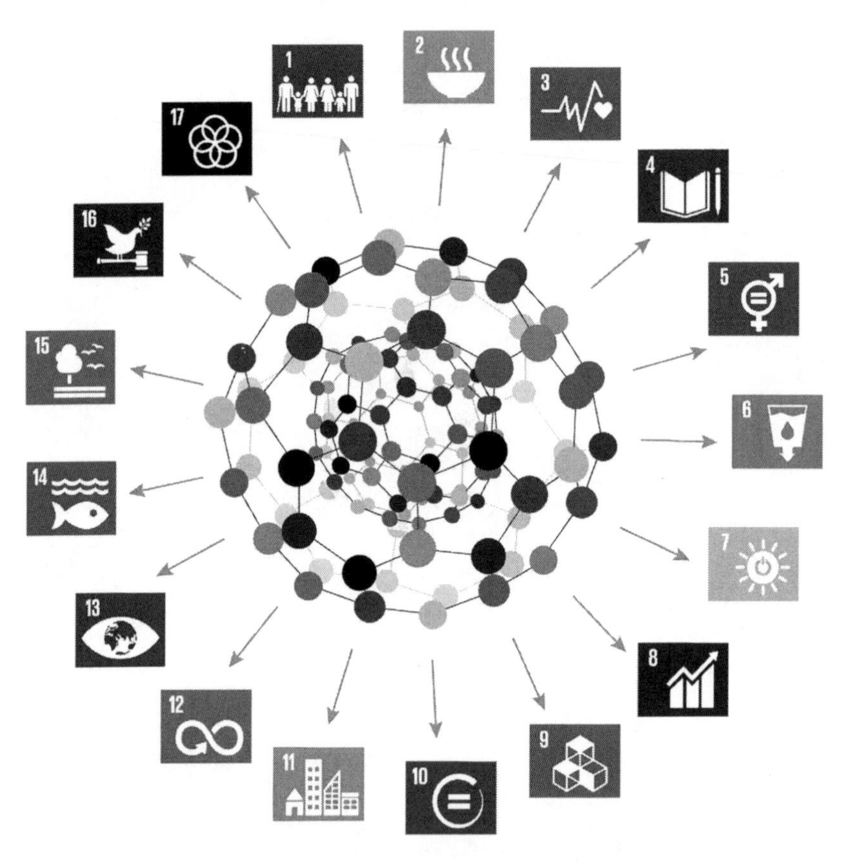

FIGURE 28.7. Achieving the full agenda of sustainable development requires the collaborative efforts of an interdisciplinary network of global and local institutions representing all areas of Agenda 2030. Design by A. Cerezo, 2015.

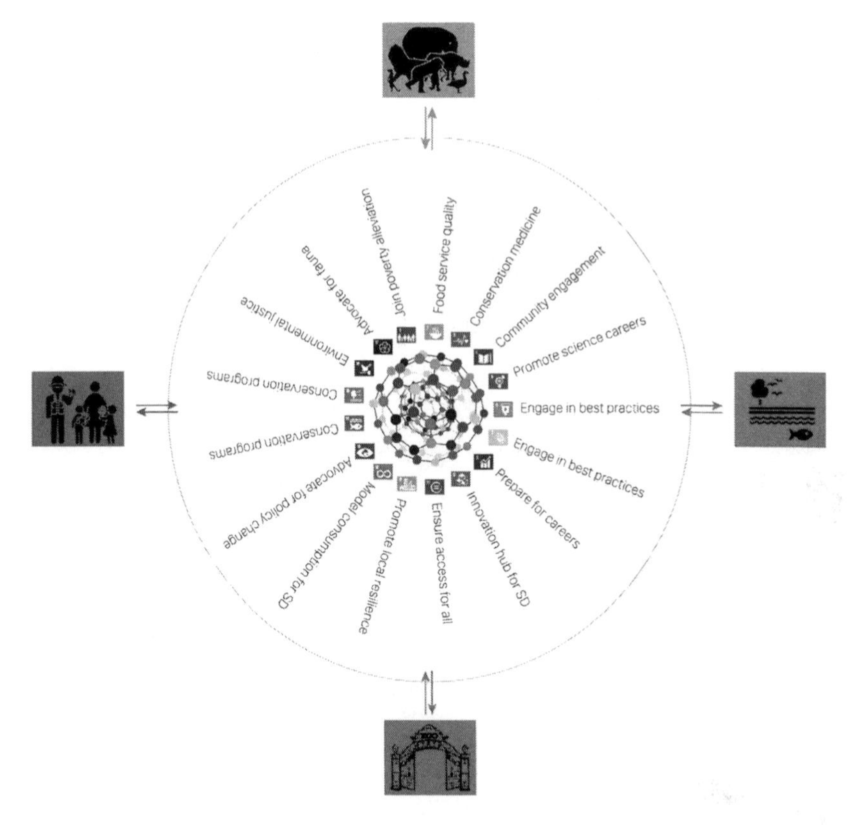

FIGURE 28.8. The sustainable development agenda (and engaging the global network) provides a platform to align the mission, programs, and practices in all areas of the zoo. This alignment increases the quality and effectiveness of science and public programs. Design by A. Cerezo, 2015.

dimensional sustainable development programs at the local, national, or global level. By relying on the expertise and resources of partners for the design and delivery of field programs, zoological institutions also reduce the time and funding devoted to taking on these responsibilities on their own. These resources can then be directed to addressing the challenges inherent in zoo-based conservation and increasing the quality of the specialized expertise and insight they contribute to the network (fig. 28.8).

On an institutional level, engaging in sustainable development also provides a platform to align the mission, policies, and practices of zoological institutions. In building collaborations, partners will promote reflection on how zoological institutions think about and address specific aspects of sustainable development. They will also provide knowledge and support in shifting toward more sustainable approaches. Eventually most aspects of the zoological institution (collections, infrastructure, and public programs) will be consistent with the conservation mission. This level of alignment will increase quality and effectiveness at all levels of endeavor. More important, every dollar spent in the zoological institution will serve the dual purpose of supporting the institution's conservation agenda and projecting the sustainable development agenda at the local, national, and international levels (fig. 28.9).

Finally, by reframing zoo-based conservation in the larger context of sustainable development, the discussion shifts from one of overwhelming limitations (of funding, resources, time, and capacity) toward a conversation about abundance (of specialized capacity, commitment, potential partners, and hope). This is why we believe it is time to retire the "zoo as ark" metaphor and look for metaphors that relate better to this expanded, more complex and optimistic paradigm (fig. 28.10).

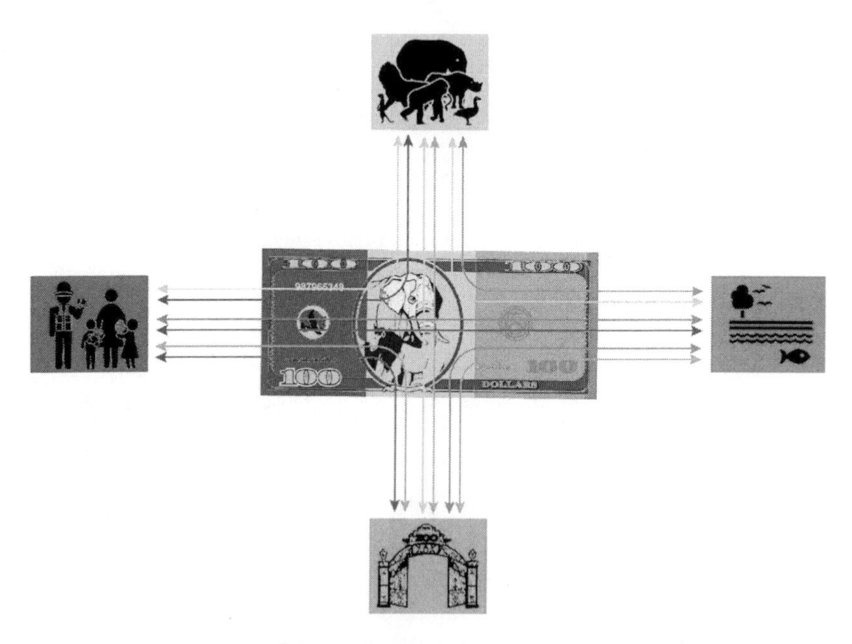

FIGURE 28.9. As areas of the zoological institutions align their missions, policies, and programs, their budgets also integrate, align, and start serving multiple purposes. This approach allows them to virtually increase the budget devoted to conservation. Design by A. Cerezo, 2015.

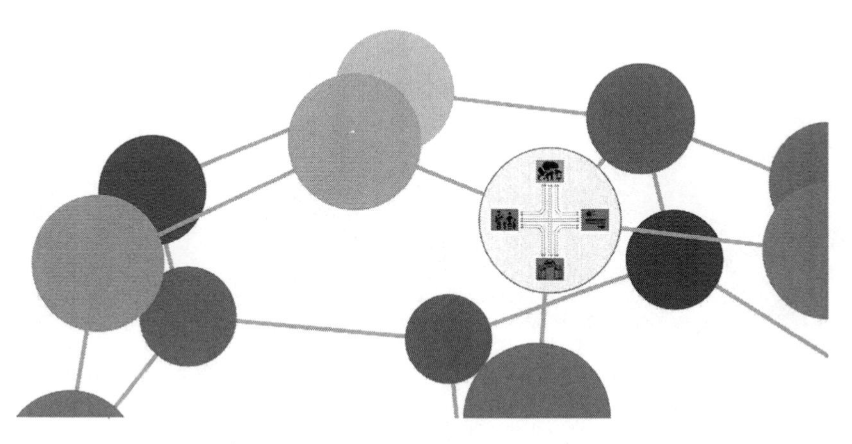

FIGURE 28.10. The future of zoo-based conservation requires the engagement and active participation of zoological institutions in the global network of sustainable development. Design by A. Cerezo, 2015.

THE WORLD BEYOND THE ARK METAPHOR: CONCLUSION AND EXHORTATION

The idea of zoos as a modern-day Noah's ark originally served to explain the practice that defines the identity of zoological institutions: collecting and displaying living animals. More recently, the metaphor has provided a weak platform to explain the role of these institutions (and captive animal populations) in global biodiversity conservation. In this narrative, zoological institutions are granted license to collect living animals because they are the refuge, one ship afloat in an endless ocean, outfitted to save fauna from a world engulfed in an apocalyptic deluge of biodiversity extinction. While we acknowledge the magnitude of the threat we face in global biodiversity loss, we believe that relying on the ark metaphor is counterproductive in imagining the future of wildlife conservation.

Ecological research demonstrates that natural systems (and their associated social systems) are complex, interconnected, and entropic (Rohde 2005). The health and productivity of these systems is a reflection of the diversity of elements, the richness of the environment, and the quality of the connections between the elements. These principles are reflected in the sustainable development agenda, where the process and the outcomes are intended to reflect and promote diversity, richness, interconnection, and complexity. The idea of one group, with a very specialized skill set, tasked with safeguarding all species from extinction is contrary to the principles

FIGURE 28.11. The ark metaphor proposes zoological institutions as a ship outfitted to save fauna from a world engulfed by an apocalyptic deluge. Design by A. Cerezo, 2015.

FIGURE 28.12. Disembarking from the ark, zoological institutions will find a global network of partners ready to support biodiversity conservation as we work together to build a sustainable future. Design by A. Cerezo, 2015.

of nature and to the process of sustainable development. Just as important, it is also detrimental to the conservation mission of zoological institutions. If zoos and aquariums are to contribute to sustainable development and successfully address biodiversity loss, they would be well served by moving beyond the ark metaphor (fig. 28.11).

It is time to find a metaphor that embodies the complex, diverse, collaborative, and optimistic nature of sustainable development and describes the role of zoological institutions in achieving it. Doing so allows zoological institutions space and resources to embrace their passion for the living world, a passion that is not just welcome but a fundamental element of true sustainable development.

The detailed consideration for human issues in Agenda 2030 (e.g., economy, health, education, justice, food, urbanization, and poverty) tends to overshadow the fact that nonhuman nature is the foundation of sustainable development. In the dialogue about the future of sustainable development, wildlife conservation institutions speak for the animals. To make this voice of wildlife loud and clear, we need to come ashore, disembark, and join the global community as active and productive participants.

Rather than ending this chapter by proposing a new metaphor for zoo-based conservation, we provide three principles that should be represented in the new metaphor: first, the passion we feel for wildlife; second, a sense of optimism about the future of the biosphere; and third, a commitment to engage the larger mission of sustainable development. The new metaphor for zoo-based conservation will be revealed as our footsteps move ever closer to a sustainable world (fig. 28.12).

CHAPTER TWENTY-NINE

Rewilding the Lifeboats

Harry W. Greene

INTRODUCTION

Controversies about captive animals ultimately concern values, so let's begin by stipulating that zoos—however imperfect, however much they also entertain—play admirable roles in conservation, research, and education. Plenty of published evidence supports that assertion, as do other contributors to this volume. More personally, as a young child I came to love nature while visiting zoos in Denver, San Francisco, and Tucson, then as a graduate student gathered dissertation data on snake behavior at zoos in Atlanta, Dallas, Fort Worth, Houston, and Knoxville. Let's also admit to financial, logistical (e.g., animal health, keeper safety), and political constraints, themselves reflecting cultural norms. The more philosophically challenging problem, then, is, To what extent do politics and culture constrain zoo design and management? Might the answer be "too much," and could zoos better achieve their important core goals through greater attention to natural processes, including those that entail mortality?

This chapter confronts these questions by channeling ethologist Gordon Burghardt's writings on "critical anthropomorphism," "private experiences" of nonhuman animals, and "controlled deprivation" to underscore the relevance to zoos of ecology and evolution (see Burghardt 1996, 1997, 2013). My original title for this chapter was "Rewilding the Ark," but Chad Peeling's (2015) portrayal of zoos as "lifeboats with very limited seating" is germane to culling, a central issue here—thus the change. In brief, I argue that "wild" can usefully be understood in terms of ecological and evolutionary processes rather than minimal human impact, and this distinction bears on captive animal management; that, thereby defined, the wild lives of even "simple" organisms are far more complex than those we typically afford them in captivity; that how zoo animals are kept often is so divorced from the wild as to diminish their applicability to conservation, research, and education; and that, informed by ecological and evolutionary perspectives, zoos can do better.

My remarks also echo Conway's (1973) classic "How to Exhibit a Bullfrog," first published half a century ago, in which he emphasized the need to convey a sense of wildness (see also Sigler 2015). Here I expand on this theme with discussions of rattlesnake natural history, the death of a Danish zoo giraffe, and displays of giant lizards and vultures. Rather than winning arguments, my aim is to provoke more nuanced, *open* debate in the service of preserving as much of wild nature as possible, including in zoos.

WHAT ARE WILD LIVES LIKE?

"Wild" has various meanings, but in the United States the word usually implies places or things minimally influenced by humans—the opposite of tame or domestic. Accordingly, wild places are those where, to employ a common aphorism, we "take only photos and leave only footprints"; wild landscape attributes (e.g., rivers) and wild organisms are "self-willed" in the sense of not being subject to human control (for diverse explorations of "wild," see Wuerthner, Crist, and Butler 2014; Minteer and Pyne 2015). In practice these notions are ever more problematic owing to our near-ubiquitous, enormously destructive presence, and elsewhere I've built on earlier claims that they also can be misleading (Greene 2013, 2015). Truly "minimal influence" would have applied to Africa, for example, only before we *became* human, more than two million years ago (e.g., Werdelin and Lewis 2013), and to the New World only before humans *arrived*, at least thirteen thousand years ago (e.g., Rick et al. 2014; for a global perspective, see Boivin et al. 2016).

Instead, I regard places and lifestyles as wilder to the extent that they entail locale-appropriate ecological and evolutionary processes, including predation, competition, selection, disturbance, and nutrient cycling. Maximally wild ecosystems therefore must contain organisms that enable those processes, including apex predators, megaherbivores, and scavengers—among which a few species can kill us, many more are regarded as ugly or otherwise unpopular, and most are increasingly imperiled by our global domination of ecosystems (Estes et al. 2011). It's worth emphasizing that some naturalists and philosophers are uncomfortable with the harsher realities of ecology and evolution, an implicitly and at times astonishingly *antiwild* mind-set (e.g., a renowned ornithologist barely stopped short of advocating extermination of all predators [Skutch 1998], and McMahan [2016] offered an ethical justification for that goal).

Animals in zoos typically are kept in ways considered healthy and humane as measured by veterinary principles of physiological and behavioral well-being and, when desirable, reproduction (see, e.g., chapters by Maple and Segura and by Palmer, Kasperbauer, and Peter Sandøe, this volume). To highlight the disparity between those criteria and an ecological and evolutionary concept of wild, consider a lifestyle that even many animal-loving biologists would regard as simple. Under generous contemporary standards, a captive black-tailed rattlesnake (about a yard long and weighing a bit under a pound) occupies nine square feet of cage floor space. Her enclosure is landscaped with rocks and plants, but such "cage props" are minimized in the interest of hygiene. She is disturbed as little as possible, and unless there are specific plans for breeding, she never interacts with other nonhuman animals; once weekly she is handed prekilled lab mice weighing perhaps 10 percent of her mass, and this schedule is not interrupted seasonally. Most zoo visitors won't see that rattler so much as flick her tongue, and they reasonably might regard her—and by implication her wild counterparts—as exceedingly boring. For years one Texas zoo actually exhibited a plastic venomous coral snake, not labeled as such.

A *wild* black-tailed rattler, however, can within weeks of birth find, kill, eat, and digest adult mice more than equaling her own mass. Once mature, she ambushes mainly wood rats and rabbits. She may track a bitten, dying prey many yards over rugged terrain, then after consuming it hide from predators under a ledge while protruding a coil containing the food lump into the hot midday sun. Her home range encompasses dozens of acres of sparsely wooded canyon, throughout which she reuses specific rock shelters, cactus patches, and so forth—even returning every couple of years to the same squirrel burrow to gestate and give birth, after which she remains with her pups for about ten days, until they shed their natal skins. During a seven-month active season she might not change hunting or basking sites for a week, then abruptly travel hundreds of yards to a new one. Her social scene encompasses offspring, closely related females, potential mates, and distantly related conspecifics. Copulation can last almost twenty-four hours, during which the female controls the pair's movements. She reacts adaptively to occasional danger and, despite a gauntlet of local predators on snakes, might well survive several decades (Greene et al. 2002; Persons, Feldner, and Repp 2016; Schuett et al. 2017).

How much more complex, then, might be the wild lives of elephants, lions, and other mammals? What would it take to provide them with even a modicum of wildness in captivity, and why might wilder conditions matter?

SHOULD WE REWILD THE LIFEBOATS?

The avowed zoo goals of conservation, research, and education might each be diminished by lack of wildness and consequent shifts in the population genetics and phenotypes of captive animals. Abundant evidence testifies to inadvertent selection in captivity, despite absence of accidents, predation, deaths during birthing, and other forms of mortality (reviewed in Schulte-Hostedde and Mastromonaco 2015). Probably less widely appreciated beyond evolutionary developmental biology is the role of experience in shaping phenotypes (e.g., Hall 2001; Schulte-Hostedde and Mastromonaco 2015), although a century-old comparison of wild-shot and zoo-raised lions is illustrative: differences between their skulls (including smaller braincases in the captives) are of the sort associated with species-level distinctions in mammals and evidently resulted from lack of behavioral effects on bone growth in juveniles (Hollister 1918; fig. 29.1). Because of inadvertent selection and developmental effects from unnatural diets (Hartstone-Rose et al. 2014; Schulte-Hostedde and Mastromonaco 2015; Wines et al. 2015), captive-bred animals could be less suited for release into wild situations, and research on them might well be misleading with regard to the morphology, physiology, and behavior of wild animals.

Compelling reasons to enhance wildness in zoos also involve welfare (e.g., effects of controlled deprivation on the private experiences of captives; see especially Burghardt 1996), as well as ways animals are portrayed in exhibits and educational materials. Because the former topic has been so extensively addressed elsewhere, I'll underscore here that zoos generally represent nature as simple and benign—and we should be asking whether that blatantly unrealistic vision undermines appreciating species that are neither, especially those that might kill and devour us. Let me now further exemplify both points with a controversial example.

MARIUS THE GIRAFFE: IT'S ALL ABOUT VALUES

Marius was a young male giraffe in the Copenhagen Zoo, an institution whose explicit population control policies include culling individuals deemed genetically inappropriate for breeding. Accordingly, out of public view, in 2014 he was killed with a rifle shot to the head, then dissected in

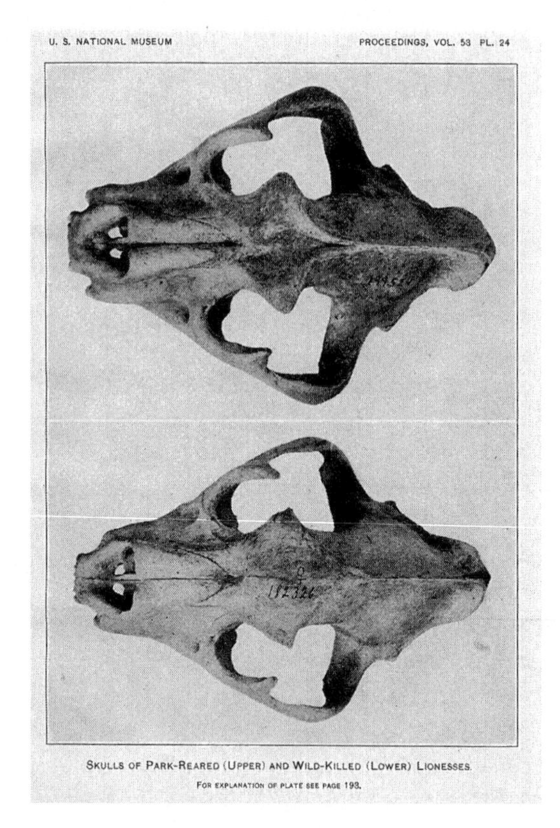

FIGURE 29.1. "Skulls of the captive [lionesses] . . . are broader and shorter, more massive and bulky. . . . The obvious reason for these great differences is that the principal muscles operating the jaws and neck (those muscles used by a wild lion in mauling and killing game, biting, gripping, and shaking) have had little influence on the shape of the bones during development. In a wild-reared lion these powerful muscles naturally and in a normal way mold the growing skull, particularly in the regions of their attachment" (Hollister 1918, 184).

front of invited zoogoers and fed to lions, as is also institutional policy. There followed an international uproar (the zoo received more than thirty thousand overwhelmingly critical e-mails), fueled by inaccurate, sensationalist reporting—one piece claimed Marius was "slaughtered" with a shotgun, another wrongly implied that children were coerced to watch or unexpectedly viewed his necropsy. Respected science journalists editorialized against the Danish practices (e.g., King 2015; Morrell 2014), as did at least one prominent former US zoo director (Maple 2014). As it happens, the Copenhagen Zoo had had those policies for decades, and polls within

Denmark revealed more than 90 percent support for them (Bertelsen 2014). In fact, my correspondence and conversations while preparing this chapter, including with some contributors to this volume, revealed a sharp dichotomy within the US zoo community, with those supporting the Danish approach generally unwilling to express their views publicly or even in e-mails. I also was told repeatedly that culling is widespread in US zoos, though more prevalent with reptiles than with mammals.

That reactions to this incident reflect international cultural differences is evident in Bertelsen's (2014, 65) report from Copenhagen Zoo on "Marius-gate":

> A zoo has three choices: 1. Pretend that animals are only born [i.e., they never die], only euthanize geriatric or very sick animals . . . and feed carnivores minced meat off exhibit, joining everybody else in the belief that it really wasn't animal derived. 2. Manage populations rationally including culling of certain individuals, feeding carnivores real meat etc. but not publicizing details of meat supply, surplus animals, population control, etc., or 3. Take on the responsibility of showing people how things actually are: let the public see lions eating a recognizable piece of horse, and ultimately also stand by the fact that, e.g., while 100 percent of [wild] female impalas can find a new home, perhaps 5 percent of the males can. Most U.S. zoos live by option 1, most European zoos by option 2 (although an increasing number of countries are leaning toward 1), and a few zoos, primarily Scandinavian, support option 3.

Discussions of how animals should be treated often are cast in the form of "How would you like it if such and such were done to you?" Herzog (2010), for example, detailed the lives of gamecocks and industrial poultry, then concluded he'd prefer to be the former but that cockfighting should be illegal and he'll keep on eating chicken. In the spirit of such thought experiments, let's imagine a cosmic dice roll such that I'm a male zoo giraffe, my only choice being to be born in the United States or Denmark. In the former case, odds are that, thanks to isolation or hormonal suppression, I'll rarely if ever experience sexual behavior. I will live as long as the best veterinary care can support, and once I'm finally severely enfeebled, my death by toxic chemical injection will be announced after private carcass burial or incineration. By contrast, in Denmark I might end up a stud bull, with all that entails psychologically and socially; maybe I'll even tour Scandinavia, depending on where my genes are needed. Of course, instead I might be born Marius, such that after a short but relatively rich life, my light will abruptly go out. Either way, once killed I'll be dissected for research and education, then my body will be fed to the very species that likely would have eventually eaten me had I been born on an African savanna.

That my cosmic wager is based on a false analogy—I'm not a giraffe—is relevant only to the extent one excludes anthropomorphic perspectives on how other animals are treated. In fact, just how we judge Marius's fate might reflect personal tendencies for empathy as well as our particular ethical beliefs and relevant cultural norms. My choice of the Danish option no doubt will appall some fellow animal lovers, but then I've spent a career promoting conservation of natural-born killers and generally regard the loss of an entire species as more tragic than that of an individual animal. Moreover, I'd vote for death with dignity laws; my parents donated their remains to science, and I'd rather be dumped in the backcountry for scavengers than perfused with formalin and planted in the ground or belched into the sky as gas and cinders. One might argue all that is trumped by Marius's having no choice in the timing and manner of his death, but neither did my wonderful dog Riley when I took him for one last walk, fed him a venison steak, and paid a veterinarian to end his terminal decline.

REWILDED DRAGONS AND VULTURES: WHAT MIGHT "BETTER" MEAN?

Komodo monitors ("dragons") are the largest extant lizards, with males topping out at about two hundred pounds and nine feet in length. They ambush or chase down deer, pigs, water buffalo, and rarely humans, killing them by slashing bites to the limbs or abdomen. By all accounts dragons are behaviorally complex, sometimes hunting and feeding communally. Their home ranges encompass hundreds of acres, many aspects of which they know well. In captivity they play with objects in the manner of puppies (see Murphy et al. 2002 for details of dragon biology), but given a naturalistic context they can erupt with astonishing ferocity:

> Years ago [the US] National Zoo routinely fed dead animals from the collection after necropsy to reptiles . . . [a keeper] acquired the carcass of a muntjac . . . and fed it to the Komodo dragon during visiting hours. The dragon seized the muntjac by the belly and tossed its head in a violent shaking motion, which eviscerated the deer, splashing the glass front of the enclosure with blood, metabolic by-products, and entrails. Predictably, the incident was reported in record time to higher authorities and . . . [the keeper] was warned to never let it happen again. (Murphy 2015, 678)

Times are indeed changing, and Komodo monitor exhibits exemplify prospects for rewilding zoos. At Clyde Peeling's Reptiland, two adult dragons occupy an enclosure thirteen hundred feet square (fig. 29.2), within which four basking hot spots (programmed to move every few hours) and a spacious pool provide for thermoregulation and drinking. The lizards are often seen strolling about, flicking their long forked tongues, and during public feeding times they bolt prekilled lab rats and chunks of pork. Overhead screens feature a graphic four-minute video of wild dragons killing and eating a water buffalo. At Reptiland I've noted onlookers lingering in front of the display, their faces brimming with curiosity and fascination rather than horror—as surely also occurs at the Los Angeles and San Antonio zoos, where visitors can watch these giant reptiles ripping apart preslaughtered, eviscerated pigs, the dragons' jaws bloody with dangling bones, muscle, and connective tissue (fig. 29.3).

Just as monitor lizards and other predators are essential components of healthy ecosystems, scavengers provide an ecologically significant postscript to the lives of prey animals, yet one that is rarely portrayed in educational materials (Moleón and Sánchez-Zapata 2015). Zoo Atlanta's lappet-winged vultures, however, exemplify a version of rewilding that is sensitive to viewers' concerns while heralding the importance of nature's cleanup crews for nutrient cycling. These giant birds are displayed in an area of carnivore enclosures, their food provisioned within an artificial zebra carcass. Placing a piece of tree limb over the simulated dead mammal's head so its eyes couldn't be seen allayed visitors' complaints that the exhibit was disturbing (fig. 29.4).

FIGURE 29.2. Komodo monitors at Clyde Peeling's Reptiland. Photo by H. Greene.

FIGURE 29.3. A Komodo monitor at the Los Angeles Zoo feeds on a pig carcass. Photos by Ian Recchio, courtesy of the Los Angeles Zoo.

FIGURE 29.4. Lappet-winged vultures at Zoo Atlanta and the artificial zebra carcass within which their food is provided. Photo by H. Greene.

NOW WHAT?

Twenty-five years ago Herzog (1991) examined why it might be considered ethical to feed lab rats but not domestic cats to captive boas, a question we can now reframe as an opening punch line for my conclusion: Why should it be acceptable to breed mice for snakes and raise swine for dragons but not supply giraffes, born in the larger context of conservation genetics and killed humanely, as food for lions? Or put more broadly, why is it tolerable to raise domestic but not wild species of hoofed stock to feed carnivores and scavengers in zoos? Rattlesnake natural history, the saga of Marius, and forward-thinking dragon and vulture exhibits underscore that our disparate responses to such controversial propositions likely have more to do with values and their murky, often unexamined underpinnings—with competing ethical and aesthetic preferences—than with biological or logistical concerns.

And so I'll close with still more questions: Wouldn't rewilded zoos better serve conservation, research, and education through more realistic and *interesting* portrayals of ecological and evolutionary processes? Wouldn't better public understanding of how nature really works, including that life *and death* frame the same eternal cycles, be a good thing? Could we begin rewilding zoological garden lifeboats with innovations that are naturalistic, yet acceptable to educable urban sensibilities—perhaps offer large carnivores and scavengers road-killed deer, eviscerated but not skinned, with heads still attached? Maybe advise visitors to watch for the pop-up plastic python that occasionally will startle baboons near one of their food trays? Because zoos might shape our attitudes, including those toward dangerous and otherwise unpopular species, I believe the dilemma of how captive animals are managed and displayed bears on the future of wild nature.[1]

ACKNOWLEDGMENTS

The influences on my thinking of grad school pal Hal Herzog and our adviser, Gordon Burghardt, are obvious. I also appreciate feedback from Doug Armstrong, Cheri Asa, Mads Bertelsen, Matt Brooks, Janine Brown, Jim Collins, Dante Fenolio, Nancy Greig, Darragh Hare, Mats Höggren, Lynne Isbell, Barbara King, Jonathan Losos, Joe Mendelson, Ben Minteer, Steve Monfort, Jim Murphy, Chad and Clyde Peeling, Jenny Pramuk, Aleta Quinn, Ian Recchio, Gordon Schuett, Kurt Schwenk, Steve Thoms, Paul Weldon, Chris Wemmer, Dave Wildt, and William Xanten. None of them necessarily agree with anything I've written here.

CHAPTER THIRTY

The Parallax Zoo

Ben A. Minteer

THE CALL OF THE (NOT SO) WILD AT YOSEMITE

John Muir probably would have been shocked at the sight of it: a caged collection of mountain lions, deer, and bears on public display in the heart of his beloved Yosemite National Park. Yet had the iconic naturalist and wilderness advocate lived only a few years more (he died in 1914), he would surely have encountered Yosemite's unlikely—and today mostly forgotten—"zoo."

By all accounts a misfire in the park's (and the National Park Service's) history of wildlife management, the Yosemite Valley zoo is nevertheless a fascinating episode in the evolution of one of the nation's iconic wilderness parks. The "menagerie," as park naturalist Ansel Hall called it in his 1920 guide to Yosemite, was about a third of a mile from the center of Yosemite village, part of a cluster of structures that included the park schoolhouse, barns, and various utility buildings (fig. 30.1). It opened in 1918 following the donation of several orphaned mountain lion cubs to the park by Jay C. Bruce, a lion hunter for the state of California (Janiskee 2009).

The motivation behind putting a zoo in Yosemite seems to have been to ensure that tourists would have a chance to see some of the park's most popular wildlife during a visit (Sellars 2009, 78). The commitment to the display of native animals was fairly relaxed, however. Although naturalist Hall described it as containing "several wild animals captured in the region" (Hall 1920, 28), not all the animals in the Yosemite zoo were, in fact, plucked from native populations. Two of the zoo's mountain lions, it turns out, were actually Rocky Mountain varieties taken from Yellowstone National Park, making them "exotic" to the northern California landscape (Grinnell and Storer 1924, 97).

FIGURE 30.1. The Yosemite Valley "zoo" in 1922. Yosemite National Park Research Library, used by permission.

The Yosemite Valley zoo never amounted to anything more than a small, opportunistic, ad hoc animal collection. Still, it was apparently enough to draw the ire of one of the nation's most prominent biologists: Joseph Grinnell, director of the Museum of Vertebrate Zoology at the University of California at Berkeley. Grinnell, like Muir before him, was a staunch supporter of the national parks (and of nearby Yosemite in particular). More significant, he was a pivotal figure in early efforts to put wildlife protection in the parks on solid scientific footing in the formative decades of the National Park Service (Runte 1990a).

"I recommend the elimination of the 'zoo,'" Grinnell wrote in an open letter to the park superintendent after a Yosemite Valley visit in 1927, a missive he summarized and published in the *Journal of Mammalogy* the following year (Grinnell 1928). The park resources currently employed in supporting the menagerie would be much better spent, Grinnell thought, on the Yosemite Museum, which he suggested was a far more suitable vehicle for educating park visitors on the diversity and natural history of the area's wildlife. Although Grinnell rather reluctantly admitted that zoos were appropriate institutions "in a crowded city, for benefit of people who cannot reach the open spaces," he emphasized that a national park should be a completely different kind of place: an area maintained as far as possible as a natural landscape.

Grinnell's vision of park management at times bordered on the Edenic, however, elevating an aesthetic and a historical ideal of wild country while simultaneously discounting the record of human influence on the landscape, especially the activities of Native Americans. A dozen years earlier, Grinnell (with his Berkeley colleague Tracy Storer) wrote in *Science* that the national parks, as those lands "kept fairly immune from human influence," offered the rare opportunity for visitors to experience nature as it was before "the advent of the white man" (Grinnell and Storer 1916, 377). Grinnell's celebration of the untrammeled character of the great parks of the American West was, of course, not an unusual sentiment during the early years of the National Park Service; indeed, as an aesthetic expectation and environmental ethic for US park and wilderness management it would persist throughout most of the twentieth century (see, e.g., Cronon 1995; Nash 2014).

For Grinnell and similarly minded wildlife biologists it thus seemed an obvious question: Why have a zoo in a magnificent place like Yosemite when it already served as a "zoological park in the widest and best sense"? (Grinnell, quoted in Runte 1990b, 133). An "artificial" zoo just didn't belong in a national park, especially a flagship wildlife and wilderness park like Yosemite. Despite their shared interest in attracting and satisfying a curious public, biologists and park administrators clearly saw the park and the zoo as fundamentally different entities, reflecting disparate aesthetic and ecological values and priorities. Transplanting a zoo into a place like Yosemite (or Yellowstone, which in 1925 opened its own menagerie of bison, bears, coyotes—and a badger [Sellars 2009, 78])—must therefore result in the institutional equivalent of tissue rejection. The message was clear. In the midst of a truly spectacular wild landscape, a zoo could only be a pitiable spectacle.

ZOO NATURALISM AND ITS DISCONTENTS

Grinnell's lobbying worked: Yosemite's zoo was shuttered in 1933. But the dissatisfaction with the perceived inauthenticity and artificiality of the zoo, especially compared with what was deemed "truly" wild nature, would prove a difficult narrative for zoos to shake over the twentieth century. Half a century after Grinnell's public plea to close the zoo at Yosemite, the distinguished art critic and novelist John Berger would still be able to make the sweeping statement that, as an institution, "the zoo cannot but disappoint" (Berger 1991, 28). Once again, the manifest artificiality of the zoo was the primary culprit. But Berger went further. He argued that the zoological artifice led to a deeper and more profound ethos of "separation": species were segregated from one another (and from their natural habitats); animals were cordoned off from people by layers of glass, concrete, and steel. It all reinforced an ethic and a visual culture, he wrote, that promoted the greater marginalizing of animals in modern society.

Although some of Berger's punches may have landed back in the early 1980s when his critique was originally published, the design standards and environmental aesthetics of professional zoos have evolved in the past several decades, in many cases dramatically. In particular, the development of what came to be known as "immersive" animal exhibits and the growth of a more ecological philosophy in zoo landscape architecture, which many observers link to the redesign of the Woodland Park Zoo in Seattle beginning in the late 1970s, have done much to reduce the patent artificiality of the zoo landscape (Hanson 2002). Today many zoological parks contain large barrier-free and naturalistic enclosures (including mixed-species exhibits) that offer not only more biotic diversity and space—and less separation—but more opportunities for animals to engage in a fuller suite of natural and social behaviors, a widely acknowledged (though not unambiguous) component of zoo animal welfare (Maple and Perdue 2013; Maple and Segura, this volume; though see also Kawata 2012 and Fábregas et al. 2012). The Arizona-Sonora Desert Museum in Tucson (see, e.g., Ivanyi and Colodner, this volume) is one of the more distinctive institutions employing this approach, a "regional biopark" with naturalistic grottoes that often seem to well up out of—and melt into—the Arizona high desert.

This isn't to say that all zoos are equally innovative and professional when it comes to exhibit naturalism and animal care, or that the divisive ethical issues surrounding some of the more controversial cases of keeping zoo animals such as elephants have been settled once and for all (see, e.g., Blau and Rothfels 2015). But in general terms it would be difficult to deny that as an institution the professional, accredited zoo has changed in significant, and in some cases revolutionary, ways with respect to many of Berger's earlier objections. The expansion of the zoo agenda to engage more significantly in conservation breeding and reintroduction during this same period is a significant marker of that evolution, a movement that may also be seen as a return to some of the scientific, educational, and wildlife protectionist roots of the modern zoo (Minteer, Collins, and Raschke, forthcoming). It's an effort that dovetails in many ways (though not always perfectly) with trends promoting heightened naturalism and improvements in zoo animal welfare (Fa, Funk, and O'Connell 2011; see also Palmer, Kasperbauer and Sandøe, this volume).

Clearly, though, these reforms haven't been enough to mollify many zoo critics. In a recent article in *Outside* magazine ("The Case for Closing Zoos"), journalist Tim Zimmermann (2015) posed the question, "Are there any good arguments for keeping animals in artificial enclosures that, at best, are only a fraction of the size of their natural habitats?" He ended up concluding that there weren't any, dismissing the educational and conservation claims of public zoos and arguing that the move to large animal sanctuaries and reserves (with limited public access) was the only defensible future of the zoological facility.

Writing a few months later in *New York* magazine, journalist and self-described "zoo lover" Benjamin Wallace-Wells similarly argued that the mainstream zoo was playing a game with naturalism and the wild that it simply could not win. Its residents, moreover, were aware of the score. "It is hard to avoid the conclusion," Wallace-Wells wrote, "that in some way the animals understand that the world around them is an artificial one . . . the central illusion of the zoo is no longer holding. The animals know" (Wallace-Wells 2014).

A RADICAL REBOOT?

Into this long-standing debate over zoos, artificiality, and naturalism (which goes back at least to Grinnell's screeds against the Yosemite Valley zoo in the 1920s) steps "Zootopia," a three-hundred-acre expansion of Denmark's Givskud Zoo proposed by iconoclastic architect Bjarke Ingels and his firm, BIG (http://www.big.dk/#projects-zoo). A creative mash-up of immersive zoo and safari park, Zootopia has been promoted as a radical reboot of the tired zoo concept: a nearly wall-less and cage-free zoo landscape in which the animals roam relatively freely in multispecies habitats. Ingels's innovative plan has even been described as proposing a dramatic reversal of "captor and captive" (Wainwright 2014). The first phase of the new park is planned to open in 2019.

Zootopia's design philosophy is clearly driven by the goal of minimizing, and in some cases completely concealing, the human presence. Visitors to the new zoo, for example, will be sequestered in hidden viewing galleries and transported through the air in mirrored pods (fig. 30.2). Elsewhere they'll use bicycles and boats to get up close and personal with the zoo's elephants and zebras, which will be separated from zoogoers by an ingenious array of natural and mostly undetectable barriers (strategic placement of log piles, water, bamboo stalks, etc.). The largest and most discernible human artifice in the park will be the dramatic bowl-shaped "arrival crater," a resplendent entry pavilion that serves as a gateway to the zoo's different "continents" (Africa, Asia, the Americas) and a symbolic entry point to the "wild" environs awaiting visitors beyond the threshold (fig. 30.3).

FIGURE 30.2. Out of sight (and out of mind)? Zootopia's airborne pod transportation system. Bjarke Ingels Group, used by permission.

FIGURE 30.3. A visualization of Zootopia's "arrival crater." Bjarke Ingels Group, used by permission.

Perhaps not surprisingly, commentary on Ingels's plan has tended to evoke film analogies, including the false-reality conceit of *The Truman Show*, rendered here as a simulated wilderness that "fools" the animals into thinking they're on the savanna or in the North American woods rather than in a three-hundred-acre Danish zoological park (e.g., Siebert 2014). And although the Zootopia design was unveiled well before 2015's *Jurassic World*, its mirrored transport pods bear more than a passing resemblance to the gyrospheres the film's characters employ (with less than happy results) to move through the resurrected dinosaur exhibits in the summer blockbuster.

These cinematic qualities, furthermore, are not all accidental. Ingels has admitted that the park's grand arrival crater—the liminal structure partitioning "civilization" and the zoo's "wilderness" habitats—was partly inspired by the jungle gate protecting (spoiler alert: not very well) the villagers from the rampaging wild beast in the 1933 classic *King Kong* (Ingels 2015; fig. 30.4). It's an amusing and perhaps also a disconcerting confession for a zoo design that has already raised some concerns about visitor safety (see, e.g., Schultz 2014). But it's clearly all part of the desire to create a zoo that—sensu Berger—will do anything but disappoint.

RETHINKING WILDNESS IN THE AGE OF HUMANS

Although Zootopia has received mostly positive (at times even fawning) press coverage, not everyone is sold on Ingels's reimagining of the modern zoological park. In "The Dark Side of Zootopia," *New York Times Magazine* writer Charles Siebert describes it as instead auguring a rather bleak outlook for wildlife and wilderness in the twenty-first century. "Ultimately," Siebert writes, "Zootopia is not a reinvention of the zoo as much as a prefigurement of its inhabitants' only possible future . . . a wilderness with us lurking at its very heart." The new zoo project at Givskud, he concludes, is the manifestation of a wider and more depressing trend: the eclipse of the wild in the human age. Zootopia, in fact, "could well be one of the singular achievements of the [A]nthropocene, a time when human representations of the wild threaten to become the wild's reality" (Siebert 2014).

I share Siebert's worry about the environmental ethos courted by the conceit that we are living in the "age of humans," an idea some believe compels us to loosen our moral and political commitments to traditional nature preservation (see, e.g., Minteer and Pyne 2015). But the notion that wilderness *should* exist apart from human culture and experience—and that Zootopia somehow violates the integrity of this relationship by offering

FIGURE 30.4. The beast breaks through the village gate in a French promotional poster for the 1933 RKO Pictures film King Kong. Wikimedia commons, public domain image.

an illicitly anthropocentric and contrived vision of "wild" nature—is also problematic. Although Siebert acknowledges that the wild is increasingly subject to the forces of human alteration and control, he nevertheless still seems to be in the grip of a classical ideal of the wilderness, a version of that older, dualistic image of nature and culture that droves of archaeologists, paleobotanists, ethnohistorians, and others have increasingly called into question by documenting the deeper narrative of human modification of the wilder corners of Earth (Mann 2005). John Muir's and Joseph Grinnell's wild Yosemite, for instance, reflects the historical activities of California Indians, who through their "harvesting, tilling, sowing, pruning, and burning" shaped, at least to some degree, the modern landscape of the Sierra Nevada (Anderson 2005). But we don't need to go digging into pre-Columbian soil to find evidence of this influence.

In the case of the national parks, we're talking about sites that since the early decades of the twentieth century have been subjected to extensive scenic and recreational development, "natural areas" that have nevertheless been shaped by generations of landscape architects, planners, and engineers seeking to encourage mass tourism and accommodate growing visitor access by building road systems, bridges, trails, campsites, lodges, and park villages (Carr 1999). The management of park wildlife, too, took many years to conform to the naturalistic and ecological principles pushed by Grinnell and other scientists in the early years of the National Park Service. For most of the twentieth century, in fact, the Park Service groped for a coherent philosophy to guide its wildlife policy as it confronted a host of controversial and challenging issues, from decisions about culling wildlife herds and introducing nonnative species to the suitability of "unnatural" zoolike animal entertainments such as roadside feeding of bears and the popular "bear shows" at park garbage dumps in Yosemite and Yellowstone (Sellars 2009).

The point is that the "real" wilderness values that are supposed to represent such a stark contrast to Zootopia's simulation of the wild are not nearly as ecologically pure or as historically tidy as we might think. And of course this isn't just an American story. As more than a few observers have noted (e.g., Greene 2015), there are fences even in South Africa's Kruger National Park, barriers that artificially hem in the elephants, lions, and rhinos at one of the most iconic wildlife reserves in the world.

But this line of argument can at times be pushed too far. Science journalist Emma Marris (2011) has suggested, for instance, that the wilderness can really only ever be "half wild" given the narrative of human influence and ecological change. In making so much of the "altered" and anthropogenic character of what we consider wild, however, there's been an unfortunate tendency to overcorrect, to swing the rhetorical pendulum too far in the other direction. Pushing against the wilderness orthodoxy of American environmentalism, some scientists have even argued that since nothing is truly or fully wild anymore (if indeed it ever was), we should back away from the wilderness as a core concept in conservation and environmental thought and focus more seriously on meeting human needs, wants, and interests (e.g., Kareiva, Marvier, and Lalasz 2012; Ellis 2015).

Accepting a more nuanced cultural and technological narrative about the wilderness, though, doesn't require rejecting the idea that a meaningful sense of the wild is available to us, even in this age of accelerating human influence and control. A remote and roadless stretch of Amazon rain forest—or of the Yosemite backcountry—is wilder than the enclosures of Zootopia ever will be or can be. Zootopia, though, may prove to be much wilder—and wilder in an important sense—than the average zoo, even if its wildness is necessarily qualified and relative rather than pure and absolute.

Furthermore, just as Zootopia is not the inevitable future of the wild, it also isn't the future of "the zoo." Or at least it likely isn't the future of most zoos. Like the national parks, which vary from small urban landscapes to millions of acres of (relatively) untrammeled wilderness, zoos have come in all shapes and sizes and will doubtless continue to do so, with an increasingly diverse range of institutional profiles and identities. The recent uptick in talking about the future of "the zoo" as if it were a singular, monolithic entity (e.g., Cohen 2013; Grazian 2015) therefore seems to me somewhat misplaced. A more likely path, I'd predict, is a kind of divergent evolution, with some zoos morphing slowly into more serious conservation organizations (as described and championed by many contributors to this volume), and others continuing to hew to the familiar recreation and entertainment path, amending at the margins rather than overhauling the core of their traditional missions.

Zootopia suggests yet another direction. Its calling card won't be conservation or recreation in the traditional sense but rather will be providing amped-up visitor excitement tied to an enhanced aesthetic of naturalism, a high-profile experiment with the experiential possibilities of augmented wildness in a zoological park. It's clear from some of Ingels's own commentary on the project that he sees it at least in part as an effort to hold the interest of an increasingly fickle and attention-scattered public that, thanks to the ubiquity of wild animal live webcams, *Animal Planet*, and the National

Geographic channel, has access to a seemingly bottomless digital well of virtual, high-definition "encounters" with wildlife. Ingels has been quite frank about his goal of embracing and upgrading the entertainment function of the zoo, stating that he wants Zootopia to provide an alternative to the typical "premeditated, prepackaged" zoo experience (Ingels 2015).

But there are also glimmers of a more progressive environmental philosophy operating within Ingels's Zootopia plan, ideas that suggest a desire to move beyond the entertainment agenda as well as the more traditional framings of the human-wild relationship. "It's almost a question of trying to find ways of actually creating successful cohabitation between humans and different species of animals," he told an interviewer for National Public Radio in 2014 (NPR 2014).

Yet as an exercise in cohabitation, Zootopia still seems curiously one-sided: it offers a journey into the wild with the human presence mostly submerged, in some case quite literally. The older romantic view of the wilderness, that is, appears alive and well in Ingels's design, a philosophy reinforced by putting us so ingeniously, but also so invisibly, into the animals' world.

SO CLOSE, SO FAR

The Zootopia plan and much of the discussion it has generated remind us that zoos today are caught on the horns of a dilemma regarding their relationship with the wilderness. On one side, they continue to be criticized for being too artificial and too contrived, an indictment not all that different from the one Joseph Grinnell leveled at the Yosemite Valley zoo back in the 1920s. Yet when zoos actually try to become more parklike and "wild" they're often pilloried for falling well short of the mark and for trying to simulate something—wilderness—that some believe simply can't be replicated. Furthermore, as Harry Greene (this volume) reminds us, if zoos do become appreciably wilder in their animal management practices they'll likely run up against the aesthetic and ethical predispositions of a public conditioned to expect a peaceable kingdom rather than a collection of animals behaving naturally, as their wilder counterparts do.

Ultimately I think how we view Zootopia and other future efforts to push the limits of wildness and naturalism in zoological parks depends not only on how we see the prospect of constructing an authentic version of the wild in meticulously designed and managed landscapes, but on how optimistic we are about our ability to maintain respect for what's taken to be the "real" wilderness on a human-dominated planet. From one vantage point, Ingels's project is simply another effort to conceal the inherent unnaturalness of the zoo with the latest architectural wizardry, a vision that only lowers our expectations for the wild. Yet from another angle it's an innovative and exhilarating attempt to inspire different ways of seeing and valuing wildlife and wild places in the twenty-first century. It is, in short, the parallax zoo.

In the end, radically immersive projects like Zootopia embody a difficult and probably inescapable moral friction, one that exists even if we accept a more pragmatic and nuanced view of the wild in the Anthropocene. It's the growing recognition that we're both coinhabitants with other animals and (increasingly) creators of their worlds, including those beyond the zoo walls. And it's a tension reinforced by the recognition that our attempts to get closer to other species often end up only reminding us of our distance, and our difference. As Carl Sandburg put it in the closing lines of his 1918 poem "Wilderness," "For I am the keeper of the zoo: I say yes and no: I sing and kill and work: I am a pal of the world: I came from the wilderness."

ACKNOWLEDGMENTS

A shorter version of this chapter appeared in *Slate* ("The Real Zootopia," March 3, 2016). Accessible at http://www.slate.com/articles/technology/future_tense/2016/03/the_real_zootopia_in_denmark_has_experts_divided_on_the_state_of_the_zoo.html.

Acknowledgments

This book owes its origins to the appointment, in 2013, of lead editor Ben Minteer to the Arizona Zoological Society Endowed Chair at Arizona State University (ASU), a position originally created by the Robert Maytag family about the time it founded what is now the Phoenix Zoo. The appointment was part of the new partnership between ASU's School of Life Sciences (SOLS) and the Phoenix Zoo, a collaboration expressly focused on conservation. We thank Brian Smith, former director of ASU's School of Life Sciences, and Bert Castro, president/CEO of the Arizona Center for Nature Conservation/Phoenix Zoo, for their vision in building this program. Our Phoenix Zoo colleagues have brought great enthusiasm, collegiality, and support to the partnership, as well as to this book.

We are also grateful for the support provided by the National Science Foundation (#SES 1430514) and the Carnegie Humanities Investment Fund (awarded by Dr. George Justice, dean of humanities in the College of Liberal Arts and Sciences [CLAS] at ASU). These resources made it possible to fully realize our plan for a multidisciplinary, multiyear, and multivenue project. Likewise, we want to recognize Bill Kavan and Linda Raish in the CLAS Development Office for their efforts on behalf of the new ASU-Phoenix Zoo program, especially their support of the endowed chair.

The *Ark and Beyond* was launched at the Marine Biological Laboratory (MBL) in Woods Hole, Massachusetts, as part of the ASU-MBL History of Biology Seminar Series. The seminar began in 1987 and offers different topics each year, with a focus on the history of zoo and aquarium conservation in May 2014. The discussions and presentations there provided a foundation for the historical perspective of this volume.

Our ASU colleagues Bertram Jacobs (director, SOLS) and Ferran Garcia-Pichel (dean, Division of Natural Sciences, CLAS) have been enthusiastic supporters of this project, and we especially appreciate the warm welcome they gave at our *Ark and Beyond* symposium in Tempe in fall 2015. We also gratefully acknowledge the efforts of Andrea Cottrell, program coordinator in ASU's Center for Biology and Society (CBS), who assisted us in planning and running the project events at the MBL and at ASU, and Jessica Ranney, manager of the CBS, for her administrative support over the course of the project.

Two doctoral students in biology and society in SOLS helped us at various points along the way, and we recognize them here. Aireona Bonnie Raschke provided key background research for the project in the early stages. Michelle Sullivan helped wrestle a large manuscript into submission. Her extensive work with the *Ark and Beyond* events and the volume is greatly appreciated.

Our editor at the University of Chicago Press, Christie Henry, was a source of great encouragement and wisdom throughout this process. She encouraged us to turn ideas into a book and to be creative in including multiple perspectives and approaches. We count ourselves lucky to have been in such good hands.

We thank all the participants who jumped in and added their voices and contributions. Our project has been complex, with different parts playing out at different times, in different locations, and in different ways. We especially value the energy and thinking everybody put into helping us integrate those parts into a coherent whole.

Finally, as we were going to press in the summer of 2017 we received news that George Rabb, an icon in zoo conservation and author of this volume's foreword, passed away. It is impossible to tell the full history of zoo conservation without George, and hard to envision a future for zoos without him in it. His legacy lives on in the places and programs he helped to shape, and in the scores of people he influenced and inspired along the way, including many of the contributors to this book.

Notes

INTRODUCTION

1. Just to be clear, we are not claiming that SeaWorld is the typical zoo (or, more accurately, aquarium), nor are we implying that there aren't important differences between for-profit and nonprofit institutions. Still, the SeaWorld example has laid bare the fundamental tension between entertainment, conservation, and animal welfare that has dominated much of the current debate about the place of zoos in the twenty-first century.

2. You will search in vain for references to zoos and aquariums in the classic texts of conservation and environmental history (e.g., Samuel B. Hays's *Conservation and the Gospel of Efficiency*, Rod Nash's *Wilderness and the American Mind*, Don Worster's *Wealth of Nature*, and Stephen Fox's *The American Conservation Movement*, among others). These works ignore the zoo narrative, presumably because it is focused on a landscape (built) and animals (captive) that depart from the dominant emphasis on wildlands and wild species. Historians of the life sciences and human-animal relations, including some of the contributors to this volume, have been much more interested in zoos, though not always emphasizing the conservation theme as they do here.

CHAPTER ONE

1 . A timeline of the development of zoos and related institutions can be found in Hoage and Deiss 1996, ix.

2. On these issues see Guerrini 2003, 23–47.

3. These and other society activities are enumerated at the website of the Société nationale de protection de la nature, http://www.snpn.com. See especially the pages "La SNPN en quelques dates."

CHAPTER FOUR

1. This chapter revises and updates findings I first reported in Barrow 2009.

2. On the impact of wildlife conservation initiatives on local residents, see Warren 1999 and Jacoby 2001.

3. The American Museum of Natural History in New York, the Museum of Comparative Zool-

ogy at Harvard, and the Museum of Vertebrate Zoology at Berkeley were among the most active early centers of wildlife conservation in the United States. Even William T. Hornaday, whose pioneering efforts to make zoos more conservation-centered are detailed below, first acquired his protective impulse while working as a taxidermist for the US National Museum. For more on the role of naturalists and natural history museums in the development of American wildlife conservation, see Barrow 2009.

4. In 1963 a fire destroyed the administrative records of the organization, which might have shed light on the confusion surrounding the institution's holdings of the species. On the history of the Cincinnati Zoo, see Ehrlinger 1993.

5. Four years after Martha's death, the last well-documented Carolina parakeet (*Conuropsis carolinensis*) would also die in the Cincinnati Zoo. See Barrow 2009, 127–32, and Snyder 2004; Snyder offers evidence that the species continued to survive in the wild after 1918.

6. A central though contentious figure in the turn-of-the-century wildlife conservation movement in America, Hornaday has recently been receiving the biographical attention he deserves. See, for example, Dehler 2013 and Bechtel 2012, as well as two key earlier studies: Dolph 1975 and Fox 1986.

7. Since it proved a key turning point in his life, biographies of Hornaday invariably include details on his Montana bison expedition. In addition, he wrote two accounts of the expedition himself: Hornaday [1889] 2002, and Hornaday 1887b.

8. The account of the founding and early years of the NYZS that follows is from Bridges 1974; Horowitz 1975; and Dehler 2013, 74–85. The society was an outgrowth of the Boone and Crockett Club, an organization of patrician sportsmen founded in 1887 by George B. Grinnell (who also established the first Audubon Society) and Theodore Roosevelt. Many of the leaders of both organizations were also active in the eugenics movement. Fearful of the transformations brought on by industrialization, urbanization, and mass immigration, they called for better management of natural resources and the human gene pool. On the links between conservation and eugenics in America, see Powell 2016; Allen 2013; and Brechin 1996.

9. Hornaday's struggles to obtain and then maintain the bison and other threatened mammals are chronicled in Bridges 1974, 71–74, 265–66.

10. Ed Hewins had maintained buffalo at his home in Cedarville, Kansas, before moving to Oklahoma around 1895. A newspaper report from the period (Untitled 1897) notes that soon he had obtained two buffalo at his new ranch west of Fort Supply, Oklahoma.

11. On the high mortality rates wild animals once experienced during capture, transport, and confinement at zoos, see Baratay and Hardouin-Fugier 2002, 117–24, 272–73; Bender 2016; see also Fox 1923, which analyzes the cause of death for nearly six thousand animals in the Philadelphia Zoo over a twenty-year period, based on their autopsy results.

12. Roosevelt established the first federal wildlife refuge, the Pelican Island Bird Reservation, in 1903 and then established fifty-four more federal bird reservations and game refuges during his presidency. See Cutright 1985; Brinkley 2009.

13. The layers of irony surrounding the transplant of bison from a zoo in the eastern United States to land in the West that had once belonged to Native Americans are detailed in Barrow 2009, 122.

14. Alicia Conrad was the widow of Charles Conrad, an entrepreneur and rancher who settled in and helped to found Kalispell, Montana, in 1891. The Conrad bison herd began with about fifty animals (Devlin 2014). Austin Corbin II was a wealthy banker and railroad mogul who, before his death in 1896, established the twenty-six-thousand-acre Blue Mountain Forest and Game Preserve in eastern New Hampshire, stocking it with a variety of native and exotic game species including the bison (Billin 2004).

15. On the five nucleus herds of most surviving bison, see Coder 1975; on DNA analysis of bison, see Robbins 2007; Halbert and Derr 2007; and Marris 2009.

16. See the many examples of bison commemoration in Dary 1974, 279–85.

17. Although the Cincinnati Zoo had little success breeding the passenger pigeon, according to Ehrlinger (1993, 40, 58), it had one of the largest herds of bison in American zoos and was exporting its surplus animals to European zoos.

CHAPTER SEVEN

1. The lack of a clear identification of what type or size of institution one means when referring to a public aquarium leads to difficulties in analysis. There are several types of institutions dedicated to public display of aquatic organisms and their environments. The most common form is the *public aquarium*. The term public does not necessarily refer to the funding for these institutions: some are privately funded, while others receive support from state or federal agencies. The term signifies that the institution is open to the general public and has a mission to educate the public about aquatic science and conservation. Many are closely associated with research institutions and function as the public arm of those institutions. In addition to public aquariums, there are other forms of aquatic parks in the aquatic conservation network. The most common is the *marine park*. Marineland and SeaWorld are examples of private institutions focused on public entertainment through the captive breeding and display of marine organisms, particularly marine mammals. While these institutions work within the conservation network and claim that their main functions are research and conservation, they commonly elevate public amusement over conservation and education. Not all public aquariums belong to the Association of Zoos and Aquariums (AZA) (Grow, Luke, and Ogden, this volume); many marine parks do belong. Therefore AZA accreditation cannot be the identifier of public aquariums. In this chapter I refer to both public aquariums and marine parks simply as aquariums.

2. The narrative seems to skip over the middle of the twentieth century. At the turn of the century, most aquariums were founded on fisheries principles. They were literally built out of fisheries exhibits and staffed by fisheries-trained people. This led to a conservation narrative similar to that of the US Bureau of Fisheries: shore up native stocks and introduce hardy, edible species to new locations. This conservation concept faded by the 1930s, overtaken by an influx of new aquarists trained both in university laboratories and in the aquarium setting itself. Ties to fisheries concerns intertwined with experimental laboratory methods and increased emphasis on aquarium husbandry and realistic displays (Muka 2014). The period between the late 1930s and late 1960s was characterized by an emphasis on laboratory work and development of aquarium craft rather than a particular conversation ethos. When conservation concerns and environmental degradation became a public concern in the 1970s and 1980s, aquariums used this combination of laboratory and husbandry research, and their links to a wide network of marine practitioners, to develop a diverse range of conservation programs. While this era is interesting and deserves attention, it is beyond the scope of this chapter, and I note it to explain the absence in the timeline.

3. While we might look askance at calling this type of activity "conservation," we can think of this as similar to the link between hunting, gamekeeping, and early zoological parks in the United States (Dehler 2013; see also Barrow, this volume; Henson, this volume; and Ritvo, this volume).

4. New York Aquarium Director's Office, C. H. Townsend, NYA Daybook, 30 October 1913–20 August 1918. Charles H. Townsend Collection, Wildlife Conservation Society Archives, Bronx, New York.

5. Marine mammals did not engender the same ethical concerns for animal welfare that mammals in zoos seemed to provoke; while there are some examples of concerned visitors' complaining about the treatment of animals on display, I have not run across any early complaints about marine mammals in captivity from daily visitors or from larger organizations such as the ASPCA. It is not until the second half of the twentieth century that we see public concern about the ethics of maintaining these species in captivity. For more information on these organizations, see Beers 2006.

6. 98.3 Specimen Register 1912–1915, Fairmount Park Aquarium Records, Philadelphia City Archives, Philadelphia, PA.

7. Marine parks developed more robust mammal populations in captivity, but initial focus was not on conservation but on entertainment. The earliest marine parks, including Marineland Florida (1938) and Marineland of the Pacific (1954), were in warmer climates that made it possible to maintain marine mammals in outdoor tanks throughout the year. The first trained dolphin, Flippy, was reported in *Parks and Recreation* in 1951. Throughout the second half of the twentieth century marine parks focused on establishing secure captive breeding and training programs for cetaceans such as bottlenose dolphins and orcas. The success of these programs contributed to behavioral studies on cetacean intelligence, as well as to an understanding of their reproduction. However, while couched in conservation terms since the 1970s, these institutions do not breed threatened or endangered species to serve as assurance populations or for eventual reintroduction into the wild. In fact, reintroductions of captive-bred or captive-reared cetaceans have historically failed. In addition to cetaceans, marine parks also successfully breed and display seals and manatees. These parks participate in MMSN. More work needs to be done to understand the place of these institutions in scientific and conservation constellations in the marine community (Davis 1997; Mitman 1996).

CHAPTER NINETEEN

1. In some cases, such as with highly social animals like elephants, individual care may require providing appropriate intraspecific contacts.

2. Alternatively, one might argue that nonhuman individuals are persons in the appropriate ethical sense.

3. Admittedly, some economists insist that sustainability can be calculated within a system of identifying, measuring, and aggregating individual welfare. They do so only by ignoring arguments that protection of ecological processes cannot be fully supported without recognizing the importance of resilient ecosystems (Walker and Salt 2006, especially, chap. 1; Norton 2015).

CHAPTER TWENTY-ONE

1. On this romantic notion of wildlife, see also Mitman 1996, Oelschlaeger 1991, Takacs 1996.

2. It is important to note that species conservation and preservation are different. Where one focuses on sustaining the conditions through which species, habitats, or both can continue to evolve—and thus change—into the future, the other focuses on preserving them as they existed at a specific moment in time. The point here, however, is to explore how both these practices are changed through de-extinction.

CHAPTER TWENTY-SEVEN

1. Lück 2007 offers a helpful survey of the positions on captive marine mammals. See also Weston 2005, Fennell 2015, Rose, Farinato, and S. Sherwin 2006, and Jiang, Lück, and E. C. M. Parsons 2008. Sally Kestin (2004a, 2004b), an investigative reporter at the *Sun Sentinel* in Florida, uncovered deaths of approximately 7,120 captive marine mammals over a thirty-year period. She notes that a quarter of these animals die before age one, and half die before age seven.

CHAPTER TWENTY-NINE

1. I greatly regret not crediting Cottle et al.'s 2009 study of zoo visitors' reponses to feeding live prey, as well as other publications cited therein regarding effects of that practice on zoo animals.

References

Abarca, J., G. Chaves, A. García-Rodríguez, and R. Vargas. 2010. "Reconsidering Extinction: Rediscovery of *Incilius holdridgei* (Anura: Bufonidae) in Costa Rica after 25 Years." *Herpetological Review* 41:150–52.

Acclimatisation Society of Victoria. 1861. *The Rules and Objects of the Acclimatisation Society of Victoria*. Melbourne: William Goodhugh.

———. 1864. *Answers, Furnished by the Acclimatisation Society of Victoria*. Melbourne: Wilson and Mackinnon.

———. n.d. [1864?] "Acclimatisation Society of Victoria." Circular. Victoria State Library SLT 285.2945.M24.

"Acclimatisation Society's Dinner Held at Scott's Hotel, Collins Street West, on Wednesday, July 6th, 1864." 1864. Reprinted from the *Yeoman*.

Adams, N. R. 1995. "Organizational and Activational Effects of Phytoestrogens on the Reproductive Tract of the Ewe." *Proceedings of the Society for Experimental Biology and Medicine* 208 (1): 87–91.

Adams, W. M., R. Aveling, D. Brockington, B. Dickson, J. Elliott, J. Hutton, D. Roe, B. Vira, and W. Wolmer. 2004. "Biodiversity Conservation and the Eradication of Poverty." *Science* 306 (5699): 1146–49. doi:10.1126/science.1097920.

Ahmad, A., and S. Grow. 2015. *AZA Annual Report on Conservation and Science*. Silver Spring, MD: Association of Zoos and Aquariums.

Albertus Magnus. 1999. *Albertus Magnus On Animals: A Medieval Summa Zoologica*. Translated by K. F. Kitchell and I. M. Resnick. Baltimore: Johns Hopkins University Press.

Alexander-Bloch, B. 2015. "Seafood Watch Removes Louisiana Shrimp from 'Avoid' List." *Times-Picayune*, July 2. http://www.nola.com/environment/index.ssf/2015/07/seafood_watch_removes_louisian.html#incart_river.

Al Jahdhami, M., S. Al-Mahdhoury, and H. Al Amri. 2011. "The Re-introduction of Arabian Oryx to the Al Wusta Wildlife Reserve in Oman: 30 years On." In *Global Re-introduction Perspectives, 2011: More Case Studies from around the Globe*, edited by P. S. Soorae, 194–98. Gland, Switzerland: IUCN/SSC Re-introduction Specialist Group and Abu Dhabi; UAE: Environment Agency-Abu Dhabi.

Allard, D. C., Jr. 1978. *Spencer Fullerton Baird and the U.S. Fish Commission*. New York: Arno Press.

Allard, R. A. 2005. "Partnering for the Pe'e Pe'e Maka 'Ole: AZA's North American Regional Conservation Consortium Concept." In *2005 Invertebrates in Captivity Conference Proceedings, Rio Rico, Arizona, July 27–31, 2005*, 13–16. Tucson, AZ: Sonoran Arthropod Studies Institute.

Allen, G. 2013. "'Culling the Herd': Eugenics and the Conservation Movement in the United States, 1900–1940." *Journal of the History of Biology* 46 (1): 31–72.

Allen, J. A. 1876a. "The American Bisons, Living and Extinct." *Memoirs of the Museum of Comparative Zoology* 4 (10). Cambridge, MA: Welch, Bigelow.

———. 1876b. "The Extirpation of the Large Indigenous Mammals of the United States." *Penn Monthly* 7:794–806.

———. 1876c. "The North American Bison and Its Extermination." *Penn Monthly* 7:214–24.

Allen, M. W., M. Hunstone, J. Waerstad, E. Foy, T. Hobbins, B. Wikner, and J. Wirrel. 2002. "Human-to-Animal Similarity and Participant Mood Influence Punishment Recommendations for Animal Abusers." *Society and Animals* 10:267–84.

Allendorf, F. W., P. A. Hohenlohe, and G. Luikart. 2010. "Genomics and the Future of Conservation Genetics." *Nature Reviews Genetics* 11 (10): 697–709.

Amphibian Ark (AArk). 2012. *AArk Amphibian Conservation Needs Assessment Process*. Apple Valley, MN: Amphibian Ark.

———. 2016a. "Welcome to the Amphibian Ark." Accessed June 30. http://www.amphibianark.org/.

———. 2016b. "Progress on Programs." Accessed June 30. http://progress.amphibianark.org/progress-of-programs.

———. 2016c. "Amphibian Conservation Needs Assessments." Accessed June 30. http://conservationneeds.org/AssessmentSearch.aspx.

Anderson, M. K. 2005. *Tending the Wild: Native American Knowledge and the Management of California's Natural Resources*. Berkeley: University of California Press.

Andrade, K., C. Corbin, S. Diver, M. V. Eitzel, J. Williamson, J. Brashares, and L. Fortmann. 2014. "Finding Your Way in the Interdisciplinary Forest: Notes on Educating Future Conservation Practitioners." *Biodiversity and Conservation* 23:3405–23.

Appleby, M., and P. Sandøe. 2002. "Philosophical Debate on the Nature of Well-Being: Implications for Animal Welfare." *Animal Welfare* 11 (3): 283–94.

Aragón, S. 2005. "Le rayonnement international de la Société zoologique d'acclimatation: Participation de l'Espagne entre 1854 et 1861." *Revue d'Histoire des Sciences* 58:169–206.

Ardoin N. M., and J. E. Heimlich. 2013. "Views from the Field: Conservation Educators' and Practitioners' Perceptions of Education as a Strategy for Achieving Conservation Outcomes." *Journal of Environmental Education* 44 (2): 97–115.

Aristotle. 1968. *Parts of Animals*. Translated by A. L. Peck and E. S. Forster. Cambridge, MA: Harvard University Press.

Armitage, D. R., R. Plummer, F. Berkes, R. I. Arthur, A. T. Charles, et al. 2009. "Adaptive Co-management for Social-Ecological Complexity." *Frontiers in Ecology and the Environment* 7 (2): 95–102. doi:10.1890/070089.

Ash, D. 2004. "How Families Use Questions at Dioramas: Ideas for Exhibit Design." *Curator* 47 (1): 84–100.

Association of Zoos and Aquariums (AZA). 2012. "Regional Collection Plan Handbook." Accessed December 8, 2015. https://www.aza.org/regional-collection-plans/.

———. 2013a. "Defining Field Conservation for the AZA Community." Accessed October 19, 2015. https://www.aza.org/uploadedFiles/Conservation/The%20Definition%20of%20Conservation_FCC2011.pdf.

———. 2013b. "Defining Education Programs for the AZA Community." Accessed October 19, 2015. https://www.aza.org/assets/2332/aza_arcshighlights_2014_web.pdf).

——. 2013c. *AZA Green Guide: Introduction to Building Zoo and Aquarium Sustainability Plans.* Silver Spring, MD: Association of Zoos and Aquariums.

——. 2013d. “AZA Board Resolution on Conservation.” Accessed October 19, 2015. https://www.aza.org/uploadedFiles/Conservation/July2013BoardResolution.pdf.

——. 2014a. “Ocean Conservation.” Accessed April 4, 2016. www.aza.org/ocean-conservation.

——. 2014b. *2014 Annual Report on Conservation and Science: Highlights*, compiled by S. Grow, N. Pletcher, and A. Ahmed. Available from https://www.aza.org/uploadedFiles/Conservation/Commitments_and_Impacts/AZA_ARCSHighlights_2014_web.pdf.

——. 2014c. About AZA. Accessed September 15, 2015. http://www.aza.org/about-aza.

——. 2014d. *International Studbook for White Rhinoceros*. Silver Spring, MD: Association of Zoos and Aquariums.

——. 2014e. *Taxon Advisory Group (TAG) Handbook* Silver Spring, MD: Association of Zoos and Aquariums.

——. 2014f. “Zoos and Aquariums Confront the Extinction Crisis.” Accessed December 8, 2015. https://www.aza.org/Membership/detail.aspx?id=35702.

——. 2014g. “AZA Strategic Plan.” Accessed December 8, 2015. https://www.aza.org/StrategicPlan/.

——. 2015a. “Conservation Programs Database.” Accessed October 19, 2015. https://www.aza.org/.

——. 2015b. “2015 Accreditation Standards and Related Policies.” Accessed October 19, 2015. https://www.aza.org/uploadedFiles/Accreditation/AZA-Accreditation-Standards.pdf.

——. 2015c. “The Accredited Zoos And Aquariums in Your Community Are Bringing the Risk Of Wildlife Extinction to the Forefront on May 15, Endangered Species Day.” Accessed April 4, 2016. https://www.aza.org/PressRoom/detail.aspx?id=38486.

——. 2016a. SAFE: Saving Animals from Extinction.” Accessed June 29, 2016. https://www.aza.org/aza-safe.

——. 2016b. “Conservation Funding.” Accessed June 30, 2016. https://www.aza.org/conservation-funding.

Association of Zoos and Aquariums Field Conservation Committee (AZA/FCC). 2016. *March 2016 AZA Field Conservation Committee Board Report*. Silver Spring, MD: M. Penning, Association of Zoos and Aquariums.

Babbage, C. 1830. *Reflections on the Decline of Science in England and on Some of Its Causes*. London: B. Fellowes.

Bae, J. 2015. Interview by Anne Clay. Gwacheon, South Korea. April.

Baird, S. 1873. *Report of the Condition of the Sea Fisheries of the South Coast of New England in 1871 and 1872*. Washington, DC: Government Printing Office.

Ballantyne, R., J. Packer, K. Hughes, and L. Dierking. 2007. “Conservation Learning in Wildlife Tourism Settings: Lessons from Research in Zoos and Aquariums.” *Environmental Education Research* 13 (3): 367–83.

Ballou, J. D., D. G. Kleiman, J. J. C. Mallinson, A. B. Rylands, C. B. Valladares-Padua, and K. Leus. 2002. “History, Management, and Conservation Role of the Captive Lion Tamarin Populations.” In *Lion Tamarins: Biology and Conservation*, edited by M. Hutchins, T. L. Maple, and C. Andrews, 95–114. Washington, DC: Smithsonian Institution Press.

Balmford, A. 2000. “Separating Fact from Artifact in Analyses of Zoo Visitor Preferences.” *Conservation Biology* 14:1193–95.

Balmford, A., J. Kroshko, N. Leader-Williams, and G. Mason. 2011. “Zoos and Captive Breeding.” *Science* 332 (6034): 1149–50.

Balmford, A., N. Leader-Williams, and M. J. B. Green. 1995. “Parks or Arks: Where to Conserve Threatened Mammals?” *Biodiversity and Conservation* 4 (6): 595–607.

Balmford, A., G. M. Mace, and N. Leader-Williams. 1996. "Designing the Ark: Setting Priorities for Captive Breeding." *Conservation Biology* 10 (3): 719–27.

Bamberg, S., and G. Möser. 2007. "Twenty Years after Hines, Hungerford, and Tomera: A New Meta-analysis of Psycho-social Determinants of Pro-Environmental Behavior." *Journal of Environmental Psychology* 27 (1): 14–25.

Baratay, E., and E. Hardouin-Fugier, eds. 2002. *Zoo: A History of Zoological Gardens in the West*. London: Reaktion Books.

Barber, D. 2008. "Building Conservation Partnerships with Zoos." *Endangered Species Bulletin* 33 (1): 18–19.

Barber, D., and V. Poole. 2014. *Association of Zoos and Aquariums Amphibian Taxon Advisory Group Regional Collection Plan*, 3rd ed. Silver Spring, MD: Association of Zoos and Aquariums.

Barnosky, A. D., N. Matzke, S. Tomiya, G. O. U Wogan, B. Swartz, et al. 2011. "Has the Earth's Sixth Mass Extinction Already Arrived?" *Nature* 471:51–57.

Barongi, R., F. A. Fisken, M. Parker, and M. Gusset, eds. 2015. *Committing to Conservation: The World Zoo and Aquarium Conservation Strategy*. Gland, Switzerland: World Association of Zoos and Aquariums Executive Office.

Barrow, Mark V., Jr. 1988. *A Passion for Birds: American Ornithology after Audubon*. Princeton, NJ: Princeton University Press.

———. 2009. *Nature's Ghosts: Confronting Extinction from the Age of Jefferson to the Age of Ecology*. Chicago: University of Chicago Press.

Barsness, Larry. 1985. *Heads, Hides, and Horns: The Compleat Buffalo Book*. Fort Worth: Texas Christian University Press.

Bashaw, M. J., M. A. Bloomsmith, M. J. Marr, and T. L. Maple. 2003. "To Hunt or Not to Hunt? A Feeding Enrichment Experiment with Captive Large Felids." *Zoo Biology* 22 (2): 189–98.

Bastian, B., Costello, K., Loughnan, S., and Hodson, G. 2012. "When Closing the Human-Animal Divide Expands Moral Concern: The Importance of Framing." *Social Psychological and Personality Science* 3:421–29.

Batson, C. D., and Shaw, L. L. 1991. "Evidence for Altruism: Toward a Pluralism of Prosocial Motives." *Psychological Inquiry* 2:107–22.

Batson, C. D., C. Turk, L. Shaw, and T. Klein. 1995. "Information Function of Empathic Emotion: Learning That We Value the Other's Welfare." *Journal of Personality and Social Psychology* 68:300–313.

Bean, M. J. 1983. *The Evolution of National Wildlife Law*. New York: Praeger.

Beardsworth, A., and A. Bryman. 2001. "The Wild Animal in Late Modernity: The Case of the Disneyization of Zoos." *Tourist Studies* 1:83–104.

Beauclerc, K. B., B. Johnson, and B. N. White. 2010. "Genetic Rescue of an Inbred Captive Population of the Critically Endangered Puerto Rican Crested Toad (*Peltophryne lemur*) by Mixing Lineages." *Conservation Genetics* 11:21–32.

Bechtel, S. 2012. *Mr. Hornaday's War: How a Peculiar Victorian Zookeeper Waged a Lonely Crusade for Wildlife That Changed the World*. Boston: Beacon Press.

Beers, D. L., 2006. *For the Prevention of Cruelty: The History and Legacy of Animal Rights Activism in the United States*. Athens: Ohio University Press.

Bekoff, M. 2015. "Compassionate Conservation: More Than 'Welfarism Gone Wild.'" *Huffington Post*, February 9.

———. 2016. "Why Was Harambe the Gorilla in a Zoo in the First Place?" *Scientific American*, May 31, 2016. Accessible at https://blogs.scientificamerican.com/guest-blog/why-was-harambe-the-gorilla-in-a-zoo-in-the-first-place/.

Bell, C. E. 2001. *Encyclopedia of the World's Zoos*. Chicago: Fitzroy Dearborn.

Bender, D. E. 2016. *The Animal Game: Searching for Wildness at the American Zoo*. Cambridge, MA: Harvard University Press.

Bendiner, R. 1981. *The Fall of the Wild, the Rise of the Zoo*. New York: Dutton.

Bennett, G. 1862. *Acclimatisation: Its Eminent Adaptation to Australia.* Melbourne: William Goodhugh.

Ben-Nun, I. F., S. C. Montague, M. L. Houck, H. T. Tran, I. Garitaonandia, et al. 2011. "Induced Pluripotent Stem Cells from Highly Endangered Species." *Natural Methods* 8 (10): 829–31. doi: 10.1038/nmeth.1706.

Berger, J. 1991. *About Looking.* New York: Vintage Books.

Berger, L., A. A. Roberts, J. L. Voyles, J. E. Longcore, K. A. Murray, and L. F. Skerratt. 2016. "History and Recent Progress on Chytridiomycosis in Amphibians." *Fungal Ecology* 19:89–99.

Berger, L., R. Speare, P. Daszak, D. E. Green, A. A. Cunningham, et al. 1998. "Chytridiomycosis Causes Amphibian Mortality Associated with Population Declines in the Rainforests of Australia and Central America." *Proceedings of the National Academy of Sciences* 95:9031–36.

Berger-Tal, O., D. T. Blumstein, S. Carroll, R. N. Fisher, S. L. Mesnick, et al. 2015. "A Systematic Survey of the Integration of Behavior into Wildlife Conservation and Management." *Conservation Biol*ogy, advance online publication. doi: 10.1111/cobi.12654.

Bergl, R. A., A. Dunn, I. Imong, A. Fowler, and R. Ikfuingei. 2014. "Use of Mobile Device and Database Technology Demonstrates Conservation Successes and Challenges across the Range of the Critically Endangered Cross River Gorilla (*Gorilla gorilla diehli*)." Paper presented at the Twenty-Fifth Congress of the International Primatological Society, Hanoi, Vietnam, August 11–16.

Berkes, F. 2007. "Community-Based Conservation in a Globalized World." *Proceedings of the National Academy of Sciences* 104 (39): 15188–93. doi:10.1073/pnas.0702098104.

Bertelsen, M. F. 2014. "On the Euthanasia of Surplus Animals, and the Role of Zoos in Alleviating Biological Illiteracy—Lessons Learned." In *Proceedings of the International Conference on Diseases of Zoo and Wild Animals*, 65–67. Warsaw, May 28–31.

Billin, D. 2004. "Private Game Preserve Has Storied History." *Valley News*, January 28. http://www.meyette.us/DanBillinCorbinParkArticle.htm.

Black, S. A., H. M. R. Meredith, and J. J. Groombridge. 2011. "Biodiversity Conservation: Applying New Criteria to Assess Excellence." *Total Quality Management* 22:1165–78.

Blanckaert, C. 1992. "Les animaux utiles chez Isidore Geoffroy Saint-Hilaire: La mission sociale de la zootechnie." *Revue de Synthèse* 113:347–82.

Blau, D., and N. Rothfels. 2015. *Elephant House.* University Park: Pennsylvania State University Press.

Blockstein, D. E. 2002. "Passenger Pigeon." In *The Birds of North America Online*, edited by A. Poole. Ithaca, NY: Cornell Lab of Ornithology. http://bna.birds.cornell.edu.ezproxy.lib.vt.edu/bna/species/611.

Blockstein, D. E., and H. B. Tordoff. 1985. "A Contemporary Look at the Extinction of the Passenger Pigeon." *American Birds* 39:845–52.

Bloomsmith, M. A., M. Jackson Marr, and T. L. Maple. 2007. "Addressing Nonhuman Primate Behavioral Problems through the Application of Operant Conditioning: Is the Human Treatment Approach a Useful Model?" *Applied Animal Behaviour Science* 102:205–22.

Bluefin Futures. 2015. "Bluefin Futures Symposium." Accessed April 4, 2016. http://bluefinfutures2016.org.

Bode, M., K. A. Wilson, T. M. Brooks, W. R. Turner, R. A. Mittermeier, et al. 2008. "Cost-Effective Global Conservation Spending Is Robust to Taxonomic Group." *Proceedings of the National Academy of Sciences* 105:6498–6501.

Bode, M., W. Probert, W. R. Turner, K. A. Wilson, and O. Venter. 2010. "Conservation Planning with Multiple Organizations and Objectives." *Conservation Biology* 25:295–304.

Boivin, N. L., M. A. Zeder, D. Q. Fuller, A. Crowther, G. Larson, J. M. Erlandson, T. Denham, and M. D. Petraglia. 2016. "Ecological Consequences of Human Niche Construction: Examining Long-Term Anthropogenic Shaping of Global Species Distributions." *Proceedings of the National Academy of Sciences* 113:6388–96.

Bonner, M. G. 1926. "New Fishes Join the Aquarium Family: This Is the Recruiting Season for One of New York's Most Popular Shows." *New York Times*, July 25.

Bottrill, M. C., L. Joseph, J. Carwardine, M. Bode, C. Cook, E. Game, H. Grantham, S. Kark, S. Linke, E. McDonald-Madden, R. Pressey, S. Walker, K. Wilson, and H. Possingham. 2008. "Is Conservation Triage Just Smart Decision Making?" *Trends in Ecology and Evolution* 23:649–54.

Boughton, D. A., and A. S. Pike. 2013. "Floodplain Rehabilitation as a Hedge against Hydroclimatic Uncertainty in a Migration Corridor of Threatened Steelhead." *Conservation Biology* 27:1158–68.

Bouman, I., and J. Bouman. 1994. "The History of the Przewalski's Horse." In *Przewalski's Horse: The History and Biology of an Endangered Species*, edited by L. Boyd and K. Houpt, 5–38. Albany: State University of New York Press.

Bourdieu, P. 1984. *Distinction: A Social Critique of the Judgment of Taste*. Translated by R. Nice. Cambridge, MA: Harvard University Press.

Bourgeois, M. 1854. "Sur les chèvres d'angora et à duvet et les avantages qui pourraient résulter du croisement de ces races." *Bulletin de la Société Zoologique d'Acclimatation* 1:268–71.

Bowkett, A. E. 2014. "Ex Situ Conservation Planning Is More Complicated Than Prioritizing the Keeping of Threatened Species in Zoos." *Animal Conservation* 17:101–3.

Bowling, A. T., W. Zimmerman, O. Ryder, C. Penado, S. Peto, L. Chemnick, N. Yasinetskaya, and T. Zharkikh. 2003. "Genetic Variation in Przewalski's Horses, with Special Focus on the Last Wild Caught Mare, 231 Orlitza III." *Cytogenetic and Genome Research* 102:226–34.

Boyd, L., and K. Houpt, eds. 1994. *Przewalski's Horse: The History and Biology of an Endangered Species*. Albany: State University of New York Press.

Boyes, S. J., and M. Elliott. 2014. "Marine Legislation–the Ultimate 'Horrendogram': International Law, European Directives and National Implementation." *Marine Pollution Bulletin* 86 (1): 39–47.

Braun, J., M. Schrenzel, C. Witte, L. Gokool, J. Burchell, and B. A. Rideout. 2014. "Molecular Methods to Detect *Mycoplasma* spp. and Testudinid Herpesvirus 2 in Desert Tortoises and Implications for Disease Management." *Journal of Wildlife Diseases* 50 (4): 757–66.

Braverman, I. 2012. *Zooland: The Institution of Captivity*. Palo Alto, CA: Stanford University Press.

Brechin, Gray. 1996. "Conserving the Race: Natural Aristocracies, Eugenics, and the U.S. Conservation Movement." *Antipode* 28 (3): 229–45.

Breckheimer, I. N., M. Haddad, W. F. Morris, A. M. Trainor, W. R. Fields, et al. 2014. "Defining and Evaluating the Umbrella Species Concept for Conserving and Restoring Landscape Connectivity." *Conservation Biology* 28:584–93.

Bridges, W. 1974. *A Gathering of Animals: An Unconventional History of the New York Zoological Society*. New York: Harper and Row.

Brinkley, D. 2009. *Wilderness Warrior: Theodore Roosevelt and the Crusade for America*. New York: HarperCollins.

Briseño-Garzón A. D., D. Anderson, and A. Anderson. 2007. "Adult Learning Experiences from an Aquarium Visit: The Role of Social Interactions in Family Groups." *Curator* 50 (3): 299–318.

Brunner, B. 2005. *The Ocean at Home: An Illustrated History of the Aquarium*. Princeton, NJ: Princeton Architectural Press.

Burghardt, G. M. 1996. "Environmental Enrichment or Controlled Deprivation?" In *The Well-Being of Animals in Zoo and Aquarium Sponsored Research*, edited by G. M. Burghardt, J. T. Blelltski, J. R. Boyce, and D. O. Schaeffer, 91–101. Greenbelt, MD: Scientists Center for Animal Welfare.

———. 1997. "Amending Tinbergen: A Fifth Aim for Ethology." In *Anthropomorphism, Anecdotes, and Animals*, edited by R. W. Mitchell, N. S. Thompson, and H. L. Miles, 254–76. Albany: SUNY Press.

———. 2013. "Environmental Enrichment and Cognitive Complexity in Reptiles and Amphibians: Concepts, Review, and Implications for Captive Populations." *Applied Animal Behaviour Science* 147:286–98.

Burke, E. 1910. *Reflections on the French Revolution and Other Essays*. London: J. M. Dent.

Burkhardt, R. W., Jr. 1997. "La ménagerie et la vie du muséum." In *Le muséum au premier siècle de son histoire*, edited by C. Blankaert, C. Cohen, P. Corsi, and J. Fischer, 481–508. Paris: Éditions du Muséum National d'Histoire Naturelle.

Burnett, D. G. 2012. *The Sounding of the Whale: Science and Cetaceans in the Twentieth Century*. Chicago: University of Chicago Press.

Burton, J. 2003. "The Context of Red Data Books, with a Complete Bibliography of the IUCN Publications." In *The Harmonization of Red Lists for Threatened Species in Europe, Leiden (The Netherlands): The Netherlands Commission for International Nature Protection*, edited by H. H. de Iongh, O. S. Bánki, W. Bergmans and M. J. van der Werfften Bosch, 291–300. Proceedings of an International Seminar November 27–28, 2002. Leiden: Netherlands Commission for International Protection, Mededelingen 38.

Butchart, S. H. M., M. Walpole, B. Collen, A. van Strien, J. P. W. Scharlemann, E. A. Almond, et al. 2010. "Global Biodiversity: Indicators of Recent Declines." *Science* 328:1164–68.

Butterfield, M. E., Hill, S. E., and Lord, C. G. 2012. "Mangy Mutt or Furry Friend? Anthropomorphism Promotes Animal Welfare." *Journal of Experimental Social Psychology* 48:957–60.

Byers, O., C. Lees, J. Wilcken, and C. Schwitzer. 2013. "The One Plan Approach: The Philosophy and Implementation of CBSG's Approach to Integrated Species Conservation Planning." *WAZA Magazine* 14:2–5.

Callahan, D. 1998. "Cloning: Then and Now." *Cambridge Quarterly of Healthcare Ethics* 7:141–44.

Callicott, J. B. 1989. "Animal Liberation: A Triangular Affair." In *In Defense of the Land Ethic: Essays in Environmental Philosophy*, 15–38. Albany: State University of New York Press.

———. 1999. *Beyond the Land Ethic: More Essays in Environmental Philosophy*. Albany: State University of New York Press.

Conservation Breeding Specialist Group (CBSG). 2006. *Okinawa Rail Population Viability Analysis Workshop Final Report* (Japanese). Apple Valley, MN: IUCN/SSC Conservation Breeding Specialist Group.

———. 2011. *Intensively Managed Populations for Conservation Workshop Report*. Apple Valley, MN: IUCN/SSC Conservation Breeding Specialist Group.

———. 2016. *Poweshiek Skipperling and Dakota Skipper: An Ex Situ Assessment and Planning Workshop Report*. Apple Valley, MN: IUCN/SSC Conservation Breeding Specialist Group.

Carlin, N. F., I. Wurman, and T. Zakin. 2013. "How to Permit Your Mammoth: Some Legal Implications of 'De-extinction.'" *Stanford Enviornmental Law Journal* 33 (1): 3–57.

Carmi, N., S. Arnon, and N. Orion. 2015. "Transforming Environmental Knowledge into Behavior: The Mediating Role of Environmental Emotions." *Journal of Environmental Education* 46:183–201.

Carr, E. 1999. *Wilderness by Design: Landscape Architecture and the National Park Service*. Lincoln: University of Nebraska Press.

Carrillo, L., K. Johnson, and J. R. Mendelson III. 2015. "Principles of Program Development and Management for Amphibian Conservation Captive Breeding Programs." *International Zoo News* 62 (2): 96–107.

Carroll, C., M. Patterson, S. Wood, A. Booth, J. Rick, and S. Balain. 2007. "A Conceptual Framework for Implementation Fidelity." *Implementation Science* 2:40. doi:10.1186/1748-5908-2-40.

Carson, R. 1962. *Silent Spring*. Boston: Houghton Mifflin.

Carter, N. H., A. Viña, V. Hull, W. J. McConnell, W. Axinn, D. Ghimire, and J. Liu. 2014. "Coupled Human and Natural Systems Approach to Wildlife Research and Conservation." *Ecology and Society* 19 (3): 43. doi:10.5751/ES-06881-190343.

Cater, C. 2010. "Any Closer and You'd Be Lunch! Interspecies Interactions as Nature Tourism in Marine Aquaria." *Journal of Ecotourism* 9:133–48.

Ceballos, G., M. M. Vale, C. Bonacic, J. Calvo-Alvarado, R. List, N. Bynum, R. Medellin, A. Simonetti, and J. P. Rodriguez. 2009. "Conservation Challenges for the Austral and Neotropical America Section." *Conservation Biology* 23:811–17.

Chang, T. R., D. L. Forthman, and T. L. Maple. 1999. "Comparison of Captive Mandrills in Traditional and Ecologically Representative Exhibits." *Zoo Biology* 18:163–76.

Cheng, T. L., S. M. Rovito, D. B. Wake, and V. T. Vredenburg. 2011. "Coincident Mass Extirpation of Neotropical Amphibians with the Emergence of the Infectious Fungal Pathogen *Batrachochytrium dendrobatidis*." *Proceedings of the National Academy of Sciences* 108:9502–7.

Chermayeff, P. 1992. "The Age of Aquariums." *World Monitor* 5:54–60.

Cheyenne Mountain Zoo. 2016. "Palm Oil Crisis." Accessed December 22, 2015. http://www.cmzoo.org/index.php/conservation-matters/palm-oil-crisis/.

Cho, K. U. 2007. "A Study on the Development of Animal Welfare in Korean Zoos." Thesis, Konkuk University Graduate School.

Cho, K. U., B. I. Choe, H. Y. Kim, J. S. Han, and J. S. Kim. 2009. "A Basic Study on the Animal Welfare Evaluation in Korean Zoos." *Korean Journal of Veterinary Research* 49 (1): 91–99.

Choi, J. Y. 2013a. *해외사례 통해 본 동물원 관리제도 개선방안* [Improving the management of zoos through foreign examples]. Seoul, South Korea: Korean National Assembly Research Service Issue and Point.

———. 2014. Interview by Anne Clay. Seoul, South Korea. October.

Cioc, M. 2009. *The Game of Conservation: International Treaties to Protect the World's Migratory Animals*. Athens: Ohio University Press.

Clark, N. 1999. "Wild Life: Ferality and the Frontier with Chaos." In *Quicksands: Foundational Histories in Australia and Aotearoa New Zealand*, edited by K. Neumann, N. Thomas, and H. Ericksen, 133–52. Sydney: University of New South Wales Press.

Clarke, A. E. 1998. *Disciplining Reproduction: Modernity, American Life Sciences, and the Problems of Sex*. Berkeley: University of California Press.

Clay, A. W., M. A. Bloomsmith, M. Jackson Marr, and T. L. Maple. 2009. "Habituation and Desensitization as Methods for Reducing Fearful Behavior in Singly Housed Rhesus Macaques." *American Journal of Primatology* 71:30–39.

Clayton, S. 2003. "Environmental Identity: A Conceptual and an Operational Definition." In *Identity and the Natural Environment*, edited by S. Clayton and S. Opotow, 45–65. Cambridge, MA: MIT Press.

Clayton, S., J. Fraser, and C. Burgess. 2011. "The Role of Zoos in Fostering Environmental Identity." *Ecopsychology* 3:87–96.

Clayton, S., J. Fraser, and C. D. Saunders. 2009. "Zoo Experiences: Conversations, Connections, and Concern for Animals." *Zoo Biology* 28 (5): 377–97.

Clayton, S., J. Luebke, C. Saunders, J. Matiasek, and A. Grajal. 2014. "Connecting to Nature at the Zoo: Implications for Responding to Climate Change." *Environmental Education Research* 20:460–75. doi:10.1080/13504622.2013.816267.

Coder, G. D. 1975. "The National Movement to Preserve the American Buffalo in the United States and Canada between 1880 and 1920." PhD diss., Ohio State University.

Coffman, D. 2013. *Reflecting the Sublime: The Rebirth of an American Icon*. Fort Benton, MT: River and Plains Society.

Cohen, J. 2013. "Zoo Futures." *Conservation*, March 8. Available online at http://conservationmagazine.org/2013/03/zoo-futures/.

Cohen, M. P. 1988. *The History of the Sierra Club: 1892–1970*. San Francisco: Sierra Club Books.

Cohen, S. 2014. "The Ethics of De-extinction." *Nanoethics* 8:165–78.

Cokinos, C. 2000. *Hope Is the Thing with Feathers: A Personal Chronicle of Vanished Birds*. New York: Jeremy B. Tarcher.

Colborn, T., F. S. vom Saal, and A. M. Soto. 1993. "Developmental Effects of Endocrine-Disrupting Chemicals in Wildlife and Humans." *Environmental Health Perspectives* 101:378–84.

Colen, S. 1995. "'Like a Mother to Them': Stratified Reproduction and West Indian Childcare

Workers and Employers in New York." In *Conceiving the New World Order: The Global Politics of Reproduction*, edited by F. D. Ginsburg and R. Rapp, 78–102. Berkeley: University of California Press.

Collar, N. J., and H. M. Butchart. 2014. "Conservation Breeding and Avian Diversity: Chances and Challenges." *International Zoo Yearbook* 48:7–28.

Collins, J. P., and M. L. Crump. 2009. *Extinction in Our Times: Global Amphibian Decline*. Oxford: Oxford University Press.

Conde, D. A., F. Colchero, M. Gusset, P. Pearce-Kelly, O. Byers, et al. 2013. "Zoos through the Lens of the IUCN Red List: A Global Metapopulation Approach to Support Conservation Breeding Programs." *PLoS ONE* 8 (12): e80311. doi:10.1371/journal.pone.0080311.

Conde, D. A., F. Colchero, B. Güneralp, M. Gusset, B. Skolnik, et al. 2015. "Opportunities and Costs for Preventing Vertebrate Extinctions." *Current Biology* 25 (6): R219–21. doi:10.1016/j.cub.2015.01.048.

Conde, O., D. Amor, N. Flesness, F. Colchero, O. R. Jones, and A. Scheuerlein. 2011a. "An Emerging Role of Zoos to Conserve Biodiversity." *Science* 331:1390–91.

——. 2011b. "Zoos and Captive Breeding—Response." *Science* 332 (3): 1150–51.

Conservation Alliance for Sustainable Seafood. 2014. *A Common Vision for Sustainable Seafood*. Accessed April 4, 2016. http://www.solutionsforseafood.org/wp-content/uploads/2014/10/A-Common-Vision-for-Sustainable-Seafood.pdf.

Conservation International. 2016. "Nature Is Speaking." Accessed June 29, 2016. http://www.conservation.org/nature-is-speaking/Pages/default.aspx.

Conservation Measures Partnership. 2015. "Open Standards for the Practice of Conservation." Version 3.0. Accessed October 19, 2015. http://www.conservationmeasures.org/.

Convention on Biological Diversity (CBD). 2010a. "Decision Adopted by the Conference of the Parties to the Convention on Biological Diversity at Its Tenth Meeting." https://sustainabledevelopment.un.org/content/documents/733FutureWeWant.pdf.

——. 2010b. *Strategic Plan for Biodiversity 2011–2020 and the Aichi Targets: "Living in Harmony with Nature."* Montreal: Secretariat of the Convention on Biological Diversity.

Conway, W. 1973. "How to Exhibit a Bullfrog: A Bed-time Story for Zoo Men." *International Zoo Yearbook* 13:221–26.

——. 1995a. "The Conservation Park: A New Zoo Synthesis for a Changed World." In *The Ark Evolving: Zoos and Aquariums in Transition*, edited by C. M. Wemmer, 259–76. Front Royal, VA: Smithsonian Institution Conservation and Research Center.

——. 1995b. "Zoo Conservation and Ethical Paradoxes." In *Ethics on the Ark: Zoos, Animal Welfare, and Wildlife Conservation*, edited by B. G. Norton, M. Hutchins, E. F. Stevens, and T. L. Maple, 1–9. Washington, DC: Smithsonian Institution Press.

——. 1982. "The Species Survival Plan: Tailoring Long-Term Propagation Species by Species." Paper presented at the annual meeting of the American Association of Zoological Parks and Aquariums, Phoenix, Arizona, September 19–23.

——. 2011. "Buying Time for Wild Animals with Zoos." *Zoo Biology* 30 (1): 1–8.

Conway, W. G., and M. Hutchins. 2001. "Introduction." In *AZA Field Conservation Resource Guide*, edited by W. G. Conway, M. Hutchins, M. Souza, Y. Kapetanakos, and E. Paul, 1-7. Atlanta, GA: Wildlife Conservation Society and Zoo Atlanta.

Costello, K., and G. Hodson. 2010. "Exploring the Roots of Dehumanization: The Role of Animal—Human Similarity in Promoting Immigrant Humanization." *Group Processes and Intergroup Relations* 13:3–22.

Cottle, L., D. Tamir, M. Hyseni, D. Bühler, and P. Lindemann-Matthies. 2009. "Feeding Live Prey to Zoo Animals: Response of Zoo Visitors in Switzerland." *Zoo Biology* 28:1–7.

Coutancier, B., and C. Barthe. 2002. "Exhibition et médiatisation de l'autre: Le Jardin zoologique d'acclimatation (1877–1890)." In *Zoos humains, XIXᵉ et XXᵉ siècles*, edited by N. Bancel, P. Blanchard, G. Boëtsch, E. Deroo, and S. Lemaire, 306–14. Paris: Éditions la Découverte.

Croke, V. 1997. *The Modern Ark: The Story of Zoos, Past, Present and Future*. New York: Scribner.

Cronon, W. 1995. "The Trouble with Wilderness, or Getting Back to the Wrong Nature." In *Uncommon Ground: Rethinking the Human Place in Nature*, ed. W. Cronon. New York: Norton.

Crowley, P. 2003. "Aquarium Seeking $4.5M Expansion." *Cincinnati Enquirer*, May 31. http://enquirer.com/editions/2003/05/31/loc_aquarium.html.

Cutright, P. 1985. *Theodore Roosevelt: The Making of a Conservationist*. Urbana: University of Illinois Press.

Daly, H., and J. Cobb. 1989. *For the Common Good: Redirecting the Economy toward Community, the Environment and a Sustainable Future*. Boston: Beacon Press.

Danz, H. P. 1997. *Of Bison and Man*. Niwot: University Press of Colorado.

Dary, D. A. 1974. *The Buffalo Book: The Full Saga of the American Animal*. Chicago: Sage Books.

Daubenton, L. 1801. *Instruction pour les bergers et pour les propriétaires de troupeaux*. Paris: Imprimerie de la République.

David, O. T., and D. Bar-Tal. 2009. "A Sociopsychological Conception of Collective Identity: The Case of National Identity as an Example." *Personality and Social Psychology Review* 13 (4):354–79. doi:10.1177/1088868309344412.

Davis, S. G. 1997. *Spectacular Nature: Corporate Culture and the Sea World Experience*. Berkeley: University of California Press.

Dawson, J., F. Patel, R. A. Griffiths, and R. P. Young. 2015. "Assessing the Global Zoo Response to the Amphibian Crisis through 20-Year Trends in Captive Collections." *Conservation Biology* 30 (1): 82–91. doi:10.1111/cobi.12563.

Deane, R. 1896. "Notes on the Passenger Pigeon (*Ectopistes migratorius*) in Confinement." *Auk* 13 (3): 234–37.

——. 1908. "The Passenger Pigeon (*Ectopistes migratorius*) in Confinement." *Auk* 25 (2): 181–83.

DeBlieu, J. 1991. *Meant to Be Wild: The Struggle to Save Endangered Species through Captive Breeding*. Golden, CO: Fulcrum.

Defenders of Wildlife. n.d. "Fact Sheet: Bison," Accessed June 6, 2016. http://www.defenders.org/bison/basic-facts.

Dehler, G. J. 2013. *The Most Defiant Devil: William Temple Hornaday and His Controversial Crusade to Save American Wildlife*. Charlottesville: University of Virginia Press.

DePinho, J. R., C. Grilo, R. B. Boone, K. A. Galvin, and J. G. Snodgrasss. 2014. "Influence of Aesthetic Appreciation of Wildlife Species on Attitudes towards Their Conservation in Kenyan Agropastoralist Communities." *PLoS ONE* 9 (2): e88842.

Der Sarkissian, C., L. Ermini, M. Schubert, M. A. Yang, P. Librado, et al. 2015. "Evolutionary Genomics and Conservation of the Endangered Przewalski's Horse." *Current Biology* 25 (19): 2577–83.

Desiderio, F. 2000. "Raising the Bars: The Transformation of Atlanta's Zoo, 1989–2000." *Atlanta History: A Journal of Georgia and the South* 43 (4): 7–43.

Desmond, A. 1985. "The Making of Institutional Zoology in London, 1822–1836." *History of Science* 23:153–85, 223–50.

Devenney, R. 2011. "Waterfront Revitalization: Bridgeport Aquarium and Waterfront Promenade." Thesis, Roger Williams University. http://docs.rwu.edu/archthese/70.

Devlin, V. 2014. "Conrad Mansion Important Part of Kalispell Landscape." *Billings Gazette*, May 16. http://billingsgazette.com/special-section/travel-tourism/rediscover-montana/conrad-mansion-important-part-of-kalispell-landscape/article_36643d13-16f9-5d09-b7a5-fdb05e96266c.html.

De Vos, J. M., L. N. Joppa, J. L. Gittleman, P. R. Stephens, and S. L. Pimm. 2015. "Estimating the Normal Background Rate of Species Extinction." *Conservation Biology* 29:452–62.

Dickie, Lesley A. 2009. "The Sustainable Zoo: An Introduction." *International Zoo Yearbook* 43 (1): 1–5. http://doi.wiley.com/10.1111/j.1748-1090.2008.00086.x.

Dietz, T., G. Gardner, J. Gilligan, P. Stern, and M. Vandenberg. 2009. "Household Actions Can Create a Behavioral Wedge to Rapidly Reduce U.S. Carbon Emissions." *Proceedings of the National Academy of Sciences* 106:18452–56. doi:10.1073/pnas.0908738106.

DiRenzo, G., P. F Langhammer, K. R Zamudio, and K. R Lips. 2014. "Fungal Infection Intensity and Zoospore Output of *Atelopus zeteki*, a Potential Chytrid Supershedder." *PLoS ONE* 9 (3): e93356. doi:10.1371/journal.pone.0093356.

Dirzo, R., et al. 2014. "Defaunation in the Anthropocene." *Science* 345:401-6.

Dobson, A., and A. Lyles. 2000. "Black-Footed Ferret Recovery." *Science* 288 (5468): 985–88.

Dolph, J. A. 1975. "Bringing Wildlife to Millions: William Temple Hornaday, the Early Years, 1854–1896." PhD diss., University of Massachusetts.

Dorsey, K. 1998. *The Dawn of Conservation Diplomacy: U.S.-Canadian Wildlife Protection Treaties in the Progressive Era*. Seattle: University of Washington Press.

Doughty, R. W. 1975. *Feather Fashions and Bird Preservation: A Study in Nature Protection*. Berkeley: University of California Press.

Dropkin, L., S. Tipton, and L. Gutekunst. 2015. *American Millennials: Cultivating the Next Generation of Ocean Conservationists*. Arlington, VA: Edge Research and David and Lucile Packard Foundation. Available at https://www.packard.org/what-were-learning/resource/american-millennials-cultivating-the-next-generation-of-ocean-conservationists/.

Drouin, J. 1994. "Baudiment, Émile (1816–1863)." In *Les professeurs du Conservatoire national des arts et métiers, dictionnaire biographique, 1794–1955*, edited by C. Fontanon and A. Grelon, 2:140–47. Paris: Institut National de Recherche Pédagogique and Conservatoire National des Arts et Métiers.

Dubois, S., and D. Fraser. 2013. "Rating Harms to Wildlife." *Animal Welfare* 22:49–55.

Dudley, J. P., J. R. Ginsberg, A. J. Plumptre, J. A. Hart, and L. C. Campos. 2002. "Effects of War and Civil Strife on Wildlife and Wildlife Habitats." *Conservation Biology* 16 (2): 319–29. doi:10.1046/j.1523-1739.2002.00306.x.

Dugatkin, L. A. 2009. *Mr. Jefferson and the Giant Moose*. Chicago: University of Chicago Press.

Durrant, B. 2008. "The Importance and Potential of Artificial Insemination in CANDES (Companion Animals, Non-domestic, Endangered Species)." *Theriogenology* 71:113–21.

Durrell, G. 1976. *The Stationary Ark*. New York: Simon and Schuster.

Economist. 2016. "World Ocean Summit 2017." Accessed April 4, 2016. http://www.economist.com/events-conferences/asia/ocean-summit-2017.

Ehrlich, P. R. 2014. "The Case against De-extinction: It's a Fascinating but Dumb Idea." Accessed October 25, 2015. http://e360.yale.edu/feature/the_case_against_de-extinction_its_a_fascinating_but_dumb_idea/2726/.

Ehrlinger, D. 1993. *The Cincinnati Zoo and Botanical Garden: From Past to Present*. Cincinnati: Cincinnati Zoo and Botanical Garden.

Ellis, E. 2015. "Too Big for Nature." In *After Preservation: Saving American Nature in the Age of Humans*, edited by B. A. Minteer and S. J. Pyne, 24–31. Chicago: University of Chicago Press.

Emslie, R., and M. Brooks. 1999. *African Rhino: Status Survey and Conservation Action Plan*. Gland, Switzerland: IUCN/SSC African Rhino Specialist Group.

Eo, G. 2015. Interview by Anne Clay. Gwacheon, South Korea. November.

Estes, J. A., J. Terborgh, J. S. Brasheres, M. E. Power, J. Berger, et al. 2011. "Trophic Down-Grading of Planet Earth." *Science* 333:301–6.

Estrada, A., B. Gratwicke, A. Benedetti, G. Della Togna, D. Garrelle, et al., eds. 2014. *The Golden Frogs of Panama (Atelopus zeteki, A. varius): A Conservation Planning Workshop*. Final Report. Apple Valley, MN: IUSN/SSC Conservation Breeding Specialist Group.

European Association of Zoos and Aquaria (EAZA). 2015. *Code of Ethics*. Amsterdam: EAZA. Accessed December 8, 2015. http://www.eaza.net/assets/Uploads/Standards-and-policies/EAZA-Code-of-Ethics-2015.pdf.

Evely, A. C., M. Pinard, M. S. Reed, and I. Fazey. 2011. "High Levels of Participation in Conservation Projects Enhance Learning." *Conservation Letters* 4:116–26.

Eves, H., M. Hutchins, and N. Bailey. 2008. "The Bushmeat Crisis Task Force (BCTF)." In *Conservation in the 21st Century: Gorillas as a Case Study*, edited by T. S. Stoinski, H. Dieter Skeklis, and P. T. Mehlman, 327–44. New York: Springer.

Fa, J. E., S. M. Funk, and D. M. O'Connell. 2011. *Zoo Conservation Biology*. Cambridge: Cambridge University Press.

Fa, J. E., M. Gusset, N. Flesness, and D. A. Conde. 2014. "Zoos Have Yet to Unveil Their Full Conservation Potential." *Animal Conservation* 17:97–100.

Fábregas, M. C., F. Guillén-Salazar, and C. Garcés-Narro. 2012. "Do Naturalistic Enclosures Provide Suitable Environments for Zoo Animals?" *Zoo Biology* 31:362–73.

Fairbank, Maslin, Maullin, Metz and Associates (FM3). 2013. "Coastal Protection Assembly." Unpublished Polling Data. Monterey Bay Aquarium, April 18.

Falk, J. H., and L. D. Dierking. 2000. *Learning from Museums: Visitor Experiences and the Making of Meaning*. Walnut Creek, CA: AltaMira Press.

Falk, J. H., J. E. Heimlich, and K. Bronnenkant. 2008. "Using Identity-Related Visit Motivations as a Tool for Understanding Adult Zoo and Aquarium Visitors' Meaning-Making." *Curator* 51:55–80.

Falk, J. H., E. M. Reinhard, C. L. Vernon, K. Bronnenkant, N. L. Deans, and J. E. Heimlich. 2007. *Why Zoos and Aquariums Matter: Assessing the Impact of a Visit*. Silver Spring, MD: Association of Zoos and Aquariums. https://www.aza.org/uploadedFiles/Education/why_zoos_matter.pdf.

Fennell, D. A. 2015. "Contesting the Zoo as a Setting for Ecotourism, and the Design of a First Principle." *Journal of Ecotourism* 12:1-14.

Ferraro, P. J., and S. K. Pattanayak. 2006. "Money for Nothing? A Call for Empirical Evaluation of Biodiversity Conservation Investments." *PLoS Biology* 4:482–88.

Ferreira, S. M., J. M. Botha, and M. C. Emmett. 2012. "Anthropogenic Influences on Conservation Values of White Rhinoceros." *PLoS ONE* 7 (9): e45989.

Field, D., L. Chemnick, M. Robbins, K. Garner, and O. Ryder. 1998. "Paternity Determination in Captive Lowland Gorillas and Orangutans and Wild Mountain Gorillas by Microsatellite Analysis." *Primates* 39:199–209.

Finlay, T., L. James, and T. Maple. 1988. "People's Perceptions of Animals: The Influence of Zoo Environment." *Environment and Behavior* 20:508–28.

Fisher, B., and T. Christopher. 2007. "Poverty and Biodiversity: Measuring the Overlap of Human Poverty and the Biodiversity Hotspots." *Ecological Economics* 62 (1): 93–101. doi:10.1016/j.ecolecon.2006.05.020.

Fisher, C., ed. 2002. *A Passion for Natural History: The Life and Legacy of the 13th Earl of Derby*. Liverpool, UK: National Museums and Galleries on Merseyside.

Fisher, M. C., D. A. Henk, C. J. Briggs, J. S. Brownstein, L. C. Madoff, S. L. McCraw, and S. J. Gurr. 2012. "Emerging Fungal Threats to Animal, Plant and Ecosystem Health." *Nature* 484:186–94.

Flesness, N. R. 2003. "International Species Information System (ISIS): Over 25 years of Compiling Global Animal Data to Facilitate Collection and Population Management." *International Zoo Yearbook* 38 (1): 53–61.

Floyd, M. 1999. "Race, Ethnicity and Use of the National Park System." *NPS Social Science Research Review* 1 (2): 1–24. Available at http://digitalcommons.usu.edu/govdocs/427/.

Folke, C., S. Carpenter, T. Elmqvist, L. Gunderson, C. S. Holling, and B. Walker. 2002. "Resilience and Sustainable Development: Building Adaptive Capacity in a World of Transformations." *AMBIO: A Journal of the Human Environment* 31 (5): 437–40. doi:10.1579/0044-7447-31.5.437.

Food and Agriculture Organization of the United Nations (FAO). 2014. *The State of World Fisheries and Aquaculture: Opportunities and Challenges*. Rome: Food and Agricultural Organization of the United Nations.

Foster, S. J., and A. C. J. Vincent. 2004. "Life History and Ecology of Seahorses: Implications for Conservation and Management." *Journal of Fish Biology* 65 (1): 1–61.

Fox, H. 1923. *Disease in Captive Wild Mammals and Birds: Incidence, Description, Comparison*. Philadelphia: Lippincott.

Fox, S. 1986. *The American Conservation Movement: John Muir and His Legacy*. Madison: University of Wisconsin Press.

Franklin, A. 2002. *Nature and Social Theory*. London: Sage.

Franklin, S. 1999. "Review Essay: What We Know and What We Don't about Cloning and Society." *New Genetics and Soceity* 18:111–20.

——. 2007. *Dolly Mixtures: The Remaking of Genealogy*. Durham, NC: Duke University Press.

——. 2013. *Biological Relatives: IVF, Stem Cells, and the Future of Kinship*. Durham, NC: Duke University Press.

Franklin, S., and H. Ragone, eds. 1998. *Reproducing Reproduction: Kinship, Power, and Technological Innovation*. Philadelphia: University of Pennsylvania Press.

Fraser, J., and J. Sickler. 2009. "Measuring the Cultural Impact of Zoos and Aquariums." *International Zoo Yearbook* 43:103–12.

Fravel, L. 2003. "Critics Question Zoos' Commitment to Conservation." *National Geographic News*, November 13.

Friese, C. 2013. *Cloning Wild Life: Zoos, Captivity, and the Future of Endangered Animals*. New York: New York University Press.

Friese, C., and C. Marris. 2014. "Making De-extinction Mundane?" *PLoS Biology* 12 (3): e1001825. doi:1001810.1001371/journal.pbio.1001825.

Frynta, D., S. Lisková, S. Bültmann, and H. Burda. 2010. "Being Attractive Brings Advantages: The Case of Parrot Species in Captivity." *PLoS ONE* 5:e12568.

Frynta, D., O. Simková, S. Lisková, and E. Landová. 2013. "Mammalian Collection on Noah's Ark: The Effects of Beauty, Brain and Body Size." *PLoS ONE* 8:e63110.

Fuller, S. 2011. "The Economic Impact of Spending for Operations and Construction by AZA-Accredited Zoos and Aquariums." Accessed August 30, 2015. www.aza.org/uploadedFiles/Press_Room/News_Releases/AZA%20Impacts%202011.pdf.

Gagliardo, R., P. Crump, E. Griffith, J. R. Mendelson III, H. Ross, and K. C. Zippel. 2008. "The Principles of Rapid Response for Amphibian Conservation, Using the Programmes in Panama as an Example." *International Zoo Yearbook* 42:125–35.

Gärdenfors, U. 2001. "Classifying Threatened Species at National versus Global Levels." *Trends in Ecology and Evolution* 16 (9): 511–16.

Garelle, D., P. Marinari, and C. Lynch. 2013. *Population Analysis and Breeding and Transfer Plan Ferret-Black Footed (Mustela nigripes)*. Chicago: Association of Zoos and Aquariums.

——. 2015. *Black-Footed Ferret (Mustela nigripes) AZA Species Survival Plan Yellow Program Population Analysis and Breeding and Transfer Plan*. Silver Spring, MD: Association of Zoos and Aquariums.

Gascon, C., J. P. Collins, R. D. Moore, D. R. Church, J. E. McKay, and J. R. Mendelson III, eds. 2007. *Amphibian Conservation Action Plan*. Gland, Switzerland: IUCN/SSC Amphibian Specialist Group.

Geist, V. 1995. "Noah's Ark II: Rescuing Species and Ecosystems." In *Ethics on the Ark: Zoos, Animal Welfare, and Wildlife Conservation*, edited by B. G. Norton, M. Hutchins, E. F Stevens, and T. L. Maple, 93–101. Washington, DC: Smithsonian Institution Press.

George, A. L., M. T. Hamilton, and K. F. Alford. 2013. "We All Live Downstream: Engaging Partners and Visitors in Freshwater Fish Reintroduction Programmes." *International Zoo Yearbook* 47 (1): 140–50.

George, A. L., B. R. Kuhajda, J. D. Williams, M. A. Cantrell, P. L. Rakes, and J. R. Shute. 2009. "Guidelines for Propagation and Translocation for Freshwater Fish Conservation." *Fisheries* 34:529–45.

Gessner, K. 1551–58. *Historiae Animalium*. Vol. 4. Zurich: Christian Froschauer.

Geyer, C. J., and E. A. Thompson. 1988. "Gene Survival in the Asian Wild Horse (*Equus przewalskii*): 1. Dependence of Gene Survival in the Calgary Breeding Group Pedigree." *Zoo Biology* 7:313–27.

Gilpin, M. E., and M. E. Soulé. 1986. "Minimum Viable Populations: Processes of Species Extinction." In *Conservation Biology: The Science of Scarcity and Diversity*, edited by M. E. Soulé, 19–34. Sunderland, MA: Sinauer.

Ginsburg, F. D., and R. Rapp, eds. 1995. *Conceiving the New World Order: The Global Politics of Reproduction*. Berkeley: University of California Press.

González-Maya, J. F., J. L. Belant, S. A. Wyatt, J. Schipper, J. Cardenal, et al. 2013. "Renewing Hope: The Rediscovery of *Atelopus varius* in Costa Rica." *Amphibia-Reptilia* 34:573–78.

Goode, G. B. 1884–87. *The Fisheries and Fishery Industries of the United States*. 8 vols. Washington, DC: Government Printing Office.

Goodpaster, K. 1978. "On Being Morally Considerable." *Journal of Philosophy* 75:308–25.

Graham, F., Jr. 1990. *The Audubon Ark: A History of the National Audubon Society*. New York: Knopf.

Grajal, A., J. F. Luebke, L.-A. D. Kelly, J. Matiasek, S. Clayton, B. T. Karazsia, C. D. Saunders, S. R. Goldman, M. E. Mann, and R. Stanoss. 2017. "The Complex Relationship between Personal Sense of Connection to Animals and Self-Reported Proenvironmental Behaviors by Zoo Visitors." *Conservation Biology* 31 (2): 322-30.

Grant, E. H. C., E. Muths, R. A. Katz, S. Canessa, M. J. Adams, et al. 2015. "Salamander Chytrid Fungus (*Batrachochytrium salamandrivorans*) in the United States—Developing Research, Monitoring, and Management Strategies." *U.S. Geological Survey Open-File Report* 2015–1233. Available at http://dx.doi.org/10.3133/ofr20151233.

Grant, J. B., J. D. Olden, J. J. Lawler, C. R. Nelson, and B. R. Silliman. 2007. "Academic Institutions in the United States and Canada Ranked according to Research Productivity in the Field of Conservation Biology." *Conservation Biology* 21:1139–44.

Grazian, D. 2015. *American Zoo: A Sociological Safari*. Princeton, NJ: Princeton University Press.

Greenberg, J. 2014. *A Feathered River across the Sky: The Passenger Pigeon's Flight to Extinction*. New York: Bloomsbury.

Greene, H. W. 2013. *Tracks and Shadows: Field Biology as Art*. Berkeley: University of California Press.

——. 2015. "Rewilding Our Lives." *Mind and Nature* 8 (1): 18–24.

Greene, H. W., P. May, D. L. Hardy Sr., J. Sciturro, and T. Farrell. 2002. "Parental Behavior by Vipers." In *Biology of the Vipers*, edited by G. W. Schuett, M. Höggren, M. E. Douglas, and H. W. Greene, 179–205. Eagle Mountain, UT: Eagle Mountain.

Grew, N. 1681. *Musaeum Regalis Societatis, or A Catalogue and Description of the Natural and Artificial Rarities Belonging to the Royal Society and Preserved at Gresham Colledge*. London: W. Rawlins.

Grier, K. 2006. *Pets in America: A History*. Chapel Hill: University of North Carolina Press.Grimwood, I. R. 1967. "Operation Oryx: The Three Stages of Captive Breeding." *Oryx* 9:110–22.

Grinnell, J. 1928. "Recommendations concerning the Treatment of Large Mammals in Yosemite National Park." *Journal of Mammology* 9:76.

Grinnell, J., and T. I. Storer. 1916. "Animal Life as an Asset of National Parks." *Science* 15:375-80.

——. 1924. *Animal Life in the Yosemite: An Account of the Mammals, Birds, Reptiles, and Amphibians in a Cross-Section of the Sierra Nevada*. Berkeley: University of California Press.

Grow, S., and R. Allard. 2008. "The Zoological Community Celebrates 2008: The Year of the Frog." *Herpetological Review* 39 (1): 15–22.

Grow, S., R. Allard, and D. Luke. 2015. "The Role of AZA-Accredited Zoos and Aquariums in

Butterfly Conservation." In *Butterfly Conservation in North America: Efforts to Help Save Our Charismatic Microfauna*, edited by J. C. Daneils, 23–34. Dordrecht: Springer.

Gruen, L., ed. 2014. *The Ethics of Captivity*, Oxford: Oxford University Press.

Guerrini, A. 2003. *Experimenting with Humans and Animals: From Galen to Animal Rights*. Baltimore: Johns Hopkins University Press.

——. 2012. "Perrault, Buffon, and the Natural History of Animals." *Notes and Records of the Royal Society of London* 66:393–409.

——. 2015. *The Courtiers' Anatomists: Animals and Humans in Louis XIV's Paris*. Chicago: University of Chicago Press.

Guillette, L. J., D. A. Crain, A. A. Rooney, and D. B. Pickford. 1995. "Organization versus Activation: The Role of Endocrine-Disrupting Contaminants (Edcs) during Embryonic Development in Wildlife." *Environmental Health Perspectives* 103 (Suppl. 7): 157–64.

Gusset, M., and G. Dick. 2010. "'Building a Future for Wildlife?' Evaluating the Contribution of the World Zoo and Aquarium Community to in Situ Conservation." *International Zoo Yearbook* 44 (1): 183–91. doi:10.1111/j.1748-1090.2009.00101.x.

——. 2011. "The Global Reach of Zoos and Aquariums in Visitor Numbers and Conservation Expenditures." *Zoo Biology* 30 (5): 566–69. doi:10.1002/zoo.20369.

——. 2015. "Towards Positive Animal Welfare." *WAZA Magazine* 16:1.

Hagan, W. T. 2007. *Charles Goodnight: Father of the Texas Panhandle*. Norman: University of Oklahoma Press.

Hagenbeck, C. 1902. Letter to W. T. Hornaday, December 12, 1902. William T. Hornaday and W. Reid Blair Incoming Correspondence, 1895–1939. Collection 1001. Wildlife Conservation Society Archives, New York.

——. 1910. *Beasts and Men: Being Carl Hagenbeck's Experiences for Half a Century among Wild Animals*. Translated by H. S. R. Elliot and A. G. Thacker. London: Longmans.

Hahn, D. 2003. *The Tower Menagerie*. New York: Jeremy Tarcher/Penguin.

Haho. 2004. 슬픈 동물원 [Sad zoo]. 서울대공원 동물원 보고서 [Report on the Seoul Grand Park Zoo]. Seoul, South Korea: Haho.

Halbert, N. D., and J. N. Derr. 2007. "A Comprehensive Evaluation of Cattle Introgression into US Federal Bison Herds." *Journal of Heredity* 98 (1): 1–12.

Hale, D. S. 2002. "L''indigène' mis en scène en France, entre exposition et exhibition (1880–1931)." In *Zoos humains, XIXe et XXe siècles*, edited by N. Bancel, P. Blanchard, G. Boëtsch, E. Deroo, and S. Lemaire, 315–22. Paris: Éditions la Découverte.

Hall, A. F. 1920. *Guide to Yosemite: A Handbook of the Trails and Roads of Yosemite Valley and the Adjacent Region*. Yosemite, CA: Sunset.

Hall, B. K. 2001. "Organic Selection: Proximate Environmental Effects on Evolution of Morphology and Behaviour." *Biology and Philosophy* 16:215–37.

Han, S. 2015. Interview by Anne Clay. Hwacheon, South Korea. April.

Hancocks, D. 1995. "Lions and Tigers and Bears, Oh No!" In *Ethics on the Ark: Zoos, Animal Welfare, and Wildlife Conservation*, edited by B. G. Norton, M. Hutchins, E. Stevens, and T. L. Maple, 31–37. Washington, DC: Smithsonian Institution Press.

——. 2002. *A Different Nature: The Paradoxical World of Zoos and Their Uncertain Future*. Berkeley: University of California Press.

Hanson, E. 2002. *Animal Attractions: Nature on Display in American Zoos*. Princeton, NJ: Princeton University Press.

Hanson, J. H. 2015. "Critical Factors for Sustainable Food Procurement in Zoological Collections." *Zoo Biology* 34 (5): 483–91. doi:10.1002/zoo.21230.

Harcourt, A. H. 1987. "Behaviour of Wild Gorillas *Gorilla gorilla* and Their Management in Captivity." *International Zoo Yearbook* 26:248–55.

Harding, G., R. A. Griffiths, and L. Pavajeau. 2015. "Developments in Amphibian Captive Breeding and Reintroduction Programs." *Conservation Biology* 30 (2): 340–49. doi:10.1111/cobi.12612.

Harrison, M. A., and A. E. Hall. 2010. "Anthropomorphism, Empathy, and Perceived Communicative Ability Vary with Phylogenetic Relatedness to Humans." *Journal of Social, Evolutionary, and Cultural Psychology* 4:34–48.

Hartsonte-Rose, A., H. Selvey, J. R. Villari, M. Atwell, and T. Schmidt. 2014. "The Three-Dimensional Morphological Effects of Captivity." *PLoS One* 9 (11): e113437.

Haussler, D., S. J. O'Brien, O. A. Ryder, F. K. Barker, M. Clamp, et al. 2009. "Genome 10K: A Proposal to Obtain Whole-Genome Sequence for 10,000 Vertebrate Species." *Journal of Heredity* 100:659–74.

Hayashi, K., H. Ohta, K. Kurimoto, S. Aramaki, and M. Saitou. 2011. "Reconstitution of the Mouse Germ Cell Specification Pathway in Culture by Pluripotent Stem Cells." *Cell* 146:519–32.

Hayashi, K., and M. Saitou. 2013. "Generation of Eggs from Mouse Embryonic Stem Cells and Induced Pluripotent Stem Cells." *Nature Protocols* 8:153–24.

Hediger, H. 1970. *Man and Animal in the Zoo.* New York: Delacorte Press.

Hedrick, P. W., P. E. A. Hoeck, R. C. Fleischer, S. Farabaugh, and B. M. Masuda. 2016. "The Influence of Captive Breeding Management on Founder Representation and Inbreeding in the 'Alalā, the Hawaiian Crow." *Conservation Genetics* 17:369–78.

Herman, D. J. 2001. *Hunting and the American Imagination.* Washington, DC: Smithsonian Institution Press.

Hermes, R., T. B. Hildebrandt, C. Walzer, F. Goritz, M. L. Patton, S. Silinski, M. J. Anderson, C. E. Reid, G. Wibbelt, K. Tomasova, and F. Schwarzenberger. 2006. "The Effect of Long Non-reproductive Periods on the Genital Health in Captive Female White Rhinoceroses (*Ceratotherium simum simum,* C.s. *cottoni*)." *Theriogenology* 65 (8): 1492–515.

Herzog, H. A. 1991. "Conflicts of Interests: Kittens and Boa Constrictors, Pets and Research." *American Psychologist* 46:246–49.

———. 2010. *Some We Hate, Some We Love, Some We Eat: Why It's So Hard to Think Straight about Animals.* New York: HarperCollins.

Hewson, I., J. B. Button, B. M. Gudenkauf, B. Miner, A. L. Newton, J. K. Gaydos, J. Wynne, C. L. Groves, G. Hendler, M. Murray, and S. Fradkin. 2014. "Densovirus Associated with Sea-Star Wasting Disease and Mass Mortality." *Proceedings of the National Academy of Sciences* 111:17278–83.

Hill, R. T., W. Iliff, and T. Brailsford. 2001. "The Aquarium as a Driver of Economic and Social Development." *Marketing* 9:96.

Hillard, J. M. 1995. *Aquariums of North America: A Guidebook to Appreciating North America's Aquatic Treasures.* Lanham, MD: Scarecrow Press.

Hoage, R. J., and W. A. Deiss, eds. 1996. *New Worlds, New Animals: From Menagerie to Zoological Park in the Nineteenth Century.* Baltimore: Johns Hopkins University Press.

Hobbs, R. J., E. Higgs, C. M. Hall, P. Bridgewater, F. S. Chapin III, et al. 2014. "Managing the Whole Landscape: Historical, Hybrid, and Novel Ecosystems." *Frontiers in Ecology* 12 (10): 557–64.

Hochadel, O. 2005. "Science in the 19th-Century Zoo." *Endeavour* 29:38-42.

Hodge, C. F. 1910. "A Last Effort to Find the Passenger Pigeon." *Bird-Lore* 12 (1): 52.

———. 1911. "The Passenger Pigeon Investigation." *Auk* 28 (1): 49–53.

———. 1912. "A Last Word on the Passenger Pigeon." *Auk* 29 (2): 169–75.

Hoegh-Guldberg, O., et al. 2015. *Reviving the Ocean Economy: The Case for Action—2015.* Gland, Switzerland: World Wildlife Federation International.

Hoff, M., M. Bloomsmith, and E. L. Zucker. 2014. *Celebrating the Career of Terry L. Maple: A Festschrift.* Tequesta, FL: Red Leaf Press.

Hoffmann M., C. Hilton-Taylor, A. Angulo, M. Böhm, T. M. Brooks, et al. 2010. "The Impact of Conservation on the Status of the World's Vertebrates." *Science* 330 (6010): 1503–9.

Holdsworth, E. W. H. 1860. *Handbook to the Fish-House in the Gardens of the Zoological Society of London*. London: Bradbury and Evans.

Holling, C. S., F. Berkes, and C. Folke. 2000. "Science, Sustainability, and Resource Management." In *Linking Social and Ecological Systems: Management Practices and Social Mechanisms for Building Resilience*, edited by F. Berkes and C. Folke, 342–62. Cambridge: Cambridge University Press.

Holling, C. S., and G. K. Meffe. 1996. "Command and Control and the Pathology of Natural Resource Management." *Conservation Biology* 10 (2): 328–37. doi:10.1046/j.1523-1739.1996.10020328.x.

Hollister, N. 1918. "Some Effects of Environment and Habit on Captive Lions." *Proceedings of the U.S. National Museum* 53:177–93.

Hoole, A., and F. Berkes. 2010. "Breaking Down Fences: Recoupling Social-Ecological Systems for Biodiversity Conservation in Namibia." *Geoforum* 41 (2): 304–17. doi:10.1016/j.geoforum.2009.10.009.

Hopwood, N., P. M. Jones, L. Kassell, and J. Secord. 2015. "Introduction: Communicating Reproduction." *Bulletin of the History of Medicine* 89 (3): 379–404.

Hornaday, W. T. 1886. Letter to Spencer F. Baird, December 21, 1886. Record Unit 305, Accession Records, Accession 18617, Smithsonian Institution Archives.

———. 1887a. Letter to G. Brown Goode, December 2, 1887. Record Unit 201, Assistant Secretary in charge of the United States National Museum, box 17, folder 10, Smithsonian Institution Archives.

———. 1887b. "The Passing of the Buffalo." *Cosmopolitan* 4:85–98, 231–43.

———. [1889] 2002. *The Extermination of the American Bison*. Washington, DC: Smithsonian Institution Press.

———. 1902. Letter to C. Hagenbeck, December 31, 1902. William T. Hornaday and W. Reid Blair Outgoing Correspondence, 1895–1939. Collection 1012. Wildlife Conservation Society Archives, New York.

———. 1903. Letter to C. Hagenbeck, January 20, 1903. William T. Hornaday and W. Reid Blair Outgoing Correspondence, 1895–1939. Collection 1012. Wildlife Conservation Society Archives, New York.

———. 1904. Letter to C. Hagenbeck, March 26, 1904. William T. Hornaday and W. Reid Blair Outgoing Correspondence, 1895–1939. Collection 1012. Wildlife Conservation Society Archives, New York.

———. 1905. Letter to C. Hagenbeck, January 7, 1905. William T. Hornaday and W. Reid Blair Outgoing Correspondence, 1895–1939. Collection 1012. Wildlife Conservation Society Archives, New York.

———. 1907. Letter to Charles E. Hughes, June 28, 1907. American Bison Society Papers, President's Office, 1905–12, box 1. Wildlife Conservation Society, New York.

———. 1908. "The Founding of the Wichita National Bison Herd." *Annual Report of the American Bison Society* 1:55–69.

———. 1909. *Popular Official Guide to the New York Zoological Park*. 10th ed. New York: New York Zoological Society.

———. 1910. "Report of the President on the Founding of the National Montana National Bison Herd." *Annual Report of the American Bison Society* 3:1–18.

———. 1913. *Our Vanishing Wild Life: Its Extermination and Preservation*. New York: New York Zoological Society.

Horowitz, H. L. 1973–74. "The National Zoological Park: 'City of Refuge' or Zoo?" *Records of the Columbia Historical Society* 49:405–29.

———. 1975. "Animal and Man in the New York Zoological Park." *New York History* 56 (4): 425–55.

———. 1996. "The National Zoological Park: "'City of Refuge' or Zoo?" In *New Worlds, New Animals: From Menagerie to Zoological Park in the Nineteenth Century*, edited by R. J. Hoage and W. A. Deiss, 126–35. Baltimore: Johns Hopkins University Press.

Howard, J. G., C. Lynch, R. M. Santymire, P. E. Marinari, and D. E. Wildt. 2015. "Recovery of Gene Diversity Using Long-Term Cryopreserved Spermatozoa and Artificial Insemination in the Endangered Black-Footed Ferret." *Animal Conservation* 19:102–11.

Howard, M. S. 1879. "The Philadelphia Zoo." *Harper's New Monthly Magazine* 58:699–712.

Hubbard, J. 2014. "In the Wake of Politics: The Political and Economic Construction of Fisheries Biology, 1860–1970." *Isis* 105:364–78.

Hungerford H. R., and T. Volk. 1990. "Changing Learner Behavior through Environmental Education." *Journal of Environmental Education* 21 (3): 8–21.

Hutchins, M. 2007. "The Animal Rights-Conservation Debate: Can Zoos and Aquariums Play a Role?" In *Zoos in the 21st Century: Catalysts for Conservation?*, edited by A. Zimmermann, M. Hatchwell, L. A. Dickie, and C. West, 92–109. Cambridge: Cambridge University Press.

Hutchins, M., and W. G. Conway. 1995. "Beyond Noah's Ark: The Evolving Role of Modern Zoological Parks and Aquariums in Field Conservation." *International Zoo Yearbook* 34 (1): 117–30.

Hutchins, M., B. Dresser, and C. Wemmer. 1995. "Ethical Considerations in Zoo and Aquarium Research." In *Ethics on the Ark: Zoos, Animal Welfare, and Wildlife Conservation*, edited by B. G. Norton, M. Hutchins, E. F Stevens, and T. L. Maple, 253–76. Washington, DC: Smithsonian Institution Press.

Hutchins, M., and B. Smith. 2003. "Characteristics of a World Class Zoo in the 21st Century." *International Zoo Yearbook* 38:130–41.

Hutchins, M., B. Smith, and R. Allard. 2003. "In Defense of Zoos and Aquariums: The Ethical Basis for Keeping Wild Animals in Captivity." *Journal of the American Veterinary Medical Association* 223:958–66.

Hutchins, M., and R. Wiese. 1991. "Beyond Genetic and Demographic Management: The Future of the Species Survival Plan and Related AAZPA Conservation Efforts." *Zoo Biology* 10:285–92.

Ingels, B. 2015. "Bjarke Ingels's Human Zoo in Denmark." *Icon*, March 12. Available online at http://www.iconeye.com/architecture/features/item/11665-bjarke-ingels-s-human-zoo-in-denmark.

Intergovernmental Panel on Climate Change (IPCC). 2014a. *Climate Change 2014: Synthesis Report.* Contribution of Working Groups I, II and III to the Fifth Assessment Report of the Intergovernmental Panel on Climate Change (R. K. Pachauri and L. A. Meyer, eds.) Geneva, Switzerland: IPCC. Available online at http://ipcc.ch/report/ar5/index.shtml.

———. 2014b. "Summary for Policymakers." In *Climate Change 2014: Impacts, Adaptation, and Vulnerability. Part A: Global and Sectoral Aspects.* Contribution of Working Group II to the Fifth Assessment Report of the Intergovernmental Panel on Climate Change (C. B. Field, V. R. Barros, D. J. Dokken, K. J. Mach, M. D. Mastrandrea, T. E. Bilir, M. Chatterjee, K. L. Ebi, Y. O. Estrada, R. C. Genova, B. Girma, E. S. Kissel, A. N. Levy, S. MacCracken, P. R. Mastrandrea, and L. L. White, eds.), 1–32. New York: Cambridge University Press.

International Rhino Foundation. 2014. *Annual Report 2014*. Fort Worth, TX: International Rhino Foundation. Available online at http://rhinos.org/media-kit/annual-report-financials/.

International Union of Directors of Zoological Gardens, IUCN Species Survival Commission (SSC), Captive Breeding Specialist Group (CBSG). 1993. *The World Zoo Conservation Strategy: The Role of Zoos and Aquaria of the World in Global Conservation*. Chicago: Chicago Zoological Society.

Isaac, N. J. B., S. T. Turvey, B. Collen, C. Waterman, and J. E. M. Baillie. 2007. "Mammals on the EDGE." *PLoS ONE* 2:e296. doi:210.1371/journal.pone.0000296.

Isenberg, A. C. 2000. *The Destruction of the Bison: An Environmental History, 1750–1920*. Cambridge: Cambridge University Press.

IUCN/SSC. 2008. *Strategic Planning for Species Conservation: A Handbook*. Version 1.0. Gland, Switzerland: IUCN Species Survival Commission.

———. 2013. *Guidelines for Reintroductions and Other Conservation Translocations*. Version 1.0. Gland, Switzerland: IUCN Species Survival Commission.

——. 2014. *IUCN Species Survival Commission Guidelines on the Use of Ex Situ Management for Species Conservation*. Version 2.0. Tallinn, Estonia: Steering Committee of the IUCN.

——. 2015. "IUCN Red List of Threatened Species." Version 2015-4. Accessed April 1, 2016. www.iucnredlist.org.IUCN/SSC Antelope Specialist Group. 2013. "*Oryx leucoryx*: The IUCN Red List of Threatened Species 2013: e.T15569A4824960." Accessed June 6, 2016. http://dx.doi.org/10.2305/IUCN.UK.2011.1.RLTS.T15569A4824960.en.

Ivanyi, C., and D. Colodner. 2015a. "Arizona-Sonora Desert Museum Influence Survey." Unpublished data.

——. 2015b. "Survey of AZA Directors." Unpublished data.

Ivy, J. A., A. S. Putnam, A. Y. Navarro, J. Gurr, and O. A. Ryder. 2016. "Applying SNP-Derived Molecular Coancestry Estimates to Captive Breeding Programs." *Journal of Heredity*, advance online publication. pii: esw029.

Jachowski, D. S. 2014. *Wild Again: The Struggle to Save the Black-Footed Ferret*. Berkeley: University of California Press.

Jackson, E. 2001. *Regulating Reproduction: Law, Technology, and Autonomy*. Oxford: Hart.

Jacoby, K. 2001. *Crimes against Nature: Squatters, Poachers, Thieves, and the Hidden History of American Conservation*. Berkeley: University of California Press.

Jacquet, J. L., and D. Pauly. 2007. "The Rise of Seafood Awareness Campaigns in an Era of Collapsing Fisheries." *Marine Policy* 31 (3): 308-13.

James, T. Y., L. F. Toledo, D. Rodder, D. S. Leite, A. Belasen, et al. 2015. "Disentangling Host, Pathogen, and Environmental Determinants of a Recently Emerged Wildlife Disease: Lessons from the First 15 Years of Amphibian Chytridiomycosis Research." *Ecology and Evolution* 5 (18): 4079-97. http://dx.doi.org/10.1002/ece3.1672.

Jamieson, D. 1985. "Against Zoos." In *In Defense of Animals*, edited by P. Singer, 108-17. Oxford: Basil Blackwell.

——. 1995. "Zoos Revisited." In *Ethics on the Ark: Zoos, Animal Welfare, and Wildlife Conservation*, ed. B. G. Norton, M. Hutchins, E. Stevens, and T. Maple, 52-66. Washington, DC: Smithsonian Institution Press.

Jang, Y. 2014. "고향으로 돌아간 제돌이 [Jedol's coming home]." Unpublished paper.

Janiskee, B. 2009. "Believe It or Not, Yosemite National Park Once Had a Zoo." *National Park Traveler*, February 17. Available online at http://www.nationalparkstraveler.com/2009/02/believe-it-or-not-yosemite-national-park-once-had-zoo.

Jasanoff, S. 2004. *States of Knowledge: The Co-production of Science and Social Order*. London: Routledge.

Jefferson, W. N., H. B. Patisaul, and C. J. Williams. 2012. "Reproductive Consequences of Developmental Phytoestrogen Exposure." *Reproduction* 143 (3): 247-60.

Jensen, E. 2014. "Evaluating Children's Conservation Biology Learning at the Zoo." *Conservation Biology* 28:1004-11. doi:10.1111/cobi.12263.

Jiang, Y., M. Lück, and E. C. M. Parsons. 2008. "Public Awareness, Education, and Marine Mammals in Captivity." *Tourism Review International* 11:237-49.

Johnson, L. 1992. "Towards the Moral Considerability of Species and Ecosystems." *Environmental Ethics* 14 (2): 145-57.

Johnson, R. 1990. ""Release and Translocation Strategies for the Puerto Rican Crested Toad, *Peltophryne lemur*." *Endangered Species Update Special Issue* 8 (1): 54-57.

Joseph, L. N., R. F. Maloney, and H. P. Possingham. 2009. "Optimal Allocation of Resources among Threatened Species: A Project Prioritization Protocol." *Conservation Biology* 23:328-38.

Jule, K. R., L. A. Leaver, and S. E. G. Lea. 2008. "The Effects of Captive Experience on Reintroduction Survival in Carnivores: A Review and Analysis." *Biological Conservation* 141:355-63.

Kagan, J. 1984. *The Nature of the Child*. New York: Basic Books.

Kahn, P. H. 1999. *The Human Relationship with Nature*. Cambridge, MA: MIT Press.

Kahn, P. H., Jr., B. Friedman, B. Gill, J. Hagman, R. L. Severson, N. G. Freier, E. N. Feldman, S. Carrere, and A. Stolyar. 2008. "A Plasma Display Window? The Shifting Baseline Problem in a Technologically Mediated Natural World." *Journal of Environmental Psychology* 28:192–99.

Kane, L. 2015. "Bill to Ban Captivity of Whales 'Wrong,' says Vancouver Aquarium." *Toronto Star*, June 12.

Kang, M., Y. Kim, J. Song, H. Yang, G. U, and Y. Yim, eds. 2013. *동물행동풍부화 실전백과: 서울대공원 풍부화 10년의 기록* [Types of animal enrichment: 10 years of animal enrichment at Seoul Grand Park]. Seoul, South Korea: Seoul Grand Park.

Kaplan, S. 1995. "The Restorative Benefits of Nature: Toward an Integrative Framework." *Journal of Environmental Psychology* 15:169–82.

Kareiva, P., M. Marvier, and R. Lalasz. 2012. "Conservation in the Anthropocene." *Breakthough*, Winter. Accessible online at http://thebreakthrough.org/index.php/journal/past-issues/issue-2/conservation-in-the-anthropocene.

Kasperbauer, T. J., and P. Sandøe. 2015. "Killing as a Welfare Issue." In *The Ethics of Killing Animals*, edited by T. Visak and R. Garner, 17–31. Oxford: Oxford University Press.

Kass, L. 2002. "The Wisdom of Repugnance: Why We Should Ban the Cloning of Humans." In *The Human Cloning Debate*, edited by G. McGee, 679–706. Berkeley, CA: Berkeley Hills Books.

Kaufman, L. 2012. "Intriguing Habitats, and Careful Discussions of Climate Change." *New York Times*, August 26.

Kawata, K. 2012. "Exorcising of a Cage: A Review of American Zoo Exhibits, Part III." *Zoologische Garten* 81:132–46.

Kelly, L.-A. D., J. F. Luebke, S. Clayton, C. D. Saunders, J. Matiasek, and A. Grajal. 2014. "Climate Change Attitudes of Zoo and Aquarium Visitors: Implications for Climate Literacy Education." *Journal of Geoscience Education* 62:502–10.

Kemmerly, J. D., and V. Macfarlane. 2009. "The Elements of a Consumer-Based Initiative in Contributing to Positive Environmental Change: Monterey Bay Aquarium's Seafood Watch program." *Zoo Biology* 28 (5): 398–411.

Kestin, S. 2004a. "Sickness and Death Can Plague Marine Mammals at Parks." *Sun Sentinel*, May 17. http://www.sun-sentinel.com/sfl-dolphins-conditionsdec31-story.html.

———. 2004b. "Captive Mammals Can Net Big Profits for Exhibitors." *Sun Sentinel*, May 17. http://www.sun-sentinel.com/sfl-dolphins-moneydec31-story.html.

Kete, K., ed. 2007. *A Cultural History of Animals in the Age of Empire*. Vol. 5 of *A Cultural History of Animals*, edited by Linda Kalof and Brigitte Resl. Oxford: Berg.

Keulartz, J. 2015. "Captivity for Conservation? Zoos at a Crossroads." *Journal of Agricultural and Environmental Ethics* 28:335–51.

Kieckhefer, T. R., D. Maldini, S. Reif, J. Cassidy, and J. Hoffman. 2007. "Rise and Fall (and Rise Again) of Southern Sea Otters (*Enhydra lutris nereis*) in Elkhorn Slough, California, 1994–2006." Poster presented at the 17th Biennial Conference on the Biology of Marine Mammals, Volume 29. Accessed March 30, 2016. http://soundwaves.usgs.gov/2008/05/SW200805-100.pdf.

King, B. 2015. "Why Do European Zoos Kill Healthy Animals?" National Public Radio, October 14. Accessed June 4, 2016. http://www.npr.org/sections/13.7/2015/10/14/448527516/why-do-european-zoos-kill-healthy-animals.

King, S. R. B., L. Boyd, W. Zimmerman, and B. E. Kendall. 2015. "*Equus ferus* ssp. *przewalskii*. The IUCN Red List of Threatened Species 2015: e.T7961A45172099." Accessed September 26, 2015. http://dx.doi.org/10.2305/IUCN.UK.2015-2.RLTS.T7961A45172099.en.

Kisling, V. N., Jr. 1996. "The Origin and Development of the American Zoological Park to 1899." In *New Worlds, New Animals: From Menagerie to Zoological Park in the Nineteenth Century*, edited by R. J. Hoage and W. A. Deiss, 109–25. Baltimore: Johns Hopkins University Press.

——, ed. 2000a. *Zoo and Aquarium History: Ancient Animal Collections to Zoological Gardens*. Boca Raton, FL: CRC Press.

——. 2000b. "Zoological Gardens in the United States." In *Zoo and Aquarium History: Ancient Animal Collections to Zoological Gardens*, edited by V. N. Kisling Jr., 147-180. Boca Raton, FL: CRC Press.

——. 2013. "Our Vanishing Wild Life Centennial and William T. Hornaday's Legacy." *International Zoo News* 60 (2): 127–35.

Kleiman, D. G., and J. J. C. Mallinson. 1998. "Recovery and Management Committees for Lion Tamarins: Partnerships in Conservation Planning and Implementation." *Conservation Biology* 12 (1): 27–38.

Knapp, C. R. 2004. "Ecotourism and Its Potential Impacts on Iguana Conservation in the Caribbean." In *Iguanas: Biology and Conservation*, edited by A. C. Alberts, R. L. Carter, W. K. Hayes, and E. P. Martins, 290–301. Berkeley: University of California Press.

Knapp, C. R., S. Alvarez-Clare, and C. Perez-Heydrich. 2010. "The Influence of Landscape Heterogeneity and Dispersal on Survival of Neonate Insular Iguanas." *Copeia* 2010:62–70.

Knapp, C. R., and A. K. Owens. 2005. "Home Range and Habitat Associations of a Bahamian Iguana: Implications for Conservation." *Animal Conservation* 8:269–78.

Koebner, L. 1994. *Zoobook: The Evolution of Wildlife Conservation Centers*. New York: Forge.

Köhler, J., and U. Bender. 2016. *International Studbook for the Western Lowland Gorilla (Gorilla g. gorilla)*. Frankfurt: Frankfurt Zoo.

Kolbert, E. 2014. *The Sixth Extinction: An Unnatural History*, New York: Henry Holt.

Koldewey, H. 2005. *Syngnathid Husbandry in Public Aquariums 2005 Manual*. London: Project Seahorse.

Koldewey, H. J., and K. M. Martin-Smith. 2010. "A Global Review of Seahorse Aquaculture." *Aquaculture* 302 (3): 131–52.

Korea National Park Service (KNPS) Species Restoration Technology Institute. 2014. *Hello! Asiatic Black Bear: Asiatic Black Bear Back to Nature through the Species Restoration Project*. Republic of Korea: Park Bo-hwan.

Kreger, M. D., and M. Hutchins. 2010. "Ethics of Keeping Mammals in Zoos and Aquariums." In *Wild Mammals in Captivity: Principles and Techniques for Zoo Management*, 2nd ed., ed. D. G. Kleiman, K. V. Thompson, and C. K. Baer, 3-10. Chicago: University of Chicago Press.

Kreger, M. D., and J. A. Mench. 1995. "Visitor—Animal Interactions at the Zoo." *Anthrozoös* 8 (3): 143–58.

Kusukawa, S. 2010. "The Sources of Gessner's Pictures for the *Historia Animalium*." *Annals of Science* 67:303–28.

Lachapelle, S., and H. Mistry. 2014. "From the Waters of the Empire to the Tanks of Paris: The Creation and Early Years of the Aquarium Tropical, Palais de la Porte Dorée." *Journal of the History of Biology* 47 (1): 1–27.

Lacroix, J. 1978. "L'approvisionnement des ménageries et les transports d'animaux sauvages par la Compagnie des indes au XVIIIe siècle." *Revue Française d'Histoire d'Outre-Mer* 65:153–79.

Lacy, R. C. 2013. "Achieving True Sustainability of Zoo Populations." *Zoo Biology* 32:19–26.

Lacy, R. C., J. D. Ballou, F. Princee, A. Starfield, and E. A. Thompson. 1995. "Pedigree Analysis for Population Management." In *Population Management for Survival and Recovery*, 57–75. New York: Columbia University Press.

Lacy, R. C., K. Traylor-Holzer, and J. D. Ballou. 2013. "Managing for True Sustainability of Species." *WAZA Magazine* 14:10–14.

Lambert, J. H. 1905. *The Story of Pennsylvania at the World's Fair St. Louis, 1904*. Vol. 1. Philadelphia: Pennsylvania Commission.

Lamoreux, J., H. R. Akçakaya, L. Bennun, N. J. Collar, L. Boitani, D. Brackett, A. Bräutigam, T. M.

Brooks, G. A. da Fonseca, R. A. Mittermeier, and A. B. Rylands. 2003. "Value of the IUCN Red List." *Trends in Ecology and Evolution* 18 (5): 214–15.

Landman, W., and D. Visscher. 2009. "Planning for a Sustainable Emmen Zoo." *International Zoo Yearbook* 43 (1): 64–70. doi:10.1111/j.1748-1090.2008.00078.x.

Laslett, B., and J. Brenner. 1989. "Gender and Social Reproduction: Historical Perspectives." *Annual Review of Sociology* 15:381–404.

"A Last Attempt to Locate and Save from Extinction the Passenger Pigeon." 1910. *Auk* 27 (1): 112.

Lawson, D. P., J. Ogden, and R. J. Snyder. 2008. "Maximizing the Contribution of Science in Zoos and Aquariums: Organizations, Models, and Perceptions." *Zoo Biology* 27:458–69.

Leader-Williams, N., A. Balmford, M. Linkie, G. M. Mace, R. J. Smith, et al. 2007. "Beyond the Ark: Conservation Biologists' Views." In *Zoos in the 21st Century: Catalysts for Conservation?*, edited by A. Zimmermann, M. Hatchwell, L. A. Dickie and C. West, 236–54. Cambridge: Cambridge University Press.

Leader-Williams, N., and H. T. Dublin. 2000. "Charismatic Megafauna as 'Flagship Species.'" In *Priorities for the Conservation of Mammalian Diversity: Has the Panda Had Its Day?*, edited by A. Entwistle and N. Dunstone, 53–84. Cambridge: Cambridge University Press.

Lee, H., D. Garshelis, U. S. Seal, and J. Shillcox, eds. 2001. *Asiatic Black Bears PHVA: Final Report.* Apple Valley, MN: Conservation Breeding Specialist Group.

Lee, H., B. Lee, G. Kwon, and C. Chung. 2013. "Release Strategy for the Red Fox (*Vulpes vulpes*) Restoration Project in Korea Based on Population Viability Analysis." *한국환경생태학회지* [Korean journal of environmental ecology] 27 (4): 417–28.

Lees, C. M., and J. Wilcken. 2009. "Sustaining the Ark: The Challenges Faced by Zoos in Maintaining Viable Populations." *International Zoo Yearbook* 43:6–18.

Leopold, Aldo. 1933. "The Conservation Ethic." *Journal of Forestry* 31 (6): 634–43.

Leus, K., L. B. Lackey, W. van Lint, D. de Man, S. Riewald, A. Veldkam, and J. Wijmans. 2011. "Sustainability of European Association of Zoos and Aquaria Bird and Mammal Populations." *WAZA Magazine* 12:11–14.

Lever, C. 1977. *The Naturalized Animals of the British Isles*. London: Hutchinson.

———. 1992. *They Dined on Eland: The Story of the Acclimatisation Societies*. London: Quiller.

Lewis R. J., and M. Janse, eds. 2008. *Advances in Coral Husbandry in Public Aquaria*. Vol. 2. Arnem, Netherlands: Burgers Zoo.

Liebo, S. P., and N. Songasen. 2002. "Cryopreservation of Embryos and Gametes of Non-domestic Species." *Theriogenology* 57:303-26.

Lindburg, D. G. 1988. "Improving the Feeding of Captive Felines through Application of Field Data." *Zoo Biology* 7:211–18.

———. 1999. "Zoos and the Rights of Animals." *Zoo Biology* 18 (5): 433–48.

———. 2007. "Zoo Biology: A Voice for Science in the Association of Zoos and Aquariums." *Zoo Biology* 26 (6): 437–39.

Lindburg, D., and L. Lindburg. 1995. "Success Breeds a Quandary: To Cull or Not to Cull." In *Ethics on the Ark: Zoos, Animal Welfare, and Wildlife Conservation*, edited by B. G. Norton, M. Hutchins, E. F. Stevens, and T. L. Maple, 195–208. Washington, DC: Smithsonian Institution Press.

Linklater, W. L., J. V. Gedir, P. R. Law, R. R. Swaisgood, K. Adcock, et al. 2012. "Translocations as Experiments in the Ecological Resilience of an Asocial Mega-herbivore." *PLoS ONE* 7 (1): e30664. doi:10.1371/journal.pone.0030664.

Lips, K. R., F. Brem, R. Brenes, J. D. Reeve, R. A. Alford, J. Voyles, C. Carey, A. Pessier, L. Livo, and J. P. Collins. 2006. "Infectious Disease and Global Biodiversity Loss: Pathogens and Enigmatic Amphibian Extinctions." *Proceedings of the National Academy of Science (USA)* 103:3165–70.

Lips, K. R., J. R. Mendelson III, A. Muñoz-Alonso, L. Canseco-Márquez, and D. G. Mulcahy. 2004. "Amphibian Population Declines in Montane Southern Mexico: Resurveys of Historical Localities." *Biological Conservation* 119:555–64.

Liu, J., T. Dietz, S. R. Carpenter, M. Alberti, C. Folke, et al. 2007. "Complexity of Coupled Human and Natural Systems." *Science* 317 (5844): 1513–16. doi:10.1126/science.1144004.

Lloyd, N., K. Traylor-Holzer, J. Mickelberg, T. Stephens, M. Schroeder, et al., eds. 2014. *Greater Sage-Grouse in Canada Population and Habitat Viability Assessment Workshop Final Report*. Apple Valley, MN: IUCN/SSC Conservation Breeding Specialist Group.

Loftin, Robert. 1995. "Captive Breeding of Endangered Species." In *Ethics on the Ark: Zoos, Animal Welfare, and Wildlife Conservation*, edited by B. G. Norton, M. Hutchins, E. F. Stevens, and T. L. Maple, 164–80. Washington, DC: Smithsonian Institution Press.

Loh, T. L., C. Knapp, and S. J. Foster. 2014. "iSeahorse Trends Toolkit." Version 1.1. Accessed July 7, 2016. http://www.acheron.com/tyler/iseahorse/trends/iSeahorse_Underwater_Manual_English_LowRes_1.0.pdf.

Loisel, G. 1912. *Histoire des ménageries de l'antiquité à nos jours*. 3 vols. Paris: O. Doin.

Long, S., C. Dorsey, and P. Boyle. 2011. "Status of Association of Zoos and Aquariums Cooperatively Managed Populations." *WAZA Magazine* 12:15–18.

Longcore, J. E., A. P. Pessier, and D. K. Nichols. 1999. "*Batrachochytrium dendrobatidis* gen. et sp. nov., a Chytrid Pathogenic to Amphibians." *Mycologia* 91:219–27.

Loring, J. A. 1906. "The Wichita Buffalo Range." *Annual Report of the New York Zoological Society* 10:3–22.

Louv, R. 2008. *Last Child in the Woods: Saving Our Children from Nature Deficit Disorder*. Chapel Hill, NC: Algonquin Books.

Low, B., S. R. Sundaresan, I. R. Fischhoff, and D. I. Rubenstein. 2009. "Partnering with Local Communities to Identify Conservation Priorities for Endangered Grevy's Zebra." *Biological Conservation* 142 (7): 1548–55. doi:10.1016/j.biocon.2009.02.003.

Lück, M. 2007. "Captive Marine Wildlife: Benefits and Costs of Aquaria and Marine Parks." In *Marine Wildlife and Tourism Management: Insights from the Natural and Social Sciences*, edited by J. Hingham and M. Lück, 130–41. Cambridge, MA: CAB International.

Luebke, J. F., S. Clayton, C. D. Saunders, J. Matiasek, L.-A. D. Kelly, and A. Grajal. 2012. *Global Climate Change as Seen by Zoo and Aquarium Visitors*. Brookfield, IL: Chicago Zoological Society. Available at http://www.clizen.org/files/CliZENSurveyFinalReportMay2012.pdf.

Luebke J. F., and J. Matiasek. 2013. "An Exploratory Study of Zoo Visitors' Exhibit Experiences and Reactions." *Zoo Biology* 32 (4): 407–16.

Luebke J. F., J. V. Watters, J. Packer, L. J. Miller, and D. M. Powell. 2016. "Zoo Visitors' Affective Responses to Observing Animal Behaviors." *Visitor Studies* 19 (1): 60–76.

Luglia, Rémi. 2015. *Des savants pour protéger la nature: La Société d'acclimatation (1854–1960)*. Rennes: Presses Universitaires de Rennes.

Lukas, K., R. Elsner, and S. Long. 2015. *Population Analysis and Breeding and Transfer Plan for the Western Lowland Gorilla (Gorilla gorilla gorilla), an AZA Species Survival Plan Green Program*. Silver Spring, MD: Association of Zoos and Aquariums.

Lukas, K. E., M. Jackson Marr, and T. L. Maple. 1998. "Teaching Operant Conditioning at the Zoo." *Teaching of Psychology* 25:112–16.

Lupick, T. 2014. "Vancouver Aquarium Bucks National Trend by Keeping Whales and Dolphins." *Georgia Straight*, February 12.

Luskutoff, N. M. 2003. "Role of Embryo Technologies in Genetic Management and Conservation of Wildlife." In *Reproductive Science and Integrated Conservation*, ed. W. V. Holt, A. P. Pickard, J. C. Rodger, and D. E. Wildt, 183-94. Cambridge: Cambridge University Press.

Mabille, G., and J. Pieragnoli. 2010. *La ménagerie de Versailles*. Arles: Éditions Honoré Clair.

MacDonald, C. 2006. "Change or Swim with the Fishes." *Amusement Business* 118:14–16.

Mace, G. M. 2014. "Whose Conservation?" *Science* 345:1558–60.

Mace, G. M., N. J. Collar, K. J. Gaston, C. Hilton-Taylor, H. R. Akçakaya, et al. 2008. "Quantification of Extinction Risk: IUCN's System for Classifying Threatened Species." *Conservation Biology* 22 (6): 1424–42. doi:10.1111/j.1523-1739.2008.01044.x.

MacGregor, A. 2014. "Patrons and Collectors: Contributors of Zoological Subjects to the Works of George Edwards (1694—1773)." *Journal of the History of Collections* 26:35–44.

Maienschein, J. 2001. "On Cloning: Advocating History of Biology in the Public Interest." *Journal of the History of Biology* 34:423–32.

———. 2002. "What's in a Name: Embryos, Clones, and Stem Cells." *American Journal of Bioethics* 2 (1): 12–19.

———. 2003. *Whose View of Life? Embryos, Cloning, and Stem Cells*. Cambridge, MA: Harvard University Press.

Mamo, L. 2007. *Queering Reproduction: Achieving Pregnancy in the Age of Technoscience*. Durham, NC: Duke University Press.

Mann, C. C. 2005. *1491: New Revelations of the Americas before Columbus*. New York: Vintage Books.

Mann, L. Q. 1977. Oral History Interview, June 9, 1977, 36–39. Smithsonian Institution Archives, Record Unit 9513.

Maple, T. L. 1995. "Toward a Responsible Zoo Agenda," In *Ethics on the Ark: Zoos, Animal Welfare, and Wildlife Conservation*, edited by B. G. Norton, M. Hutchins, E. Stevens, and T. L. Maple, 31-37. Washington, DC: Smithsonian Institution Press.

———. 2008. "Empirical Zoo: Opportunities and Challenges to a Scientific Zoo Biology." *Zoo Biology* 27 (6): 431-35.

———. 2014. "Copenhagen Zoo's Giraffe Killing Was Wrong and Disturbing." *San Francisco Gate*, April 12. Accessed June 4, 2016. http://www.sfgate.com/opinion/article/Copenhagen-Zoo-s-giraffe-killing-was-wrong-and-5396183.php.

———. 2015. "Four Decades of Psychological Research on Zoo Animal Welfare." *Magazine of the World Association of Zoos and Aquariums* 16:41–44.

———. 2016. *Professor in the Zoo: Designing the Future for Wildlife in Human Care*. Fernandina Beach, FL: Red Leaf Press.

Maple, T. L., and E. F. Archibald. 1993. *Zoo Man: Inside the Zoo Revolution*. Atlanta, GA: Longstreet Press.

Maple, T. L., M. A. Bloomsmith, and A. Martin. 2008. "Primates and Pachyderms: A Primate Model of Zoo Elephant Welfare." In *An Elephant in the Room: The Science and Well-Being of Elephants in Captivity*, edited by D. Forthman, 129–53. North Grafton, MA: Tufts University Cummings School of Veterinary Medicine/Center for Animals and Public Policy.

Maple, T. L., and D. Bocian. 2013. "Wellness as Welfare." *Zoo Biology* 32:363–65.

Maple, T. L., and T. W. Finlay. 1986. "Evaluating the Environments of Captive Nonhuman Primates." In *Primates: The Road to Self-Sustaining Populations*, edited by K. Benirschke, 480–88. New York: Springer.

Maple, T. L., and D. Lindburg. 2008. "Empirical Zoo: Opportunities and Impediments to Scientific Zoo Biology" Special issue of *Zoo Biology* 27.

Maple, T., R. McManamon, and E. Stevens. 1995. "Animal Care, Maintenance, and Welfare." In *Ethics on the Ark: Zoos, Animal Welfare, and Wildlife Conservation*, edited by B. G. Norton, M. Hutchins, E. F. Stevens, and T. L. Maple, 219–34. Washington, DC: Smithsonian Institution Press.

Maple, T. L., and B. M. Perdue. 2013. *Zoo Animal Welfare*. Heidelberg: Springer.

Maple, T. L., and V. D. Segura. 2014. "Advancing Behavior Analysis in Zoos and Aquariums." *Behavior Analyst* 38:77–91.

Marino, L., G. Bradshaw, and R. Malamud. 2009. "The Captivity Industry: The Reality of Zoos and Aquariums." *Best Friends Magazine*, March/April, 25–27.

Marino, L. S., O. Lilienfeld, R. Malamud, N. Nobis, and R. Broglio. 2010. "Zoos and Aquariums

Promote Attitude Change in Visitors? A Critical Evaluation of the American Zoo and Aquarium Study." *Society and Animals* 18:126–38.

Marris, E. 2009. "The Genome of the American West." *Nature* 457:950–52.

———. 2011. *Rambunctious Garden: Saving Nature in a Post-wild World*. New York: Bloomsbury.

Martel, A., A. Spitzen-van der Sluijs, M. Blooi, W. Bert, R. Ducatelle, et al. 2013. "*Batrachochytrium salamandrivorans* sp. nov. Causes Lethal Chytridiomycosis in Amphibians." *Proceedings of the National Academy of Sciences* 110:15325–29.

Martin, A. L., M. A. Bloomsmith, M. E. Kelley, M. Marr, and T. L. Maple. 2011. "Functional Analysis and Treatment of Human-Directed Undesirable Behaviors in a Captive Chimpanzee." *Journal of Applied Behavior Analysis* 1 (44): 139-43.

Martin, E. 1987. *The Woman in the Body: A Cultural Analysis of Reproduction*. Boston: Beacon Press.

Martin, T. E., H. Lurbiecki, J. B. Joy, and A. O. Mooers. 2014. "Mammal and Bird Species Held in Zoos Are Less Endemic and Less Threatened Than Their Close Relatives Not Held in Zoos." *Animal Conservation* 17 (2): 89–96.

Maslow, Abraham. 1962. *Toward a Psychology of Being*. New York: Van Nostrand.

Mason, G. 2010. "Species Differences in Responses to Captivity: Stress, Welfare, and the Comparative Method." *Trends in Ecology and Evolution* 25 (12): 713–21.

Mason, G., C. Burn, J. Dallaire, J. Kroshko, H. Kinkaid, and J. Jeschke. 2013. "Plastic Animals in Cages: Behavioural Flexibility and Responses to Captivity." *Animal Behaviour* 85 (5): 1113–26.

McCarthy, D. P., P. F. Donald, J. P. W. Scharlemann, G. M. Buchanan, A. Balmford, et al. 2012. "Financial Costs of Meeting Global Biodiversity Conservation Targets: Current Spending and Unmet Needs." *Science* 338 (6109): 946–49. doi:10.1126/science.1229803.

McDonald, M. G., S. Wearing, and J. Ponting. 2009. "The Nature of Peak Experience in Wilderness." *Humanistic Psychologist* 37 (4): 370–85.

McGee, G. 2002. *The Human Cloning Debate*, 3rd ed. Berkeley, CA: Berkeley Hills Books.

McMahan, J. 2016. "The Meat Eaters." In *The Stone Reader: Modern Philosophy in 133 Arguments*, edited by P. Catapano and S. Critchley, 538–45. New York: Norton.

McManus, K. F., J. L. Kelley, S. Song, K. R. Veeramah, A. E. Woerner, et al. 2015. "Inference of Gorilla Demographic and Selective History from Whole-Genome Sequence Data." *Molecular Biology and Evolution* 32 (3): 600–612. doi:10.1093/molbev/msu394.

McPhate, M. 2016. "Zoo's Killing of Gorilla Holding a Boy Prompts Outrage." *New York Times*, May 30, 2016. Accessible at https://www.nytimes.com/2016/05/31/us/zoos-killing-of-gorilla-holding-a-boy-prompts-outrage.html?_r=0.

McPhee, M. E. 2003. "Generations in Captivity Increases Behavioral Variance: Considerations for Captive Breeding and Reintroduction Programs." *Biological Conservation* 115:71–77.

McShane, C., and J. A. Tarr. 2007. *The Horse in the City: Living Machines in the Nineteenth Century*. Baltimore: Johns Hopkins University Press.

Meffe, G. K. 1998. "The Potential Consequences of Pollinator Declines on the Conservation of Biodiversity and Stability of Food Crop Yields." *Conservation Biology* 12 (1): 8–17.

Meine, C., M. Soulé, and R. F. Noss. 2006. "A Mission-Driven Discipline: The Growth of Conservation Biology." *Conservation Biology* 20:631–51.

Meister, B., et al. 2014. *Community Impacts: Social and Economic Benefits Created by the Non-profit Monterey Bay Aquarium*. Monterey, CA: Monterey Bay Aquarium.

Melfi, V. A., W. McCormick, and A. Gibbs. 2004. "A Preliminary Assessment of How Zoo Visitors Evaluate Animal Welfare according to Enclosure Style and the Expression of Behavior." *Anthrozoös* 17 (2): 98–108.

Mellor, D. J., S. Hunt, and M. Gusset, eds. 2015a. *Caring for Wildlife: The World Zoo and Aquarium Animal Welfare Strategy*. Gland, Switzerland: World Association of Zoos and Aquariums.

Mendelson, J. R., III. 2016. "A Frog Dies in Atlanta, and a World Vanishes with It." *New York Times*, October 10, 2016.

Mendelson, J. R., III, K. R. Lips, R. W. Gagliardo, G. B. Rabb, J. P. Collins, et al. 2006. "Policy Forum: Confronting Amphibian Declines and Extinctions." *Science* 313:48.

Mendelson, J. R., III, and G. B. Rabb. 2006. "Global Amphibian Extinctions and the Role of Living-Collections Institutions." In *World Association of Zoos and Aquariums, Proceedings WAZA Conferences: Proceedings of the 60th Annual Meeting, New York City, USA*, October 2–6, 2005, 179–81. Gland, Switzerland: World Association of Zoos and Aquariums.

Metrione, L. C., and J. D. Harder. 2011. "Fecal Corticosterone Concentrations and Reproductive Success in Captive Female Southern White Rhinoceros." *General and Comparative Endocrinology* 171 (3): 283–92.

Meyer, D., A. Isakower, and B. Mott. 2015. *An Ocean of Opportunities: Inspiring Visitors and Advancing Conservation. Research Findings*. Providence, RI: Ocean Project. http://theoceanproject.org/wp-content/uploads/2015/01/OceanOfOpportunities-SummaryReport2014.pdf.

Mihelich, P. 2005. "Big Window to the Sea." *CNN.com: Science and Space*, November 23. Accessed June 6, 2016. http://www.cnn.com/2005/TECH/science/11/21/new.ga.aquarium/.

Miller, B., W. Conway, R. P. Reading, C. Wemmer, D. Wildt, et al. 2004. "Evaluating the Conservation Mission of Zoos, Aquariums, Botanical Gardens, and Natural History Museums." *Conservation Biology* 18 (1): 86–93. doi:10.1111/j.1523-1739.2004.00181.x.

Miller, B., R. P. Reading, and S. Forrest. 1996. *Prairie Night: Black-Footed Ferrets and the Recovery of Endangered Species*. Washington, DC: Smithsonian Institution Press.

Minteer, B. A. 2006. *The Landscape of Reform: Civic Pragmatism and Environmental Thought in America*. Cambridge, MA: MIT Press.

———. 2013. "Conservation, Animal Rights, and Human Welfare: A Pragmatic View of the Bushmeat Crisis." In *Ignoring Nature No More: The Case for Compassionate Conservation*, edited by M. Bekoff, 77–93. Chicago: University of Chicago Press.

———. 2014. "Extinct Species Should Stay Extinct." *Slate*. Accessed Ocotber 15, 2015. Retrieved from Future Tense website: http://www.slate.com/articles/technology/future_tense/2014/12/de_extinction_ethics_why_extinct_species_shouldn_t_be_brought_back.html.

———. 2015. "When Extinction Is a Virtue." In *After Preservation: Saving American Nature in the Age of Humans*, edited by B. A. Minteer and S. J. Pyne, 96–104. Chicago: University of Chicago Press.

Minteer, B. A., and J. P. Collins. 2005. "Ecological Ethics: Building a New Tool Kit for Ecologists and Biodiversity Managers." *Conservation Biology* 19:1803–12.

———. 2013. "Ecological Ethics in Captivity: Balancing Zoo and Aquarium Research under Rapid Global Change." *ILAR Journal* 54 (1): 41–51.

Minteer, B. A., J. P. Collins, and A. Raschke. Forthcoming. "From the Wild to the Walled: The Evolution and Ethics of Zoo Conservation." In *Routledge Companion to Environmental Ethics*, edited by B. Hale and A. Light. London: Routledge.

Minteer, B. A., and S. J. Pyne, eds. 2015. *After Preservation: Saving American Nature in the Age of Humans*. Chicago: University of Chicago Press.

Mitchell, P. C. 1929. *Centenary History of the Zoological Society of London*. London: Zoological Society of London.

Mitman, G. 1996. "When Nature Is the Zoo: Vision and Power in the Art and Science of Natural History." *Osiris* 11:117–43.

———. 1999. *Reel Nature: America's Romance with Wildlife on Film*. Cambridge, MA: Harvard University Press.

Mohr, E. [1959] 1971. *The Asiatic Wild Horse*. Translated by D. M. Goodall. London: J. A. Allen.

Moleón, M., and J. A. Sánchez-Zapata. 2015. "The Living Dead: Time to Integrate Scavenging into Ecological Teaching." *BioScience* 65:1003–10.

Monfort, S. L. 2014. "'Mayday Mayday Mayday,' the Millennium Ark Is Sinking!" In *Reproductive Sciences in Animal Conservation: Advances in Experimental Medicine and Biology*, edited by W. V. Holt, J. L. Brown, and P. Comizzoli, 15–31. New York: Springer.

Monterey Bay Aquarium (MBA). 2015. "We're Making a Difference for the Ocean in Sacramento," *Conservation and Science at the Monterey Bay Aquarium*, October 8. Accessed April 4, 2016. https://futureoftheocean.wordpress.com/2015/10/08/were-making-a-difference-for-the-ocean-in-sacramento.

Moreau, M. A., H. J. Hall, and A. C. J. Vincent, eds. 2000. *Proceedings of the First International Workshop on the Management and Culture of Marine Species Used in Traditional Medicines, in Cebu City, Philippines*. Montreal: Project Seahorse.

Morgan, Bethan J., A. Adeleke, T. Bassey, R. Bergl, A. Dunn, et al. 2011. *Regional Action Plan for the Conservation of the Nigeria–Cameroon Chimpanzee (Pan troglodytes ellioti)*. San Diego, CA: IUCN/SSC Primate Specialist Group and Zoological Society of San Diego.

Morrell, V. 2014. "Opinion: Killing of Marius the Giraffe Exposes Myths about Zoos." *National Geographic*, February 13. Accessed June 4, 2016. http://news.nationalgeographic.com/news/2014/02/140212-giraffe-death-denmark-copenhagen-zoo-breeding-europe/.

Morrison, E. E. 1976. *Guardian of the Forest: A History of the Smokey Bear Program*. New York: Vantage Press.

Morse-Jones, S., I. J. Bateman, A. Kontoleon, S. Ferrini, N. D. Burgess, and R. K. Turner. 2012. "Stated Preferences for Tropical Wildlife Conservation amongst Distant Beneficiaries: Charisma, Endemism, Scope and Substitution Effects." *Ecological Economics* 78:9–18.

Mortensen, M. F. 2010. "Designing Immersion Exhibits as Border-Crossing Environments." *Museum Management and Curatorship* 25:323–36.

Moss, A., and M. Esson. 2010. "Visitor Interest in Zoo Animals and the Implications for Collection Planning and Zoo Education Programmes." *Zoo Biology* 29:715–31.

Moss, A., E. Jensen, and M. Gusset. 2014. *A Global Evaluation of Biodiversity Literacy in Zoo and Aquarium Visitors*. Gland, Switzerland: World Association of Zoos and Aquariums Executive Office.

———. 2015. "Evaluating the Contribution of Zoos and Aquariums to Aichi Biodiversity Target 1." *Conservation Biology* 29:537–44.

———. 2016. "Probing the Link between Biodiversity-Related Knowledge and Self-Reported Proconservation Behavior in a Global Survey of Zoo Visitors." *Conservation Letters*, March 11. doi:10.1111/conl.12233.

Muka, S. 2014. "Portrait of an Outsider: Class, Gender, and the Scientific Career of Ida M. Mellen." *Journal of the History of Biology* 47 (1): 29–61.

Muller-Wille, S., and H. J. Rheinberger, eds. 2007. *Heredity Produced: At the Crossroads of Biology, Politics, and Culture, 1500–1870*. Cambridge, MA: MIT Press.

Murdoch, W., S. Polasky, K. A. Wilson, H. P. Possingham, P. Kareiva, and R. Shaw. 2007. "Maximizing Return on Investment in Conservation." *Biological Conservation* 139:375–88.

Murphy, J. B. 2015. "Studies on Lizards and Tuataras in Zoos and Aquariums. Part II—Families Teiidae, Lacertidae, Bipedidae, Amphisbaenidae, Scincidae, Cordylidae, Xantusiidae, Anguidae, Helodermatidae, Varanidae, Lanthanotidae, Shinisauridae, Xenosauridae, and Sphenodontidae." *Herpetological Review* 46:672–85.

Murphy, J. B., C. Ciofi, C. Panouse, and T. Walsh, eds. 2002. *Komodo Dragons: Biology and Conservation*. Washington, DC: Smithsonian Institution Press.

Myers, O. E., C. D. Saunders, and A. A. Birjulin. 2004. "Emotional Dimensions of Watching Zoo Animals: An Experience Sampling Study Building on Insights from Psychology." *Curator* 47:299–321.

Naidoo, R., A. Balmford, P. Ferraro, S. Polasky, T. Ricketts, and M. Rouget. 2006. "Integrating Economic Costs into Conservation Planning." *Trends in Ecology and Evolution* 21:681–87.

Nam, K., K. Munch, A. Hobolth, J. Y. Dutheil, K. R. Veeramah, et al. 2015. "Extreme Selective Sweeps Independently Targeted the X Chromosomes of the Great Apes." *Proceedings of the National Academy of Sciences* 112 (20): 6413–18. doi: 10.1073/pnas.1419306112.

Nash R. F. 2014. *Wilderness and the American Mind*. 5th ed. New Haven, CT: Yale University Press.

"National Buffalo Herd." [1907]. "The National Buffalo Herd: A Gift from the New York Zoological Society to Start a Great Southwestern Herd." Unpublished MS. in *American Bison Society Papers*, President's Office, 1905–12, box 1. Wildlife Conservation Society, New York, NY.

National Ocean Council. 2016. *National Ocean Policy 2016 Annual Work Plan*. Washington DC: National Ocean Council. Available online at www.whitehouse.gov/sites/default/files/microsites/ostp/2016_annual_work_plan_final_-_160105.pdf.

National Oceanic and Atmospheric Administration (NOAA). 2014. "How Important Is the Ocean to Our Economy?" Accessed March 23, 2016. http://oceanservice.noaa.gov/facts/oceaneconomy.html.

National Public Radio (NPR) 2014. "Glass-Free Menagerie: New Zoo Concept Gets Rid Of Enclosures." *NPR*, August 9. Available online at http://www.npr.org/2014/08/09/339148819/glass-free-menagerie-new-zoo-concept-gets-rid-of-enclosures.

National Science Foundation (NSF) National Center for Science and Engineering Statistics. 2015. *Women, Minorities, and Persons with Disabilities in Science and Engineering: 2015*. Special Report NSF 15–311. Arlington, VA: NSF.

Newell, D. A., Goldingay, R. L., and Brooks, L. O. 2013. "Population Recovery Following Decline in an Endangered Stream-Breeding Frog (*Mixophys fleayi*) from Subtropical Australia." *PLoS ONE* 8:e58559.

New England Aquarium. 2016. "New England Aquarium." Accessed July 1. http://www.neaq.org/index.php.

New York Zoological Society. 1904. "Our Asiatic Deer Collection." *Zoological Society Bulletin* 15:173–74.

Nicholls, H. 2004. "The Conservation Business." *PLoS Biology* 2:1256–59.

Nicholson, T. E., K. A. Mayer, M. M. Staedler, and A. B. Johnson. 2007. "Effects of Rearing Methods on Survival of Released Free-Ranging Juvenile Southern Sea Otters." *Biological Conservation* 138 (3): 313–20.

"Ninety-Six Elephants." 2016. Accessed June 29, 2016. http://www.96elephants.org/.

No, J. 2015. Interview by Anne Clay. Gwacheon, South Korea. May.

Norton, B. G. 1987. *Why Preserve Natural Variety?* Princeton, NJ: Princeton University Press.

———. 2005. *Sustainability: A Philosophy of Adaptive Ecosystem Management*. Chicago: University of Chicago Press.

———. 2015. *Sustainable Values, Sustainable Change: A Guide to Environmental Decision Making*. Chicago: University of Chicago Press.

Norton, B. G., M. Hutchins, E. F. Stevens, and T. L. Maple, eds. 1995. *Ethics on the Ark: Zoos, Animal Welfare and Wildlife Conservation*. Washington DC: Smithsonian Institution Press.

"Notes and News." 1910. *Auk* 27 (2): 243.

———. 1914. *Auk* 31 (4): 566–67.

Nowak, E. M., and M. A. Santana-Bendi. 2002. *Status, Distribution, and Management Recommendations for the Narrow-Headed Gartersnake (Thamnophis rufipunctatus) in Oak Creek, Arizona*. Final Report to Arizona Game and Fish Department Heritage Grant I9900. Colorado Plateau Research Station: US Geological Survey.

Obama, B. 2010. *Executive Order 13547: Stewardship of the Ocean, Our Coasts, and the Great Lakes*. Washington, DC: White House, Office of the Press Secretary. https://www.whitehouse.gov/the-press-office/executive-order-stewardship-ocean-our-coasts-and-great-lakes.

Oelschlaeger, M. 1991. *The Idea of Wilderness: From Prehistory to the Age of Ecology*. New Haven, CT: Yale University Press.

Ogden, J., and J. E. Heimlich. 2009. "Why Focus on Zoo and Aquarium Education?" *Zoo Biology* 28:357–60.

Oksanen, M., and H. Siipi. 2014. "Introduction: Towards a Philosophy of Resurrection Science." In *The Ethics of Animal Re-creation and Modification: Reviving, Rewilding, Restoring*, edited by M. Oksanen and H. Siipi, 1–21. Basingstoke, UK: Palgrave Macmillan.

Oldfield, S., and A. C. Newton. 2012. *Integrated Conservation of Tree Species by Botanic Gardens: A Reference Manual.* Richmond, UK: Botanic Gardens Conservation International.

Olmstead, A. L., and P. W. Rhode. 2015. *Arresting Contagion: Science, Policy, and Conflicts over Animal Disease Control.* Cambridge, MA: Harvard University Press.

Olney, P. J. S. 2001. "Studbook." in *Encyclopedia of the World's Zoos, R-Z*, ed. Catherine E. Bell. Detroit: Fitzroy Dearborn.

Olson, S. 2010. Letter to U.S. Fish and Wildlife Service, on behalf of the Association of Zoos and Aquariums. Docket No. FWS-R9-FHC-2009-0093. Available online at https://federalregister.gov/a/2010-23039.

Osbaldiston R., and J. Schott. 2012. "Environmental Sustainability and Behavioral Science: Meta-analysis of Pro-Environmental Behavior Experiments." *Environment and Behavior* 44:257–99.

Osborne, M. A. 1992. "Applied Natural History and Utilitarian Ideals: "Jacobin Science" at the Muséum d'histoire naturelle, 1789–1870." In *Re-creating Authority in Revolutionary France*, edited by B. T. Ragan Jr. and E. A. Williams, 125–43. New Brunswick, NJ: Rutgers University Press.

———. 1994. *Nature, the Exotic, and the Science of French Colonialism.* Bloomington: Indiana University Press.

———. 1996. "Zoos in the Family: The Geoffroy Saint-Hilaire Clan and the Three Zoos of Paris." In *New Worlds, New Animals*, edited by R. J. Hoage and W. A. Deiss, 33–42. Baltimore: Johns Hopkins University Press.

———. 2000. "Acclimatizing the World: A History of the Paradigmatic Colonial Science." *Osiris* 15:135–51.

Ostrom, E. 2007. "A Diagnostic Approach for Going beyond Panaceas." *Proceedings of the National Academy of Sciences* 104 (39): 15181–87. doi:10.1073/pnas.0702288104.

Ostrowski, S., E. Bedin, D. M. Lenain, and A. H. Abuzinada. 1998. "Ten Years of Arabian Oryx Conservation Breeding in Saudia Arabia: Achievements and Regional Perspectives." *Oryx* 32:209–22.

Ovels, C., E. Horberg, and D. Keltner. 2010. "Compassion, Pride, and Social Intuitions of Self-Other Similarity." *Journal of Personality and Social Psychology* 98:618–30.

Packard, J. 1989. "Bringing the Ocean to the Public at the Monteray Bay Aquarium." *Oceanography* 2:44–45.

———. 2009. "Aquariums as a Force for Change: New Roles in Conservation and Social Impact." *ASTC Blog*, September/October. Available online at http://www.astc.org/astc-dimensions/aquariums-as-a-force-for-change-new-roles-in-conservation-and-social-impact/.

Palmer, T. S. 1916. "Our National Herds of Buffalo." *Annual Report of the American Bison Society* 10:40–62.

Parfit, D. 1984. *Reasons and Persons.* New York: Oxford University Press.

Parks, N. 2015. "Only 4 Northern White Rhinos Left on Earth." *CNN News online*, July 29. Accessed December 28, 2015. http://www.cnn.com/2015/07/29/world/northern-rhino-dies/.

Patrick, P. G., C. E. Matthews, D. F. Ayers, and S. D. Tunnicliffe. 2007. "Conservation and Education: Prominent Themes in Zoo Mission Statements." *Journal of Environmental Education* 38 (3): 53–60.

Pauly, P. J. 2000. *Biologists and the Promise of American Life: From Meriwether Lewis to Alfred Kinsey.* Princeton, NJ: Princeton University Press.

Pearce, J. M. 2012. "The Case for Open Source Appropriate Technology." *Environment, Development and Sustainability* 14 (3): 425–31.

Pearson, D., S. Wells, T. Sprankle, J. Sorenson, and M. Martinez. 2014. "Reproductive Seasonality and Developmental Characteristics of the Page Springsnail (*Pyrgulopsis morrisoni*)." *Journal of the Arizona-Nevada Academy of Science* 45:4–69.

Pearson, E. L., R. Lowry, J. Dorrian, and C. A. Litchfield. 2014. "Evaluating the Conservation Impact of an Innovative ZooÐBased Educational Campaign: 'Don't Palm Us Off' for Orangutan Conservation." *Zoo Biology* 33 (3): 184–96. doi:10.1002/zoo.21120.

Peart, B., and R. Kool. 1988. "Analysis of a Natural History Exhibit: Are Dioramas the Answer?" *International Journal of Museum Management and Curatorship* 7:117–28.

Peeling, C. 2015. "Facing the Real Threats to Collection Sustainability." Paper presented at the Association of Zoos and Aquariums' National Conference, Salt Lake City, Utah, September 2015.

Pekarik, A. 2004. "Eye-to-Eye with Animals and Ourselves." *Curator* 47:257–60.

Penning, M., G. McReid, H. Koldewey, G. Dick, B. Andrews, et al., eds. 2009. *Turning the Tide: A Global Aquarium Strategy for Conservation and Sustainability*. Bern, Switzerland: World Association of Zoos and Aquariums.

People for the Ethical Treatment of Animals (PETA). 2008. "Zoos: Pitiful Prisons." Accessed December 8, 2015. http://www.peta.org/issues/animals-in-entertainment/animals-used-entertainment-factsheets/zoos-pitiful-prisons/.

———. 2015. "Zoos: An Idea Whose Time Has Come and Gone." Accessed September 6. http://www.peta.org/issues/animals-in-entertainment/zoos/.

Perez, R., C. L. Richards-Zawacki, A. R. Krohn, M. Robak, E. Griffith, et al. 2014. "Field Surveys in Western Panama Indicate Populations of *Atelopus varius* Frogs Are Persisting in Reigons Where *Batrachochytrium dendrobatidis* Is Now Endemic." *Amphibian and Reptile Conservation* 8:30–35.

Perrault, Claude, ed. 1671-76. *Mémoires pour servir à l'histoire naturelle des animaux*. 2 vols. Paris: Imprimerie Royale.

Perrier, E. 1911. "Société d'acclimatation: Distribution des recompenses." Typescript, February 12. MS. 2227, Natural History Museum Archive, Paris.

Persons, T. B., M. J. Feldner, and R. A. Repp. 2016. "Black-Tailed Rattlesnake: *Crotalus molossus* (Baird and Girard 1853)." In *The Rattlesnakes of Arizona*, edited by G. W. Schuett, M. J. Feldner, C. F. Smith, and R. S. Reiserer, 1:439–93. Rodeo, NM: Eco Press.

Pessier, A. P., and J. R. Mendelson III, eds. 2010. *A Manual for Control of Infectious Diseases in Amphibian Survival Assurance Colonies and Reintroduction Programs*. Apple Valley, MN: IUCN/SSC Conservation Breeding Specialist Group.

Petchesky, R. P. 1987. "Fetal Images: The Power of Visual Culture in the Politics of Reproduction." *Feminist Studies* 13 (2): 263–92.

Pew Oceans Commission. 2003. *America's Living Oceans: Charting a Course for Sea Change. A Report to the Nation*. Arlington, VA: Pew Oceans Commission.

Phoenix Zoo Conservation and Science Department. 2015. *Narrow-Headed Gartersnake (Thamnophis rufipunctatus) Husbandry Manual*. Phoenix: Arizona Center for Nature Conservation.

Pimm, S. 2013. "Opinion: The Case against Species Revival." *National Geographic News*, March 12. http://news.nationalgeographic.com/news/2013/03/130312—deextinction-conservation-animals-science-extinction-biodiversity-habitat-environment/.

Pimm, S., C. Jenkins, R. Abell, T. Brooks, J. Gittleman, L. Joppa, P. Raven, C. Roberts, and J. Sexton. 2014. "The Biodiversity of Species and Their Rates of Extinction, Distribution, and Protection." *Science* 344 (6187): 1246752. doi:10.1126/science.1246752.

Player, I. 1973. *The White Rhino Saga*. New York: Stein and Day.

Plous, S. 1993. "Psychological Mechanisms in the Human Use of Animals." *Journal of Social Issues* 49:11–52.

Poliakof, M. 1881. "Supposed New Species of Horse from Central Asia." Translated by E. D. Morgan. *Annals and Magazine of Natural History* 8:16–26.

Poole, V. A., and S. Grow, eds. 2012. *Amphibian Husbandry Resource Guide*, ed. 2.0. Silver Spring, MD: Association of Zoos and Aquariums.

Possehl, S. 1994. "Rare Przewalski's Horse Returns to the Harsh Mongolian Steppe." *New York Times*, October 4, C4.

Powell, D. M., and E. V. W. Bullock. 2014. "Evaluation of Factors Affecting Emotional Responses in Zoo Visitors and the Impact of Emotion on Conservation Mindedness." *Anthrozoös* 27 (3): 389–405.

Powell, M. 2016. *Vanishing America: Species Extinction and Racial Peril in U.S. Conservation's Troubled First Century*. Cambridge, MA: Harvard University Press.

Prado-Martinez, J., P. H. Sudmant, J. M. Kidd, H. Li, J. L. Kelley, et al. 2013. "Great Ape Genetic Diversity and Population History." *Nature* 499 (7459): 471–75. doi:10.1038/nature12228.

President's Council on Bioethics. 2002. *Human Cloning and Human Dignity: An Ethical Inquiry*. Washington, DC: President's Council on Bioethics.

Price, Jennifer. 1996. *Flight Maps: Adventures with Nature in Modern America*. New York: Perseus Books.

Project Golden Frog. 2016. Accessed June 30, 2016. http://www.projectgoldenfrog.org/.

Pyle, R. M. 2003. "Nature Matrix: Reconnecting People and Nature." *Oryx* 37 (2): 206–14.

Rabb, G. B. 1990. "Declining Amphibian Populations." *Species* 13–14:33–34.

———. 1994. "The Changing Roles of Zoological Parks in Conserving Biological Diversity." *American Zoologist* 34 (1): 159–64.

Rabb, G. B., and C. D. Saunders. 2005. "The Future of Zoos and Aquariums: Conservation and Caring." *International Zoo Yearbook* 39:1–26.

Rader, K. A., and V. E. M. Cain. 2008. "From Natural History to Science: Display and the Transformation of American Museums of Science and Nature." *Museum and Society* 6:152–71.

Rahbek, C. 1993. "Captive Breeding—a Useful Tool in the Preservation of Biodiversity?" *Biodiversity and Conservation* 2:426–37.

Ralls, K., and J. D. Ballou. 2004. "Genetic Status and Management of California Condors." *Condor* 106 (2): 215–28.

Ralls, K., J. D. Ballou, B. A. Rideout, and R. Frankham. 2000. "Genetic Management of Chondrodystrophy in California Condors." *Animal Conservation* 3:145–53.

Ramaswamy, K., W. Y. Yik, X. M. Wang, E. N. Oliphant, W. Lu, et al. 2015. "Derivation of Induced Pluripotent Stem Cells from Orangutan Skin Fibroblasts." *BMC Research Notes* 8 (1): 577. doi: 10.1186/s13104-015-1567-0.

Ramirez, M. 2016 "Elephants Adjusting to New Life at Dallas Zoo." *Dallas Morning News*, March 13.

Reade, L. S., and N. K. Waran. 1996. "The Modern Zoo: How Do People Perceive Zoo Animals?" *Applied Animal Behaviour Science* 47 (1): 109–18.

A Record of the Collection of Foreign Animals Kept by the Duke of Bedford in Woburn Park, 1892 to July, 1905. 1905. N.p.: n.p.

Redford, K. H., G. Amato, J. Baillie, P. Beldomenico, E. L. Bennett, et al. 2011. "What Does It Mean to Successfully Conserve a (Vertebrate) Species?" *BioScience* 61 (1): 39–48.

Redford, K. H., D. B. Jensen, and J. J. Breheny. 2012. "Integrating the Captive and the Wild." *Science* 338:1157–58.

Reed, T. H. 1989. Oral History Interviews, April 14, 1989, 10–16, August 3, 1989, 103–5, and October 13, 1989, 31, 36–48, 56–57, Smithsonian Institution Archives, Record Unit 9568.

Regan, T. 1983. *The Case for Animal Rights*. Berkeley: University of California Press.

———. 1995. "Are Zoos Morally Defensible?" In *Ethics on the Ark: Zoos, Animal Welfare, and Wild-*

life Conservation, edited by B. G. Norton, M. Hutchins, E. F. Stevens, and T. L. Maple, 38–51. Washington, DC: Smithsonian Institution Press.

Reid, G. McG., A. A. Macdonald, A. L. Fidgett, B. Hiddinga, and K. Leus. 2008. *Developing the Research Potential of Zoos and Aquaria: The EAZA Research Strategy*. Amsterdam: European Association of Zoos and Aquaria, EAZA Executive Office.

Reid, G. McG., and G. Moore, eds. 2014. *History of Zoos and Aquariums: From Royal Gifts to Biodiversity Conservation*. Chester: North of England Zoological Society.

Reid, G. McG., and K. Zippel. 2008. "Can Zoos and Aquariums Ensure the Survival of Amphibians in the 21st Century?" *International Zoo Yearbook* 42:1–6.

Reiger, J. F. 2000. *American Sportsmen and the Origins of Conservation*, 3rd ed. Corvallis: Oregon State University Press.

Reiss, C., L. Olsson, and U. Bossfeld. 2015. "The History of the Oldest Self-Sustaining Laboratory Animal: 150 Years of Axolotl Research." *Journal of Experimental Zoology Part B: Molecular and Developmental Evolution* 324 (5): 393–404.

Reiss, M. J., and S. D. Tunnicliffe. 2011. "Dioramas as Depictions of Reality and Opportunities for Learning." *Curator* 54:47–59.

Renshaw, G. 1904. *Natural History Essays*. London: Sherratt and Hughes.

"Revive and Restore." N.d. Accessed June 6, 2016. http://longnow.org/revive/.

Reynolds Polymer Technologies. 2014. *Acrylic in Modern Aquarium Construction: Aquarium Window Design Types and Examples*. Grand Junction, CO: Reynolds Polymer Technologies. Available online at http://www.reynoldspolymer.com/userfiles/files/Acrylic%20in%20Modern%20Aquarium%20Exhibits.pdf.

Rich, Nathaniel. 2014. "The Mammoth Cometh." *New York Times*, February 27.

Rick, T. C., T. S. Sillett, C. K. Ghalambor, C. A. Hoffman, K. Ralls, et al. 2014. "Ecological Change on California's Channel Islands from the Pleistocene to the Anthropocene." *BioScience* 64:680–92.

Rideout, B. A., I. Stalis, R. Papendick, A. Pessier, B. Puschner, et al. 2012. "Patterns of Mortality in Free-Ranging California Condors (*Gymnogyps californianus*)." *Journal of Wildlife Diseases* 48 (1): 95–112.

Ritvo, H. 1987. *The Animal Estate: The English and Other Creatures in the Victorian Age*. Cambridge, MA: Harvard University Press.

——. 1997. The *Platypus and the Mermaid, and Other Figments of the Classifying Imagination*. Cambridge, MA: Harvard University Press.

——. 2012. "President's Lecture: Going Forth and Multiplying: Animal Acclimatization and Invasion." *Environmental History* 17 (2): 404–14.

Robbins, J. 2007. "Strands of Undesirable DNA Roam with Buffalo." *New York Times*, January 9. http://www.nytimes.com/2007/01/09/science/09bison.html.

Robbins, L. E. 2002. *Elephant Slaves and Pampered Parrots: Exotic Animals in Eighteenth-Century Paris*. Baltimore: Johns Hopkins University Press.

Robbins, M. M., M. Gray, K. A. Fawcett, F. B. Nutter, P. Uwingeli, et al. 2011. "Extreme Conservation Leads to Recovery of the Virunga Mountain Gorillas." *PLoS ONE* 6 (6): e19788. doi:10.1371/journal.pone.0019788.

Roberts, D. 1997. *Killing the Black Body: Race, Reproduction and the Meaning of Liberty*. New York: Pantheon Books.

Rodrigues, A. S. L., J. D. Pilgrim, J. F. Lamoreux, M. Hoffmann, and T. M. Brooks. 2006. "The Value of the IUCN Red List for Conservation." *Trends in Ecology and Evolution* 21 (2): 71–76.

Rodríguez, J. P., J. A. Simonetti, A. Premoli, and M. A. Marini. 2005. "Conservation in Austral and Neotropical America: Building Scientific Capacity Equal to the Challenges." *Conservation Biology* 19:969–72.

Roe, K., A. McConney, and C. F. Mansfield. 2014. "The Role of Zoos in Modern Society—a Comparison of Zoos' Reported Priorities and What Visitors Believe They Should Be." *Anthrozoos* 27 (4): 529–41.

Rogers, P. 2011. "Monterey Bay Aquarium Shows Off New Great White Shark." *Santa Cruz Sentinel*, September 2.

Rohde, K. 2005. *Introduction to Nonequilibrium Ecology*. New York: Cambridge University Press.

Roheim, C. A. 2009. "An Evaluation of Sustainable Seafood Guides: Implications for Environmental Groups and the Seafood Industry." *Marine Resource Economics* 24 (3): 301–10.

Romanov, M. N., E. M. Tuttle, M. L. Houck, W. S. Modi, L. G. Chemnick, et al. 2009. "The Value of Avian Genomics to the Conservation of Wildlife." *BMC Genomics* 10 (Suppl. 2): S10.

Rookmaaker, L. C. 1998. "The White Rhinoceros (*Ceratotherium simum*)." In *The Rhinoceros in Captivity: a List of 2439 Rhinoceroses Kept from Roman Times to 1994*, edited by L. C. Rookmaaker, 270–74. Amsterdam: SPB Academic Press.

———. 2000. "The Alleged Population Reduction of the Southern White Rhinoceros (*Ceratotherium simum simum*) and the Successful Recovery." *Saugetierkundliche Mitteilungen* 45 (2): 55–70.

Rose, N. A., R. Farinato, and S. Sherwin. 2006. *The Case against Marine Mammals in Captivity*, 3rd ed. Washington, DC: Humane Society of the United States and World Society for the Protection of Animals.

Rosenblum, E. B., T. Y. James, K. R. Zamudio, T. J. Poorten, D. Ilut, et al. 2013. "Complex History of the Amphibian-Killing Chytrid Fungus Revealed with Genome Resequencing Data." *Proceedings of the National Academy of Sciences* 110:9385–90.

Rothfels, N. 2002. *Savages and Beasts: The Birth of the Modern Zoo*. Baltimore: Johns Hopkins University Press.

Rothstein, E. 2006. "A Hundred Thousand Fish behind a Pane Two Feet Thick." *New York Times*, March 23.

Routman, E., J. Ogden, and K. Winsten. 2010. "Visitors, Conservation Learning, and the Design of Zoo and Aquarium Experiences." In *Wild Mammals in Captivity*, edited by D. G. Kleiman, K. V. Thompson, and C. K. Baer, 137–50. Chicago: University of Chicago Press.

Rowell, T. 1972. *Social Behaviour of Monkeys*. Middlesex, UK: Penguin Books.

Ruane, M. E., and T. Patel. 2013. "Missing Red Panda Found in Adams Morgan." *Washington Post*, June 24.

Runte, A. 1990a. "Joseph Grinnell and Yosemite: Rediscovering the Legacy of a California Conservationist." *California History* 69:170–81.

———. 1990b. *Yosemite: The Embattled Wilderness*. Lincoln: University of Nebraska Press.

Russow, L. M. 1981. "Why Do Species Matter?" *Environmental Ethics* 3 (2): 101–11.

Ryan, J. 2011. *The Forgotten Aquariums of Boston*. 3rd ed. rev. Boston: New England Aquarium Corporation.

Ryder, O. A. 1993. "Przewalski's Horse: Prospects for Reintroduction into the Wild." *Conservation Biology* 7 (1): 13–15.

———. 2005. "Conservation Genomics: Applying Whole Genome Studies to Species Conservation Efforts." *Cytogenetic and Genome Research* 108:6–15.

———. 2015. "Genetic Rescue and Biodiversity Banking." Retrieved from http://tedxtalks.ted.com /video/Genetic-rescue-and-biodiversity;search%3Aoliver%20ryder.

Ryder, O. A., and A. T. C. Feistner. 1995. "Research in Zoos: A Growth Area in Conservation." *Biodiversity and Conservation* 4 (6): 671–77.

Ryder, O. A., A. McLaren, S. Brenner, Y. P. Zhang, and K. Benirschke. 2000. "DNA Banks for Endangered Animal Species." *Science* 288 (5464): 275–77.

Sachs, J. D. 2012. "From Millennium Development Goals to Sustainable Development Goals." *Lancet* 379 (9832): 2206–11. doi:10.1016/S0140-6736(12)60685-0.

Sadovy, Y., and A. C. Vincent. 2002. "Ecological Issues and the Trades in Live Reef Fishes." In

Coral Reef Fishes: Dynamics and Diversity in a Complex Ecosystem, edited by P. Sales, 391–420. Amsterdam: Academic Press.

Sagoff, M. 1984. "Animal Liberation and Environmental Ethics: Bad Marriage, Quick Divorce." *Osgood Hall Law Journal* 22:306–22.

Sagra, R. de la. 1854. "Sur un projet d'acclimatation des chèvres dite d'angora," *Bulletin de la Société Zoologique d'Acclimatation* 1:23–30.

Salafsky, N., N. Margoluis, K. H. Redford, and J. G. Robinson. 2002. "Improving the Practice of Conservation: A Conceptual Framework and Research Agenda for Conservation Science." *Conservation Biology* 16:1469–79.

Salafsky, N., D. Salzer, A. J. Stattersfield, C. Hilton-Taylor, R. Neugarten, et al. 2008. "A Standard Lexicon for Biodiversity Conservation: Unified Classifications of Threats and Actions." *Conservation Biology* 22 (4): 897–911.

Sandler, R. 2013a. "Climate Change and Ecosystem Management." *Ethics, Policy and Environment* 16 (1): 1–15.

——. 2013b. "The Ethics of Reviving Long Extinct Species." *Conservation Biology* 28 (2): 354–60.

Saragusty, J., S. Diecke, M. Drukker, B. Durrant, I. Friedrich Ben-Nun, et al. 2016. "Rewinding the Process of Mammalian Extinction." *Zoo Biol*ogy, advance online publication. doi:10.1002/zoo.21284.

Save the Rhino. 2016. "Poaching Statistics." Accessed June 29, 2016. https://www.savetherhino.org/rhino_info/poaching_statistics.

Savitz, A. W. 2013. *The Triple Bottom Line: How Today's Best-Run Companies Are Achieving Economic, Social and Environmental Success—and How You Can Too*. New York: Jossey-Bass.

Schad, K., ed. 2008. "Amphibian Population Management Guidelines." Presented at Amphibian Ark Amphibian Population Management Workshop, San Diego, CA, December 10–11, 2007.

Scheele, B. C., D. A. Hunter, L. F. Grogan, L. Berger, J. E. Kolby, et al. 2014. "Interventions for Reducing Extinction Risk in Chytridiomycosis-Threatened Amphibians." *Conservation Biology* 28:1195–1205.

Scherren, H. [1826] 1905. *The Zoological Society of London: A Sketch of Its Foundation and Development, and the Story of Its Farm, Museum, Gardens, Menagerie and Library*. London: Cassell.

Schloegel, L. M., L. F. Toledo, J. E. Longcore, S. E. Greenspan, C. A. Vieira, et al. 2012. "Novel, Panzootic and Hybrid Genotypes of Amphibian Chytridiomycosis Associated with the Bullfrog Trade." *Molecular Ecology* 21:5162–77.

Schorger, A. W. 1955. *The Passenger Pigeon: Its Natural History and Extinction*. Norman: University of Oklahoma Press.

Schrenzel, M. D., T. A. Tucker, I. H. Stalis, R. A. Kagan, R. P. Burns, et al. 2011. "Pandemic (H1N1) 2009 Virus in 3 Wildlife Species, San Diego, California, USA." *Emerging Infectious Diseases* 17:747–49.

Schuett, G. W., R. W. Clark, R. A. Repp, M. Amarello, and H. W. Greene. 2017. "Social Behavior of Rattlesnakes: A Shifting Paradigm." In *The Rattlesnakes of Arizona*, edited by G. W. Schuett, M. J. Feldner, C. F. Smith, and R. S. Reiserer, 2:161–242. Rodeo, NM: Eco Press.

Schulte-Hostedde, A. I., and G. F. Mastromonaco. 2015. "Integrating Evolution in the Management of Captive Zoo Populations." *Evolutionary Applications* 8:413–22.

Schultz, C. 2014. "Is It Wise to Build a Zoo without Cages?" *Smithsonian*, August 13. http://www.smithsonianmag.com/smart-news/it-wise-build-zoo-without-cages-180952345/?no-ist.

Schultz, P. W., and F. Kaiser. 2012. "Promoting Pro-Environmental Behavior." In *Handbook of Environmental and Conservation Psychology*, edited by S. Clayton, 556–80. New York: Oxford University Press.

Schwan, S., A. Grajal, and D. Lewalter. 2014. "Understanding and Engagement in Places of Science Experience: Science Museums, Science Centers, Zoos, and Aquariums." *Educational Psychologist* 49:70–85.

Schwarzenberger, F., C. Walzer, K. Tomasova, J. Zima, F. Goritz, et al. 1999. "Can the Problems Associated with the Low Reproductive Rate in Captive White Rhinoceroses (*Ceratotherium simum*) Be Solved within the Next Five Years?" *Verhandlungsbericht über die Erkrankungen der Zoo- und Wildtiere* 39:283–89.

Schwitzer, C., R. A. Mittermeier, A. B. Rylands, F. Chiozza, E. A. Williamson, J. Wallis, and A. Cotton, eds. 2015. *Primates in Peril: The World's 25 Most Endangered Primates, 2014–2016*. Arlington, VA: IUCN/SSC Primate Specialist Group (PSG), International Primatological Society (IPS), Conservation International (CI), and Bristol Zoological Society.

Scott, J. M., D. D. Goble, J. A. Wiens, D. S. Wilcove, M. Bean, and T. Male. 2005. "Recovery of Imperiled Species under the Endangered Species Act: The Need for a New Approach." *Frontiers in Ecology and the Environment* 3:383–87.

"Scottish Wildcat Association." 2016. Accessed June 26, 2016. http://www.scottishwildcats.co.uk/.

Seafood Watch. 2013. *Press Kit*. Monterey, CA: Monterey Bay Aquarium Seafood Watch. http://www.seafoodwatch.org/-/m/sfw/pdf/press/mba-seafoodwatch-press-kit.pdf.

——. 2016. "About Us." Accessed April 4, 2016. http://www.seafoodwatch.org/about-us.

Sellars, R. W. 2009. *Preserving Nature in the National Parks: A History*. New Haven, CT: Yale University Press.

Seoul Grand Park. 1996. *한국동물원80년사 서울대공원: 전국동물원/수족관* [Korean zoo 1980s history of Seoul Grand Park: Zoo/aquarium for the whole country]. Seoul, South Korea: Seoul Metropolitan Government.

——. 2001. "Seoul Grand Park Zoo: Conservation Masterplan Concept Development." Paper presented at Seoul Grand Park, Gwacheon, South Korea, April.

——. 2002. "*서울대공원 생태동물원 조성 기본계획*" [Basic construction plan of Seoul Grand Park EcoZoo]. Paper presented at Seoul Grand Park, Gwacheon, South Korea, October.

——. 2009. *Seoul Zoo 한동산* [Handongsan]. Vol. 25. Seoul: Seoul Grand Park.

——. 2012. *Seoul Zoo 한동산* [Handongsan]. Vol. 28. Seoul: Seoul Grand Park.

——. 2013a. "서울대공" [Seoul Grand Park]. Accessed June 8, 2016. http://grandpark.seoul.go.kr/.

——. 2013b. *Seoul Zoo 한동산 (Handongsan)*. Vol. 29. Seoul: Seoul Grand Park.

Serpell, J. 2004. "Factors Influencing Human Attitudes to Animals and Their Welfare." *Animal Welfare* 13:S145–51.

Shafer, S. 2015. "Newport Aquarium Adding Shark Bridge." *Courier Journal*, January 15.

Shell, H. R. 2002. "Introduction: Finding the Soul in the Skin." In *The Extermination of the American Bison*, edited by W. T. Hornaday, viii–xxiii. Washington, DC: Smithsonian Institution Press.

Shepherd, A. 1965. *The Flight of the Unicorns: An Expedition to Preserve the Arabian Oryx*. New York: Abelard-Schuman.

Sherkow, J. S., and H. T. Greely. 2013. "What If Extinction Is Not Forever?" *Science* 5 (340): 32–33.

Shier, D. M., and R. R. Swaisgood. 2012. "Fitness Costs of Neighborhood Disruption in Translocations of a Solitary Mammal." *Conservation Biology* 26:116–23. doi:10.1111/j.1523-1739.2011.01748.x.

Shufeldt, R. W. 1915. "Anatomical and Other Notes on the Passenger Pigeon (*Ectopistes migratorius*) Lately Living in the Cincinnati Zoological Gardens." *Auk* 32 (1): 28–41.

Shumaker, R. W., A. M. Palkovich, B. B. Beck, G. A. Guagnano, and H. Morowitz. 2001. "Spontaneous Use of Magnitude Discrimination and Ordination by the Orangutan (*Pongo pygmaeus*)." *Journal of Comparative Psychology* 115 (4): 385.

Siebert, C. 2014. "The Dark Side of Zootopia." *New York Times Magazine*, November 18. Available online at http://www.nytimes.com/2014/11/23/magazine/the-dark-side-of-zootopia.html?_r=0.

Sigler, L. 2015. "The Crocodile Museum at Zoologico Regional Miguel Alvarez del Toro (Zoomat), Mexico: 'How to Exhibit a Bullfrog' Put Into Reality." *International Zoo Yearbook* 49:162–71.

Silvertown, J. 2009. "A New Dawn for Citizen Science." *Trends in Ecology and Evolution* 24:467–71.

Singer, P. 1975. *Animal Liberation: A New Ethics for Our Treatment of Animals*. New York: Random House.

Skutch, A. F. 1998. "Biocompatability: A Criterion for Conservation." *Revista de Biologia Tropical* 46:481–86.

Smith, F. R. 1994. "A New Wave of Aquariums Brings the Ocean Ashore." *Smithsonian* 25:50–60.

Smith, L. D. G., J. Curtis, J. Mair, and P. A. van Dijk. 2012. "Requests for Zoo Visitors to Undertake Pro-Wildlife Behaviour: How Many Is Too Many?" *Tourism Management* 33 (6): 1502–10. doi:10.1016/j.tourman.2012.02.004.

Smith, R. J., D. Veríssimo, N. J. B. Isaac, and K. E. Jones. 2012. "Identifying Cinderella Species: Uncovering Mammals with Conservation Flagship Appeal." *Conservation Letters* 5:205–12.

Smith, R. K., and W. J. Sutherland. 2014. *Amphibian Conservation: Global Evidence for the Effects of Interventions*. Exeter, UK: Pelagic.

Smith, T. D. 1994. *Scaling Fisheries: The Science of Measuring the Effects of Fishing, 1855–1955*. New York: Cambridge University Press.

Snyder, N. 2004. *The Carolina Parakeet: Glimpses of a Vanished Bird*. Princeton, NJ: Princeton University Press.

Snyder, N., S. Derrickson, S. Beissinger, J. Wiley, T. Smith, et al. 1996. "Limitations of Captive Breeding in Endangered Species Recovery." *Conservation Biology* 10 (2): 338–48.

Snyder, N., and H. Snyder. 2000. *The California Condor: A Saga of Natural History and Conservation*. Princeton, NJ: Princeton University Press.

Sommer, R. 1974. *Tight Spaces*. Englewood Cliffs, NJ, Prentice-Hall.

Sorensen, L. 2015. "The Best Performing CEOs in the World." *Harvard Business Review*, November, 60–63.

Soulé, M., M. Gilpin, W. Conway, and T. Foose. 1986. "How Long a Voyage, How Many Staterooms, How Many Passsengers?" *Zoo Biology* 5:101–13.

Spary, E. C. 2000. *Utopia's Garden: French Natural History from Old Regime to Revolution*. Chicago: University of Chicago Press.

Stack, L. 2015. "Endangered White Rhino Nola Dies at San Diego Zoo Safari Park." *New York Times*, November 22.

Stanley Price, M. R. 1989. *Animal Re-introductions: The Arabian Oryx in Oman*. Cambridge: Cambridge University Press.

Staples, W., and P. Cafaro. 2012. "For a Species Right to Exist." In *Life on the Brink*, edited by P. Cafaro and C. Christ, 283–300. Athens: University of Georgia Press.

Steg, L., and J. De Groot. 2012. "Environmental Values." In *The Oxford Handbook of Environmental and Conservation Psychology*, edited by S. Clayton, 81–92. Oxford: Oxford University Press.

Steiner, C. C., A. S. Putnam, P. E. Hoeck, and O. A. Ryder. 2013. "Conservation Genomics of Threatened Animal Species." *Annual Review of Animal Biosciences* 1 (1): 261–81.

Stevens, J. 2009. "What's Killing California Sea Otters?" Accessed March 30, 2016. http://www.seaotterresearch.org.

Stoinski, T. S., and B. B. Beck. 2001. "Spontaneous Tool Use in Captive, Free-Ranging Golden Lion Tamarins (*Leontopithecus rosalia rosalia*)." *Primates* 42 (4): 319–26.

Stoinski, T. S., M. P. Hoff, and T. L. Maple. 2001. "Habitat Use and Structural Preferences of Captive Lowland Gorillas: The Effect of Environmental and Social Variables." *International Journal of Primatology* 22:431–47.

Stoinski, T. S., K. E. Lukas, and M. Hutchins. 2008. "Zoos and Conservation: Moving beyond a Piecemeal Approach." In *Conservation in the 21st Century: Gorillas as a Case Study*, edited by T. S. Stoinski, H. D. Steklis, and P. T. Mehlman, 315–26. New York: Kluwer Academic.

Stott, R. J. 1981. "Historical Origins of the Zoological Park in American Thought." *Environmental History Review* 5 (2): 52–65.

Stuart, S. N., J. S. Chanson, N. A. Cox, B. E. Young, A. S. L. Rodrigues, et al. 2004. "Status and Trends of Amphibian Declines and Extinctions Worldwide." *Science* 306:1783–86.

Sundstrom, E., P. A. Bell, P. L. Busby, and C. Asmus. 1996. "Environmental Psychology, 1989–1994." *Annual Review of Psychology* 47 (1): 485-512.

Swaisgood, R. R., D. M. Dickman, and A. M. White. 2006. "A Captive Population in Crisis: Testing Hypotheses for Reproductive Failure in Captive-Born Southern White Rhinoceros Females." *Biological Conservation* 129:468–76.

Swanagan, J. 1992. "An Assessment of Factors Influencing Zoo Visitors' Conservation Attitudes and Behavior." Master's thesis, Georgia Institute of Technology.

Takacs, D. 1996. *The Idea of Biodiversity: Philosophies of Paradise*. Baltimore: Johns Hopkins University Press.

Taylor L. 1995. "The Status of the North American Public Aquariums at the End of the Century." *International Zoo Yearbook* 34:14–25.

Thomas, C. D., et al. 2004. "Extinction Risk from Climate Change." *Nature* 427:145-48.

Thomas, E. F., C. McGarty, and K. I. Mavor. 2009. "Aligning Identities, Emotions, and Beliefs to Create Commitment to Sustainable Social and Political Action." *Personality and Social Psychology Review* 13 (3): 194–218.

Thomas, K. 1983. *Man and the Natural World: A History of the Modern Sensibility*. New York: Pantheon Books.

Thometz, N. M., M. T. Tinker, M. M. Staedler, K. A. Mayer, and T. M. Williams. 2014. "Energetic Demands of Immature Sea Otters from Birth to Weaning: Implications for Maternal Costs, Reproductive Behavior and Population-Level Trends." *Journal of Experimental Biology* 217 (12): 2053–61.

Thompson [Cussins], C. 1998. "Producing Reproduction: Techniques of Normalization and Naturalization in Infertility Clinics." In *Reproducing Reproduction: Kinship, Power, and Technological Innovation*, edited by S. Franklin and H. Ragone, 66–101. Philadelphia: University of Pennsylvania Press.

———. 2005. *Making Parents: The Ontological Choreography of Reproductive Technologies*. Cambridge, MA: MIT Press.

Tlusty, M. F., A. L. Rhyne, L. Kaufman, M. Hutchins, G. M. Reid, C. Andrews, P. Boyle, J. Hemdal, F. McGilvray, and S. Dowd. 2013. "Opportunities for Public Aquariums to Increase the Sustainability of the Aquatic Animal Trade." *Zoo Biology* 32:1–12.

Tober, J. A. 1981. *Who Owns Wildlife? The Political Economy of Conservation in Nineteenth-Century America*. Westport, CT: Greenwood Press.

Townsend, C. H. 1893. *Condition of the Seal Life on the Rookeries of the Pribilof Islands, 1893–1895*. Washington, DC: Government Printing Office.

———. 1919. *Guide to the New York Aquarium*. New York: New York Zoological Society.

Traylor-Holzer, K., K. Leus, and P. McGowan. 2013. "Integrating Assessment of *Ex Situ* Management Options into Species Conservation Planning." *WAZA Magazine* 14:6–9.

Tribe, A., and R. Booth. 2003. "Assessing the Role of Zoos in Wildlife Conservation." *Human Dimensions of Wildlife* 8 (1): 65–74.

Tubbs, C. W., B. S. Durrant, and M. R. Milnes. 2017. "Reconsidering the Use of Soy and Alfalfa in Southern White Rhinoceros Diets." *Pachyderm* 461, in press.

Tubbs, C., C. E. McDonough, R. Felton, and M. R. Milnes. 2014. "Advances in Conservation Endocrinology: The Application of Molecular Approaches to the Conservation of Endangered Species." *General and Comparative Endocrinology* 203:29–34.

Tubbs, C., L. A. Moley, J. A. Ivy, L. C. Metrione, S. LaClaire, et al. 2016. "Estrogenicity of Captive Southern White Rhinoceros Diets and Their Association with Fertility." *General and Comparative Endocrinology* 238:32-38.Tubbs, C., P. Hartig, M. Cardon, N. Varga, and M. Milnes. 2012. "Activation of Southern White Rhinoceros (*Ceratotherium simum simum*) Estrogen Receptors by Phytoestrogens: Potential Role in the Reproductive Failure of Captive-Born Females?" *Endocrinology* 153 (3): 1444–52.

Tudge, C. 1992. *Last Animals at the Zoo: How Mass Extinction Can Be Stopped*. Washington, DC: Island Press.

Tullis, P. 2014. "When You Walk into a Zoo, Are You Helping Animals or Hurting Them?" *Take Part*, May 2, 2014. Retrieved from http://www.takepart.com/feature/2014/05/02/do-zoos-matter.

Turkowski, F. J., and G. C. Mohney. 1971. "History, Management and Behavior of the Phoenix Zoo Arabian Oryx Herd, 1964–1971." *Arizona Zoological Society, Special Bulletin* 2:1–36.

Turner, G. G., D. M. Reeder, and J. T. H. Coleman. 2011. "A Five-Year Assessment of Mortality and Geographic Spread of White-Nose Syndrome in North American Bats, with a Look at the Future. Update of White-Nose Syndrome in bats." *Bat Research News* 52:13–27.

Turner, J. 1980. *Reckoning with the Beast: Animals, Pain, and Humanity in the Victorian Mind*. Baltimore: Johns Hopkins University Press.

Turner, S. S. 2008. Open-Ended Stories: Extinction Narratives in Genome Time. *Literature and Medicine* 26 (1): 55–82.

United Nations (UN). 1987. Report of the World Commission on Environment and Development, *Our Common Future*, A/42/427 (October). http://undocs.org/A/42/427.

——. 1992. Conference on Environment and Development, Agenda 21, A/CONF.151/26, June 3–14. http://www.un.org/documents/ga/conf151/aconf15126-1annex1.htm.

——. 1998. "Kyoto Protocol to the United Nations Framework." doi:10.1111/1467-9388.00150.

——. 2000. General Assembly Resolution 55/2, United Nations Millennium Declaration, A/RES/55/2 (September 18). http://undocs.org/A/RES/55/2.

———. 2002. World Summit on Sustainable Development, Johannesburg Declaration on Sustainable Development, A/CONF.199/20 (September 4). http://www.un- documents.net/jburgdec.htm.

——. 2012. "The Future We Want-Outcome Document of the United Nations Conference on Sustainable Development," 41. doi:10.1126/science.202.4366.409.

——. 2014. General Assembly Resolution 69/315, Report of the Intergovernmental Committee of Experts on Sustainable Development Financing, A/69/315 (August 15). undocs.org/A/69/314.

——. 2015. General Assembly Resolution 70/1, "Transforming Our World: The 2030 Agenda for Sustainable Development," A/RES/70/1 (September 25). http://undocs.org/A/RES/70/1.

———. 2016. *The First Global Integrated Marine Assessment: World Ocean Assessment I*. New York: United Nations. http://www.un.org/Depts/los/global_reporting/WOA_RPROC/WOA Compilation.pdf.

United Nations Development Programme (UNDP). 2015. *Goal 14: Life below Water*. New York: United Nations. http://www.undp.org/content/undp/en/home/sdgoverview/post-2015-development-agenda/goal-14.html.

Untitled. 1897. *Weekly Oklahoma State Capital* 8 (36): 2nd ed., 1.

US Commission on Ocean Policy. 2004. *An Ocean Blueprint for the 21st Century*. Washington, DC: US Commission on Ocean Policy.

US Department of Interior. 2008. "Endangered and Threatened Wildlife and Plants; Determination of Threatened Status for the Polar Bear (*Ursus maritimus*) throughout Its Range; Final Rule." *Federal Register* 73, no. 95 (May 15, 2008): 28212. https://www.gpo.gov/fdsys/pkg/FR-2008-05-15/pdf/E8-11105.pdf#page=2.

US Department of State (USDS). 2015. "Our Ocean." http://www.state.gov/e/oes/ocns/opa/ourocean/index.htm.

US Fish and Wildlife Service (USFWS). 2014. "Time Line of the American Bison." Accessed October 19, 2015. http://www.fws.gov/bisonrange/timeline.htm.

——. 2016. "News Release: Service Proposes Delisting Three Fox Subspecies on Northern Channel Islands Due to Recovery, Highlighting Historic Endangered Species Act Success." Accessed June 6, 2016. http://www.fws.gov/ventura/newsroom/release.cfm?item=357.

Urban, M. C. 2015. "Accelerating Extinction Risk from Climate Change." *Science* 348:571-73.

Van Dooren, T. 2014. *Flight Ways: Life and Loss at the Edge of Extinction*. New York: Columbia University Press.

Van Dooren, T., and D. Rose. 2013. "Keeping Faith with Death: Mourning and De-extinction." Accessed May 12, 2015. http://extinctionstudies.org/2013/11/10/keeping-faith-with-death-mourning-and-de-extinction/.

Vehrs, K. L. 2015. Letter to US Fish and Wildlife Service, on Behalf of the Association of Zoos and Aquariums. Docket No. FWS-HQ-FAC-2015-0005. https://federalregister.gov/a/2016-00452.

Veríssimo, D., T. Pongiluppi, M. C. M. Santos, P. F. Develey, I. Fraser, et al. 2013. "Using a Systematic Approach to Select Flagship Species for Bird Conservation." *Conservation Biology* 28:269-77.

Vincent, A., and H. Koldewey. 2007. *Project Seahorse at 10*. Chicago: Project Seahorse.

Vining, J. 2003. "The Connection to Other Animals and Caring for Nature." *Human Ecology Review* 10:87-99.

Volf, J. 1994. "The Studbook." In *Przewalski's Horse: The History and Biology of an Endangered Species*, edited by L. Boyd and K. Houpt, 61-74. Albany: State University of New York Press.

Voyles, J., A. M. Kilpatrick, J. P. Collins, M. C. Fisher, W. F. Frick, et al. 2014. "Moving beyond Too Little, Too Late: Managing Emerging Infectious Diseases in Wild Populations Requires International Policy and Partnerships." *Ecohealth* 12 (3): 404-7. doi:10.1007/s10393-014-0980-5.

Wadnitz, C., M. Taylor, E. Green, and T. Razak. 2003. *From Ocean to Aquarium: The Global Trade in Ornamental Fish Species*. Cambridge, UK: UNEP-WCMC.

Wainwright, O. 2014. "Denmark's Cage-Free Zoo Will Put Humans in Captivity." *Guardian*, August 5. Available online at http://www.theguardian.com/artanddesign/architecture-design-blog/2014/aug/05/denmark-cage-free-zoo-will-put-humans-in-captivity.

Wake, D. B., and V. T. Vredenburg. 2008. "Are We in the Midst of the Sixth Mass Extinction? A View from the World of Amphibians." *Proceedings of the National Academy of Sciences* 105:11466-73.

Wakefield, S., J. Knowles, W. Zimmermann, and M. van Dierendonck. 2002. "Status and Action Plan for the Przewalski's Horse (*Equus ferus przewalskii*)." In *Status Survey and Conservation Action Plan Equids: Zebras, Asses and Horses*, edited by P. D. Moehlman, 82-92. IUCN/SSC Equid Specialist Group. Cambridge, UK: IUCN.

Walker, B., and D. Salt. 2006. *Resilience Thinking: Sustaining Ecosystems and People in a Changing World*. Washington, DC: Island Press.

Wallace-Wells, B. 2014. "The Case for the End of the Modern Zoo." *New York*, July 11. Available online at http://nymag.com/daily/intelligencer/2014/07/case-for-the-end-of-the-modern-zoo.html#.

Walters, J. 2016. "Cincinnati Zoo Director: Shooting Gorilla to Protect Child Was 'Right Decision'" *Guardian*, May 31, 2016. Accessible at https://www.theguardian.com/us-news/2016/may/30/gorilla-shot-cincinnati-zoo-child.

Wapner, P. 2010. *Living through the End of Nature: The Future of American Environmentalism*. Cambridge, MA: MIT Press.

Warren, Louis S. 1999. *The Hunter's Game: Poachers and Conservationists in Twentieth-Century America*. New Haven, CT: Yale University Press.

Waylen, K. A., A. Fischer, P. J. K. McGowan, S. J. Thirgood, and E. J. Milner-Gulland. 2010. "Effect of Local Cultural Context on the Success of Community-Based Conservation Interventions." *Conservation Biology* 24 (4): 1119-29. doi:10.1111/j.1523-1739.2010.01446.x.

Waytz, A., Cacioppo, J. T., and Epley, N. 2010. "Who Sees Human? The Stability and Importance of Individual Differences in Anthropomorphism." *Perspectives on Psychological Science* 5:219-32.

Wei, F., R. Swaisgood, Y. Hu, Y. Nie, L. Yan, et al. 2015. "Progress in the Ecology and Conservation of Giant Pandas." *Conservation Biology* 29 (6): 1497–1507. doi:10.1111/cobi.12582.

Weilbacher, M. 2009. "Knowing One Big Thing: 'The Role of the Nature Center in the Next Millennium.'" *Best of Clearing Magazine, vol. 5*, May 20. Available online at http://clearingmagazine.org/archives/423.

Weiler, B., and L. Smith. 2009. "Does More Interpretation Lead to Greater Outcomes? An Assessment of the Impacts of Multiple Layers of Interpretation in a Zoo Context." *Journal of Sustainable Tourism* 17:91–105.

Wells, S., D. Pearson, T. Sprankle, J. Sorenson, and M. Martinez. 2012. "*Ex Situ* Husbandry and Environmental Parameters Resulting in Reproduction of the Page Springsnail, *Pyrgulopsis morrisoni*: Implications for Conservation." *Journal of the Arizona-Nevada Academy of Science* 44:69–77.

Wellwood, J. M., ed. 1968. *Hawkes's Bay Acclimatisation Society Centenary, 1868–1968*. Hastings, NZ: H. B. Acclimatisation Society.

Wemmer, C. M. 1995. *The Ark Evolving: Zoos and Aquariums in Transition*. Washington DC: National Zoological Park, Smithsonian Institution Press.

Wemmer, C. M., M. Rodden, and C. Pickett. 1997. "Publications Trends in Zoo Biology: A Brief Analysis of the First 15 Years." *Zoo Biology* 16 (1): 3–8.

Wemmer, C. M., R. Rudran, F. Dallmeier, and D. E. Wilson. 1993. "Training Developing-Country Nationals Is the Critical Ingredient to Conserving Global Biodiversity." *BioScience* 43 (11): 762–67.

Werdelin, L., and M. E. Lewis. 2013. "Temporal Changes in Functional Richness in the Eastern African Plio-Pleistocene Carnivoran Guild." *PLoS ONE* 8 (3): e57944. doi:10.1371/journal.pone.0057944.

Wernau, J. 2013. "Shedd Aquarium Looks to Slice Energy Bill." *Chicago Tribune*, January 26.

West, C., and L. A. Dickie. 2007. "Introduction: Is There a Conservation Role for Zoos in a Natural World under Fire?" In *Zoos in the 21st Century: Catalysts for Conservation?*, edited by A. Zimmermann, M. Hatchwell, L. Dickie, and C. West, 3–11. Cambridge: Cambridge University Press.

Westbury, R. H., and Neumann, D. L. 2008. "Empathy-Related Responses to Moving Film Stimuli Depicting Human and Non-human Animal Targets in Negative Circumstances." *Biological Psychology* 78:66–74.

Weston, P. 2005. "The Dark Side of Flipper." *Sunday Mail*, April 24:36.

Wharton, D. 2007. "Research by Zoos." In *Zoos in the 21st Century: Catalysts for Conservation?*, edited by A. Zimmermann, M. Hatchwell, L. Dickie, and C. West, 178–91. Cambridge: Cambridge University Press.

Whiteley, A. R., S. W. Fitzpatrick, W. C. Funk, and D. A. Tallmon. 2015. "Genetic Rescue to the Rescue." *Trends in Ecology and Evolution* 30 (1): 42–49. doi: 10.1016/j.tree.2014.10.009.

Whitworth, A. W. 2012. "An Investigation into the Determining Factors of Zoo Visitor Attendances in UK Zoos." *PLoS ONE* 7 (1): e29839.

Wielebnowski, N. C., N. Fletchall, K. Carlstead, J. M. Busso, and J. L. Brown. 2002. "Noninvasive Assessment of Adrenal Activity Associated with Husbandry and Behavioral Factors in the North American Clouded Leopard Population." *Zoo Biology* 21 (1): 77–98.

Wiese, R. J., and M. Hutchins. 1994. *Species Survival Plans: Strategies for Wildlife Conservation*. Bethesda, MD: American Zoo and Aquarium Association Executive Office and Conservation Center.

Wildt, D. E., S. E. Ellis, D. Janssen, and J. Buff. 2003. "Toward More Effective Reproductive Science in Conservation." In *Reproductive Science and Integrated Conservation*, edited by W. V. Holt, A. R. Pickard, J. C. Rodger, and D. E. Wildt, 2–20. Cambridge: Cambridge University Press.

Wilkinson, D. 1991. *Program Review of the Marine Mammal Stranding Network*. Report to the Assistant Administrator for Fisheries. Silver Springs, MD: US Department of Commerce, National Oceanographic and Atmospheric Administrations, National Marine Fisheries Service.

Wilkinson, D., and G. A. J. Worthy. 1999. "Marine Mammal Stranding Networks." In *Conservation and Management of Marine Mammals*, vol. 2, edited by J. R. Twiss and R. R. Reeves, 396–411. Washington, DC: Smithsonian Institution Press.

Willis, E. L., D. C. Kersey, B. S. Durrant, and A. J. Kouba. 2011. "The Acute Phase Protein Ceruloplasmin as a Non-invasive Marker of Pseudopregnancy, Pregnancy, and Pregnancy Loss in the Giant Panda." *PLoS One* 6 (7): e21159. doi: 10.1371/journal.pone.0021159. Epub 2011 Jul 13. PubMed PMID: 21765892; PubMed Central PMCID: PMC3135589.

Wilshusen, P. R., S. R. Brechin, C. L. Fortwangler, and P. C. West. 2011. "Reinventing a Square Wheel: Critique of a Resurgent 'Protection Paradigm' in International Biodiversity Conservation." *Society and Natural Resources* 15 (1): 17–40. doi:10.1080/089419202317174002.

Wilson, E. 1864. Circular Letter to Parish Priests. March. Victoria State Library SLT 285.2945. M24.

Wilson, E. O. 1984. *Biophilia*. Cambridge, MA: Harvard University Press.

——. 2016. *Half-Earth: Our Planet's Fight for Life*. New York: Liverlight.

Wineman, J., and Y. K. Choi. 1991. "Spatial/Visual Properties of Zoo Exhibition." *Curator: The Museum Journal* 34:304–15.

Wines, M. P., V. M. Johnson, B. Lock, F. Antonio, J. C. Godwin, E. M. Rush, and C. Guyer. 2015. "Optimal Husbandry of Hatching Eastern Indigo Snake Snakes (*Drymarchon couperi*) during a Captive Head-Start Program." *Zoo Biology* 34:230–38.

Wisely, S. M., O. A. Ryder, R. M. Santymire, J. F. Engelhardt, and B. J. Novak. 2015. "A Road Map for 21st Century Genetic Restoration: Gene Pool Enrichment of the Black-Footed Ferret." *Journal of Heredity* 106 (5): 581–92. doi:10.1093/jhered/esv041.

Witte, C. L., L. L. Hungerford, R. Papendick, I. H. Stalis, and B. A. Rideout. 2010. "Factors Predicting Disease among Zoo Birds Exposed to Avian Mycobacteriosis." *Journal of the American Veterinary Medical Association* 236:211–18.

Wonders, K. 1993. *Habitat Dioramas: Illusions of Wilderness in Museum of Natural History*. Acta Universitatis Upsaliensis, Figura Nova Series 25. Uppsala: Universitasis Upsaliensis.

Woo, A. 2014. "Vancouver Aquarium Challenges Citation Breeding Ban in Court." *Globe and Mail*, August 27.

World Association of Zoos and Aquariums (WAZA). 2003. *Code of Ethics and Animal Welfare*. Accessed December 8, 2015. http://www.waza.org/en/site/conservation/code-of-ethics-and-animal-welfare.

——. 2005. *Building a Future for Wildlife—the World Zoo and Aquarium Conservation Strategy*. Bern, Switzerland: WAZA Executive Office.

——. 2016. "Biodiversity Is Us." Accessed June 29, 2016. http://www.waza.org/en/site/conservation/biodiversity-is-us.

World Bank. 2013. *Fish to 2030: Prospects for Fisheries and Aquaculture*. Washington, DC: World Bank.

——. 2014. *Fisheries and Aquaculture*. Washington, DC: World Bank.

World Commission on Environment and Development. 1987. *Our Common Future*. New York: United Nations.

World Wildlife Fund (WWF). 2015. *Living Planet Report*. Gland, Switzerland: World Wildlife Fund.

World Zoo Organization. 1993. *The World Zoo Conservation Strategy: The Role of Zoos and Aquaria of the World in Global Conservation*. Chicago: Chicago Zoological Society.

Wormell, D., K. Leus, M. Stevenson, E. B. Ruivo, and A. Rylands, eds. 2014. *Regional Collection Plan for Callitrichidae*, ed. 3—2014. Amsterdam: EAZA Callitrichid Taxon Advisory Group.

Worster, D. 1985. *Nature's Economy: A History of Ecological Ideas*. Cambridge: Cambridge University Press.

Wuerthner, G., E. Crist, and T. Butler, eds. 2014. *Keeping the Wild: Against the Domestication of Earth*. Washington, DC: Island Press, 2014.

Yang, H. 2014a. Interview by A. Clay. Gwacheon, South Korea. December.

——. 2014b. "동물박사가 들려주는 동물이야기: 서울대공원 개관 30년 [Animal stories told by an animal expert: The 30-year reformation of Seoul Grand Park]." *Seoul Newspaper*, May 2. http://www.seoul.co.kr//news/newsView.php?id=20140502015003.

Yates, S. 2015. "A Life-Changing Experience Studying Iguanas in the Bahamas." Accessed December 11. http://voices.nationalgeographic.com/2015/06/25/shedd-exuma-cays-rock-iguana-research-trip/.

Yeom, I. 2014. Interview by A. Clay. Gwacheon, South Korea. November.

Yi, H. 2015. Interview by A. Clay. Seoul, South Korea. March.

Yi, O. 2014. "삵 5마리 자연으로 . . . 서울대공원, 생태균형 위해 포식자 '삵' 방사" [Five Amur leopard cats released . . . Seoul Grand Park releases predators for ecological balance]. *Chosun.com*, March 21. http://news.chosun.com/site/data/html_dir/2014/03/21/2014032100625.html.

Young, C., M. Curtis, N. Ravida, F. Mazotti, and B. Durrant. 2017. "Development of a Sperm Cryopreservation Protocol for the Argentine Black and White Tegu (*Tupinambis merianae*)." *Theriogenology* 87:55-63.

Yule, J. 2002. "Cloning the Extinct: Restoration as Ecological Prostheses." *Common Ground* 1 (2): 6–9.

Zimmermann, A., M. Hatchwell, L. Dickie, and C. West. 2007. *Zoos in the 21st Century: Catalysts for Conservation?* Cambridge: Cambridge University Press.

Zimmermann, T. 2015. "The Case for Closing Zoos." *Outside*, February 13. http://www.outsideonline.com/1930141/case-closing-zoos.

Zippel, K. 2002. "Conserving the Panamanian Golden Frog: *Proyecto rana dorada*." *Herpetological Review* 33:11–12.

Zippel, K., K. Johnson, R. Gagliardo, R. Gibson, M. McFadden, R. et al. 2011. "The Amphibian Ark: A Global Community for Ex Situ Conservation of Amphibians." *Herpetological Conservation and Biology* 6 (3): 340–52.

Zippel, K., R. Lacy, and O. Byers, eds. 2006. *CBSG/WAZA Amphibian Ex Situ Conservation Planning Workshop Final Report*. Apple Valley, MN: IUCN/SSC Conservation Breeding Specialist Group.

Zippel, K. C., and J. R. Mendelson III. 2008. "The Amphibian Extinction Crisis: A Call to Action." *Herpetological Review* 39:23–29.

Zoological Society Bulletin. 1910. New York: New York Zoological Society.

Zoological Society of London. 2016. "Introducing the Modern Zoo." Accessed June 27. https://www.zsl.org/education/introducing-the-modern-zoo.

"Zoo's Bison Herd Accepted." 1906. *New York Times*, July 9, 14.

Zoos Victoria. 2016. "Don't Palm Us Off." Accessed June 29. http://www.zoo.org.au/get-involved/act-for-wildlife/dont-palm-us-off.

Contributors

Ruth A. Allard is executive vice president for conservation and education at the Arizona Center for Nature Conservation, based at the Phoenix Zoo. Her primary professional interests are applied conservation science, community-based conservation, and fostering community engagement in environmental stewardship.

Rick Barongi was director of the Houston Zoo from 2000 to 2015. He has also worked at Disney's Animal Kingdom and the San Diego Zoo and served as coeditor of the 2015 Conservation Strategy for world zoos and aquariums. He is committed to advancing the conservation efforts of zoological parks.

Mark V. Barrow Jr. is chair and professor of history and an affiliated faculty member of the Science and Technology in Society Department at Virginia Polytechnic Institute. He has written *A Passion for Birds: American Ornithology after Audubon* and *Nature's Ghosts: Confronting Extinction from the Age of Jefferson to the Age of Ecology*.

Onnie Byers is chair of the IUCN SSC Conservation Planning Specialist Group, overseeing a network of ten regional offices and over three hundred members that provide expertise and guidance in species conservation planning and integration of in situ and ex situ conservation activities.

Adrián Cerezo is a social ecologist who explores the nexus among early childhood development, conservation, and sustainable development. Through basic research, museum exhibitions, policy design, and community programs, he promotes innovation in multiple zoological institutions including the Yale Zigler Center for Early Childhood Policy and UNICEF.

Catherine A. Christen, an environmental historian, is a conservation training manager at the Smithsonian Conservation Biology Institute. She also serves on the Society for Conservation Biology's board of governors. She coedited (with Christen M. Wemmer) *Elephants and Ethics: Toward a Morality of Coexistence*, awarded the Smithsonian Secretary's 2009 Research Prize.

Anne Safiya Clay received bachelor's and master's degrees in biology and society from Arizona State University and was awarded a Boren Fellowship to South Korea, where she researched the Seoul Zoo. She is now working toward her PhD in environmental science and policy at George Mason University.

Susan Clayton is Whitmore-Williams Professor of Psychology at the College of Wooster. A social and conservation psychologist, she is an author or editor of *Conservation Psychology* (with Gene Myers) and the *Oxford Handbook of Environmental and Conservation Psychology*, among other books.

James P. Collins is Virginia M. Ullman Professor of Natural History and the Environment at Arizona State Uiversity. He studies the role of host-pathogen interactions in species decline and extinction, especially in amphibians. Other research focuses on intellectual factors that have shaped the development of ecology as a discipline and on ecological ethics. He is coauthor (with Martha L. Crump) of *Extinction in Our Times*.

Debra Colodner is director of conservation education and science at the Arizona-Sonora Desert Museum, where she oversees interpretive and educational programs, conservation science and outreach, and natural history collections.

Carrie Friese is associate professor of sociology at the London School of Economics and Political Science. She has written on the assisted reproduction of both humans and animals, including *Cloning Wild Life: Zoos, Captivity and the Future of Endangered Animals.*

Alejandro Grajal is president and CEO of the Woodland Park Zoo in Seattle. He has over twenty-five years of multidisciplinary experience in biodiversity conservation, animal welfare, social research, and the future of zoos.

Anita Guerrini is Horning Professor in the Humanities and professor of history at Oregon State University. Her most recent book is *The Courtiers' Anatomists: Animals and Humans in Louis XIV's Paris*. She blogs at http://anitaguerrini.com/anatomia-animalia/.

Harry W. Greene is professor emeritus of ecology and evolutionary biology at Cornell University and a fellow of the American Academy of Arts and Sciences. His books include *Snakes: The Evolution of Mystery in Nature* and *Tracks and Shadows: Field Biology as Art.*

Shelly Grow is vice president of conservation and science at the Association of Zoos and Aquariums (AZA) and works closely with AZA's organizational and individual members, animal programs, committees, and partner organizations to support AZA members' participation in conservation.

Pamela Henson, institutional historian, Smithsonian Institution Archives, focuses her research on the history of natural history and the use of visual information in historical research. Recent publications include "A Baseline Environmental Survey: The 1910–12 Smithsonian Biological Survey of the Panama Canal Zone," in the journal *Environmental History*.

Craig Ivanyi is executive director of the Arizona-Sonora Desert Museum. Much of his career has focused on the natural history and conservation of lower vertebrates of the Sonoran Desert region, with particular emphasis on venomous reptiles.

Kelly Kapsar is a PhD student at the Center for Systems Integration and Sustainability in the Department of Fisheries and Wildlife at Michigan State University. Her research focuses on the dynamic relation between culture, ecosystems, and science in arctic conservation policy.

T. J. Kasperbauer received his PhD in philosophy from Texas A&M University in 2014 and is currently a visiting researcher in the Department of Philosophy at George Washington University. His research interests include animal and environmental ethics, moral psychology, and the ethical, legal, and social implications of emerging technologies.

Lisa-Anne DeGregoria Kelly is an education research scientist at the Chicago Zoological Society-Brookfield Zoo, where she manages informal science education projects. Her projects focus on climate change education, environmental conservation education, and community partnerships.

Vernon Kisling is chair emeritus, Marston Science Library, University of Florida. He was a curator at the Atlanta and Miami zoos and has done wildlife research in Papua New Guinea and Chile. He is the editor of *Zoo and Aquarium History: Ancient Animal Collections to Zoological Gardens.*

Charles R. Knapp is vice president of conservation and research at the John G. Shedd Aquarium in Chicago. He is also cochair of the IUCN Species Survival Commission (SSC) Iguana Specialist Group and serves on the Association of Zoos and Aquariums Field Conservation Committee.

Khoa D. Le Nguyen is a PhD student in social psychology at the University of North Carolina, Chapel Hill. He is generally interested in emotions, pro-sociality, and well-being.

Kristin Leus works for the Copenhagen Zoo as a program officer for the European regional network of IUCN SSC Conservation Planning Specialist Group and for the European Association of Zoos and Aquaria. Her activities include in situ and ex situ population modeling and species conservation planning.

Stefan Linquist was one of the founders of the Ucluelet Aquarium, serving on its board of directors from 2005 to 2015. He is associate professor of philosophy at the University of Guelph. His research explores contemporary debates in ecology, genomics, and evolution.

Jerry F. Luebke is senior manager of audience research at the Chicago Zoological Society-Brookfield Zoo. He conducts various audience research and program evaluation studies. His current research efforts focus on informal learning experiences of zoo visitors.

Kristen E. Lukas is director of conservation and science at the Cleveland Metroparks Zoo, chair of AZA's Gorilla Species Survival Plan, and adjunct assistant professor of biology at Case Western Reserve University. She has written more than fifty scientific publications in animal behavior, animal welfare, and conservation education.

Debborah Luke is the executive director of the Society for Conservation Biology. Previously, she served as senior vice president of conservation and science at the Association of Zoos and Aquariums and managed the field conservation components of SAFE (Saving Animals from Extinction).

Jane Maienschein is university professor and director of the Center for Biology and Society at Arizona State University and also serves as fellow and directs the History Project at the Marine Biological Laboratory in Woods Hole, Massachusetts. Her research focuses on history and philosophy of science, bioethics, and biopolicy.

Terry L. Maple served as a zoo executive for twenty-five years while also working as a professor at Georgia Institute of Technology. He is currently affiliated with Florida Atlantic University and the University of North Florida and is professor in residence at the Jacksonville Zoo and Gardens. He is the author of *Professor in the Zoo: Designing the Future for Wildlife* and *Zoo Animal Welfare*.

Joseph R. Mendelson III is director of research at Zoo Atlanta and adjunct professor in biology at Georgia Institute of Technology. With research focusing on Central American amphibians, he has been intimately involved for many years in the research, conservation, and policymaking realms of the amphibian declines.

Ben A. Minteer is professor of environmental ethics and conservation at Arizona State University, where he also holds the Arizona Zoological Society endowed chair. He has published a number of books, most recently *After Preservation: Saving American Nature in the Age of Humans*.

Steve Monfort is the John and Adrienne Mars director of the Smithsonian Conservation Biology Institute (SCBI), headquartered in Front Royal, Virginia. SCBI provides leadership in the Smithsonian's global effort to use science-based approaches to conserve species and the habitats they require for survival.

Samantha Muka is assistant professor in the College of Arts and Letters at the Stevens Institute of Technology in Hoboken, New Jersey. She is currently working on a manuscript examining the role of aquarium technology in marine knowledge production.

Bryan Norton is distinguished professor emeritus in philosophy and policy in the School of Public Policy, Georgia Institute of Technology. He has written a number of books—most recently, *Sustainable Values, Sustainable Change*. He has also edited a number of collections and has published widely in academic journals of several fields.

Jackie Ogden leads the Association of Zoos and Aquariums' conservation effort SAFE (Saving Animals from Extinction) and consults with the Walt Disney Company, having recently retired after nineteen years as vice president of animals, science, and environment, Walt Disney Parks and Resorts.

Michael A. Osborne is professor of history of science at Oregon State University, where he teaches history of science and medicine. He is currently at work on a global history of yellow fever, and his most recent book is *The Emergence of Tropical Medicine in France*.

Clare Palmer is professor of philosophy at Texas A&M University. Her publications include *Animal Ethics in Context* and (with Peter Sandøe and Sandra Corr) *Companion Animal Ethics*.

Harriet Ritvo is Conner Professor of History at Massachusetts Institute of Technology. She has written widely about environmental history and the history of humans' relations with other animals. Her books include *Noble Cows and Hybrid Zebras: Essays on Animals and History* and *The Platypus and the Mermaid, and Other Figments of the Classifying Imagination*.

Nigel Rothfels is historian and administrator at the University of Wisconsin-Milwaukee. He is the author of *Savages and Beasts: The Birth of the Modern Zoo* and, most recently (with Dick Blau), *Elephant House*, a study of the lives of elephants and their keepers.

Oliver A. Ryder is director of genetics/Kleberg Chair at the San Diego Zoo Institute for Conservation Research and adjunct professor of biology at the University of California, San Diego. He has contributed broadly to the development of the fields of conservation biology and conservation genetics.

Peter Sandøe was educated in philosophy and is currently professor of bioethics at the University of Copenhagen. His research has been on ethical issues related to animals, biotechnology, and food production, often combining perspectives from natural science, social sciences, and philosophy.

Valerie Segura is an applied animal behavior analyst at the Jacksonville Zoo and Gardens, where she uses principles derived from the behavioral sciences to quantify and promote wellness for animals in human care.

Margaret Spring is vice president of conservation and science and chief conservation officer at the Monterey Bay Aquarium. She has served at the National Oceanic and Atmospheric Administration, as general counsel to Congress, and as coastal and marine program director for the Nature Conservancy.

Tara S. Stoinski is president and chief economic officer/chief scientific officer for the Dian Fossey Gorilla Fund International, chair of AZA's Ape TAG, and an executive committee member of the IUCN Primate Specialist Group's Section on Great Apes. She has written more than seventy scientific publications and books, and her work has been featured in *National Geographic Magazine*, on CNN, and on National Public Radio.

Kathy Traylor-Holzer is a senior program officer for the IUCN SSC Conservation Planning Specialist Group. She has served as population management adviser for regional and global zoo populations for twenty-five years, as well as being a population viability analysis modeler and facilitator for species conservation planning workshops worldwide.

Christopher W. Tubbs is a scientist at the San Diego Zoo Institute for Conservation Research. His research focuses on how environmental chemicals interact with the endocrine systems of endangered species to compromise reproduction.

Stuart Wells is director of conservation and science at the Arizona Center for Nature Conservation—Phoenix Zoo's Local Species Conservation Center. He is an adjunct faculty member of Arizona State University's School of Life Sciences. He studied the reproductive behavior and physiology of cheetahs as a graduate student at George Mason University.

Index